Modern Taxonomy of Bacteria and Archaea

Wen-Jun Li • Jian-Yu Jiao
Nimaichand Salam • Manik Prabhu Narsing Rao
Editors

Modern Taxonomy of Bacteria and Archaea

New Methods, Technology and Advances

Springer

Editors
Wen-Jun Li
School of Life Sciences
Sun Yat-Sen University
Guangzhou, China

Jian-Yu Jiao
School of Life Sciences
Sun Yat-Sen University
Guangzhou, China

Nimaichand Salam
School of Life Sciences
Sun Yat-Sen University
Guangzhou, China

Manik Prabhu Narsing Rao
School of Life Sciences
Sun Yat-Sen University
Guangzhou, China

ISBN 978-981-99-5719-4 ISBN 978-981-99-5720-0 (eBook)
https://doi.org/10.1007/978-981-99-5720-0

This Springer imprint is published by the registered company Springer Nature Singapore Pte Ltd.
The registered company address is: 152 Beach Road, #21-01/04 Gateway East, Singapore 189721, Singapore

Paper in this product is recyclable.

Contents

About the Editors

Wen-Jun Li received his PhD in Microbiology from Shenyang Institute of Applied Ecology, Chinese Academy of Sciences. He is currently working as Distinguished Professor in the School of Life Sciences, Sun Yat-Sen University, Guangzhou, China. His publications include six monographs, 30 authorized patents, and more than 1000 peer-reviewed research articles (https://www.researchgate.net/profile/Wen-Jun-Li). His research mainly focuses on the microbial diversity of terrestrial extremophilic environments, using culture-dependent and culture-independent methods, and on the mechanisms by which extremophilic microorganisms adapt to these unusual environments. He was awarded the WFCC (The World Federation for Culture Collections) Skerman award for microbial taxonomy in 2007, and other six provincial and ministry-level awards for his outstanding research contributions in the fields of microbial systematics and microbial ecology.

He serves as an editorial board member or Associate Editor for more than 10 international journals, and he has been a member of the International Committee on Systematics of Prokaryotes (ICSP) since 2017, and Secretary of Bergey's International Society for Microbial Systematics (BISMiS) since April 2018 and President of BISMiS since December of 2023.

Jian-Yu Jiao graduated with a degree in Biotechnology from Henan University and received his PhD in Ecology from Sun Yat-Sen University, China. Currently, he is associate research fellow at Sun Yat-Sen University, Guangzhou, China. His main research areas are microbial resources, ecology, and evolution by using multi-omics approaches. He has described a number of novel taxa, including three classes of *Actinomycetota*, and analyzed their ecology function and evolutionary history. He is appointed as an editorial board member for *iMeta* and *Frontiers in Microbiology* and acts as a reviewer for more than 5 journals. He has published more than 80 research papers in peer-reviewed journals, including *National Science Review*, *Nature Communications*, *The ISME Journal*, *mLife*, *Environmental Microbiome*, *Microbiological Research*, and *Frontiers in Microbiology*.

Nimaichand Salam graduated from North-Eastern Hill University, India, with a bachelor's degree in Biochemistry. He completed his master's degree from Manipur University (India) in Biochemistry. He received his PhD in Microbial Ecology from Manipur University. After a stint as a Post-Doctoral Fellow and Associate Research Fellow at Sun Yat-Sen University (China), he is currently working as Scientist D at National Agri-Food Biotechnology Institute (NABI), Mohali, India. His research interest is in the development of functional foods from fermented foods of India. His research areas also include the use of in silico modeling to understand the roles of microbes in food fermentation. He is involved in the description of more than 80 new species of bacteria and archaea and has 17 book chapters to his credit.

Manik Prabhu Narsing Rao received his PhD in microbiology from Gulbarga University, Kalaburgi, Karnataka, India. Currently, he is Assistant Professor at Facultad de Ingeniería, Universidad Autónoma de Chile, Talca, Chile. His research thrust areas are microbial taxonomy and understanding plant–microbes interaction. He described a number of novel taxa isolated from various habitats. He received the Young Scientist and Excellent Post-Doctorate awards. He was appointed as an editorial board member of *iMeta* and *Frontiers in Marine Science* and as a reviewer for many peer-reviewed international journals. He has published more than 88 research papers in peer-reviewed journals and 18 book chapters.

Chapter 1
Minimal Taxonomic Standards for Declaration of New Microbial Species

Nimaichand Salam, Shuai Li, and Wen-Jun Li

1.1 Introduction

Taxonomy is the science of naming, describing, and grouping organisms with similar characteristics. By grouping, organisms (or groups of organisms) are given a taxonomic status. The species concept (a key factor in systematics) is easy to define in plants and animals where there are easily observable characteristics such as morphological features. But the same concept cannot be applied to all organisms, particularly organisms in the microscopic world. In simple terms, it is difficult to observe a definable character by direct observation in microorganisms that can be considered as defining factors for characterization.

Taxonomy encompasses both nomenclature and characterization. A review of the set of rules for the nomenclature of Prokaryotes as stated in the International Code for Nomenclature for Prokaryotes (ICNP) is covered in Chap. 2. There are, however, no definitive rules for the classification of microorganisms, and to give a taxonomic status is at the discretion of the taxonomists who are dealing with the isolation and characterization of microbial isolate(s).

Prokaryotic systematics is not static but is continuously driven by evolving theoretical and methodological advances. Until a few decades ago, the universal marker

N. Salam
State Key Laboratory of Biocontrol, Guangdong Provincial Key Laboratory of Plant Resources and Southern Marine Science and Engineering Guangdong Laboratory (Zhuhai), School of Life Sciences, Sun Yat-Sen University, Guangzhou, People's Republic of China

National Agri-Food Biotechnology Institute, Mohali, Punjab, India

S. Li · W.-J. Li (✉)
State Key Laboratory of Biocontrol, Guangdong Provincial Key Laboratory of Plant Resources and Southern Marine Science and Engineering Guangdong Laboratory (Zhuhai), School of Life Sciences, Sun Yat-Sen University, Guangzhou, People's Republic of China
e-mail: liwenjun3@mail.sysu.edu.cn

W.-J. Li et al. (eds.), *Modern Taxonomy of Bacteria and Archaea*,
https://doi.org/10.1007/978-981-99-5720-0_1

that helps improve the characterization of prokaryotes was the small ribosomal subunit RNA (Fox et al. 1977). The phylogenetic relationship based on the 16S rRNA gene provides a constitutional framework for the characterization of an organism. With genome sequencing technology and its analyses becoming more affordable, the choice of genome sequence data as the phylogenetic framework for the characterization of prokaryotes is being considered mandatory (Wu and Eisen 2008; Chun et al. 2018). This chapter gives a review of the minimal features that are being used for the characterization of prokaryotes (with a particular focusses on bacteria and archaea).

1.2 History of Microbial Taxonomy

The drawings of Antonie van Leeuwenhoek from his observation of water in 1676 using his self-made microscope were the first records of the occurrence of microscopic organisms (Porter 1976). Considering the landmark discovery, his various communications with the Royal Society about his microscopic observations were published as letters by the Society. However, the first descriptive taxonomic observation of microscopic life forms was dedicated to the early work by O. F. Müller published in 1773 and 1786, which recorded the description of bacterial genera *Vibrio* and *Monas* using the Linnaean approach of naming life forms. Other notable descriptions of bacteria that follows were that of *Polyangium vitellinum* by mycologist H.F. Link published in 1809, *Serratia marcescens* by Bizio in 1823 and *Spirillum, Spirochaeata,* and *Bacterium* by Ehrenberg in 1838 (Murray and Holt 2015).

In a view to provide a formal classification system, bacteria were first recognized as plants under the class *Schizomycetes* in the phylum *Schizophyta.* In another classification system, Haeckel (1866) placed bacteria in the phylum *Monera* under two subclasses depending on the presence or absence of a structural envelope. Later, Cohn in 1872 recognized bacteria under six genera, *Micrococcus, Bacterium, Bacillus, Vibrio, Spirillum,* and *Spirochaeta* (Murray and Holt 2015). This later study was influential in rearranging the then-known microbial species into the genera *Planococcus, Streptococcus, Klebsiella, Merista, Planomerista, Neisseria, Sarcina, Planosarcina, Metabacterium, Clostridium, Serratia, Bacterium,* and *Spirillum* (Smith 1905).

The initial phase of bacterial taxonomy was rather chaotic. While there were no consensus classification systems for characterizing and classifying bacteria, the approaches adopted by Cohn (1875), Orla-Jensen (1909), Buchanan (1916), Bergey et al. (1925), Prevot (1958), and Bisset (1962) were some of the efforts to establish a formal classification system. Later on, different authors suggested several reclassifications of the genera, often due to the difference in observation of characteristic morphological features (Buchanan 1916, 1917). Several of the problems were sorted out by the publication of the first edition of the Bacteriological Code (Buchanan et al. 1948). These successive attempts were instrumental in the bringing up of the Bergey's Manuals, which provide updates on the characterizing features of the different bacterial and archaeal genera.

1.3 Journey to Polyphasic Taxonomy and on

Over the years, the processes for microbial classifications have started taking shape with the diversification of tools and techniques. For example, the classification of the early nineteenth century was merely on the basis of observed phenotypes and responses to biochemical reactions. The similarities in characteristics between a group of organisms were stored in the form of a numerical score. The scores were then arranged in the form of a matrix; on which basis, the phenetics and cladistics form of classification systems were developed (numerical taxonomy) (Sneath and Sokal 1973). The arrival of molecular techniques enables the then taxonomists with the ability to provide an objective comparison of the inter- and intra-relatedness between organisms. This was the basis for the development of techniques such as DNA-DNA hybridization (DDH) (Brenner 1973), which was considered the gold standard for delineating bacterial species. The development of sequencing technology and more particularly the ability to extract the genetic information of the housekeeping genes further revolutionized the phylogenetic evaluation of closely related microbial taxa. In this aspect, the most widely used sequence data is that of 16S rRNA gene sequence analysis.

Polyphasic taxonomy, a term proposed by Colwell (1970), referred to the taxonomic approaches that incorporate phenotypic, genotypic, and phylogenetic information. This approach used a classification framework based on the 16S rRNA gene sequence analysis. Accordingly, a group of strains having >70% DDH similarity and >97% 16S rRNA gene sequence identity along with <5% mol G + C content differences were considered to belong to the same bacterial species (Wayne et al. 1987; Stackebrandt and Goebel 1994). This definition should be supported by the fact that each of these strains should exhibit similar phenotypic characteristics, i.e., from morphology to response to biochemical reaction, and have more or less similar chemical features of the cell. All microbial taxonomists adopt the use of this approach in the declaration of a new taxon, with consideration for modification and improvement in techniques from time to time. To further provide a general perspective to microbial taxonomists, Tindall et al. (2010) published a set of minimal characteristics that should be considered for use during prokaryote systematics. For prokaryotes with specific characteristics, subcommittees are set up by International Committee on Systematics of Prokaryotes (ICSP) to help in the taxonomy of the defined set of prokaryotes. As of now, the subcommittees constituted by ICSP related to the following prokaryotic groups are presently active (https://www.the-icsp.org/taxonomic-subcommittees).

(i) Aerobic *Bacteroidetes*
(ii) *Aeromonadaceae*, *Vibrionaceae*, and related organisms

(iii) Order *Bacillales*
(iv) *Bifidobacterium*, *Lactobacillus,* and related organisms
(v) *Campylobacter* and related bacteria
(vi) *Chlamydiae*
(vii) Gram-negative anaerobic rods
(viii) *Halobacteria*
(ix) *Halomonadaceae*
(x) *Leptospiraceae*
(xi) Methanogenic archaea
(xii) *Micrococcinaceae* suborder
(xiii) *Mollicutes*
(xiv) *Pasteurellaceae*
(xv) Phototrophic Bacteria
(xvi) *Pseudonocardineae*
(xvii) Rhizobia and Agrobacteria

1.4 Minimal Taxonomic Description for Prokaryotic Systematics

There are no straightforward methods for the identification and characterization of a cultured microbial isolate as either belonging to a known species or a new microbial lineage. All the techniques and tools currently being implemented are developed over a period of time using experimental evidence. With the advancement of technology, a technique that provides more stable and reproducible data will eventually replace the existing methodology. The following is a summarization of the authors' knowledge of the currently used methods for the interpretation and inference of experimental data that all microbial taxonomists must consider in their work plan. We declare that the information provided here should not be construed as a complete guide, and neither are we in any way authority guiding the species description. The following description will however serve in planning a general experimental workflow for the description and declaration of new microbial taxa (Fig. 1.1). Besides, prokaryotic taxonomy is an ever-evolving science, and any additional information generated about the new isolates can always be emended.

1.4.1 Sequence and Phylogeny

The taxonomic status of a strain or set of strains is primarily assessed by phylogenetic analyses of a biomarker. A good practice for preliminary investigation of the new isolate is getting a good quality sequence preferentially of an evolutionarily conserved gene, such as the 16S rRNA gene, and generating a phylogeny based on

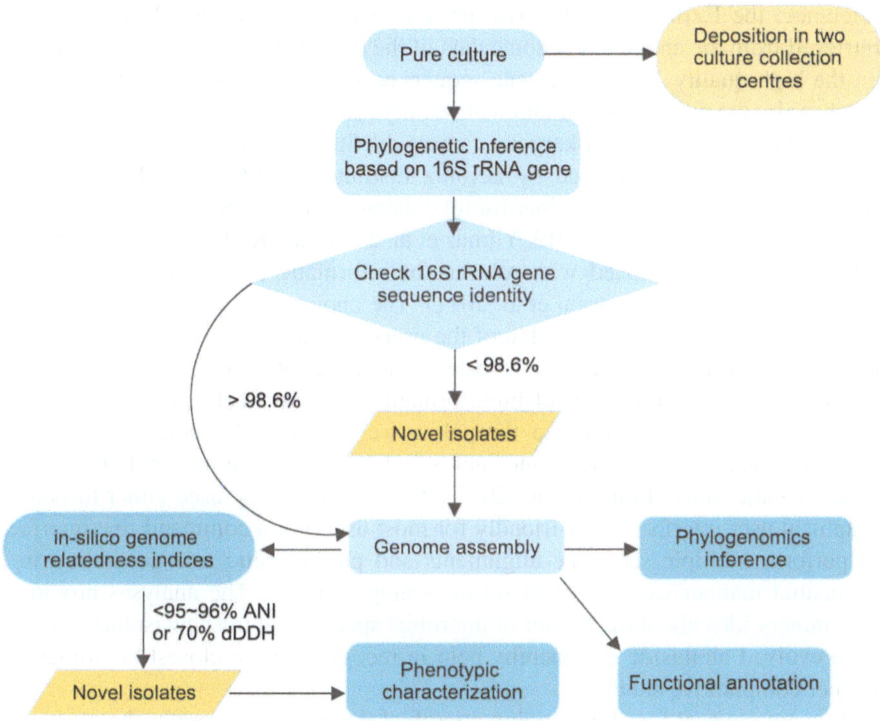

Fig. 1.1 Workflow for declaration of new microbial isolates

the sequence information. If the new isolate is shown to have a sequence identity of above 98.65% with the 16S rRNA gene sequence of a species representing a nomenclature type, the new isolate may be generally identified as belonging to the same species (Kim et al. 2014). On the other hand, strict interpretation of the result should be avoided (Rossi-Tamisier et al. 2015). In such cases, the interpretation should be accompanied by genome-based methods such as DDH (in the present context in silico model is considered) along with phenotypic characteristics.

Coming to the phylogenetic analysis, it is always better if the inference of the 16S rRNA gene is done with a well-maintained and updated 16S rRNA gene sequence database containing all the nomenclature types. There are different types of curated sequence databases maintained by scientists around the globe. Certain groups maintained them as expert-maintained aligned sequence databases that can directly be used with associated tools. While others maintained them as independently curated high-quality sequences that need to be manually analyzed and interpreted using the users' expertized tools and skills.

Most bacterial taxonomists may find the EzBioCloud server (Yoon et al. 2017) a useful resource for use in phylogenetic analysis. The database includes a regularly updated and curated 16S rRNA database of all microbial species (Bacteria and Archaea) recognized as nomenclature types. To maintain continuity and for future reference, the EzBioCloud database preferentially uses the 16S rRNA sequence retrieved from the annotated genome data of the type strains. Similarly, information on the high-quality 16S rRNA gene sequences submitted to the NCBI nucleotide database by the publishing authors with validly published nomenclature can also be retrieved from the List of Prokaryotic names with Standing in Nomenclature (LPSN) website, currently maintained by Leibniz Institute DSMZ (https://lpsn.dsmz.de) (Parte et al. 2020). There are other useful databases for 16S rRNA genes sequences such as SILVA (Quast et al. 2013; Yilmaz et al. 2014) and RDP database (Cole et al. 2014), which can be used with relevant bioinformatic tools like ARB software (Ludwig et al. 2004; Westram et al. 2011). The choice of sequence databases will however depend on the knowledge of the users and familiarity with bioinformatic tools that are necessary to analyze the available nucleotide data.

Currently, a good number of bioinformatic tools are freely available either as standalone software or for use through a web server. Molecular Evolutionary Genetics Analysis (MEGA) is one such standalone software that includes several bioinformatic tools (Tamura et al. 2021). These tools can be used either through a graphical user interface (user-friendly for most users) or a command line interface to perform multiple sequence alignments and phylogenetic tree generation in a sequential manner using a select set of treeing methods. The analyses provide a preliminary idea about the cluster of microbial species that the new isolate(s) might have evolved alongside and thereby help in recognizing the closest homologs for further comparative studies.

It is not unusual for all practicing microbial taxonomists to prefer the most easy method for inferring phylogenies. For instance, a more common practice among majority of the taxonomists is performing a multiple sequence alignment of the sequence data with retrieved sequences from any of the available sequence databases using ClustalX/ClustalW software (Larkin et al. 2007) and the preferential use of neighbor-joining phylogenetic tree regeneration based on Kimura 2-parameter substitution model. It is always advisable to use models that are more likely to provide a more accurate statistical output. For example, MUSCLE (Edgar 2004) may provide a faster and more accurate statistical value for large datasets (particularly for amino acid sequences) having unequal lengths than other methods. T-Coffee (Notredame et al. 2000) is another alternative that provides a good alignment. The multiple sequence alignment of nucleotide data for the 16S rRNA gene will still work well in Clustal programs for a small dataset of nearly complete nucleotide sequences. For phylogenetic inference, the maximum likelihood tree (Felsenstein 1981) provides more evolutionary information compared to other treeing methods based on distance matrices as this method relies on statistical modelling. In terms of speed, the neighbor-joining method (Saitou and Nei 1987) and maximum parsimony method (Fitch 1971) tend to be faster. Therefore, the choice of treeing method will depend on the need of the study.

Not all phylogenetic analyses using 16S rRNA gene sequences are unambiguous. For example, *Rhizobiales* species often tend to be polyphyletic and need to be resolved at higher levels (Dong et al. 2019). In such cases, the use of other conserved genes especially single-copy housekeeping genes should be considered. For instance, the prokaryotic systematics in archaea frequently need the phylogenetic inference from the *rpoB* gene in addition to that of the 16S rRNA gene (Liu et al. 2022). In certain cases such as in the genus *Streptomyces*, a multi-locus sequence analysis may provide a better resolution than the phylogenetic analysis of just the 16S rRNA gene (Guo et al. 2008).

1.4.2 Phylogenetic Inference Based on Genome Sequences

The rapid development of high-throughput sequencing technologies along with their easy access at lower cost has enabled the sequencing of several thousands of prokaryotic genomes. This development has made it possible to replace the experimental DDH with pairwise genome-sequence-derived similarity. Some of the tools being used for calculating the overall genome relatedness indices are average amino acid identity (AAI), average nucleotide identity (ANI), and digital DDH (dDDH) (Goris et al. 2007, Richter and Rosselló-Móra 2009, Auch et al. 2010). Provided genome sequences are of high quality and are reproducible, the use of in silico methods using genome data is more accurate in terms of reproducibility and validation and should be used in lieu of the error-prone experimental DDH method (Chun et al. 2018). Currently, the threshold values for delineating bacterial species are calculated at 95% AAI, 95–96% ANI, and 70% dDDH (Thompson et al. 2013).

A very useful web-based tool that primarily focuses on genome-based taxonomy is the Type (Strain) Genome Server (TYGS) maintained by Leibniz Institute DSMZ (https://tygs.dsmz.de) (Meier-Kolthoff and Göker 2019). An advantage of using TYGS is that the output phylogenetic inference is integrated with the types of species with validly published names (https://lpsn.dsmz.de), thereby making it more user-friendly for microbial systematics. The methods implemented in TYGS can be read in more detail from the cited publications (Meier-Kolthoff and Göker 2019; Meier-Kolthoff et al. 2022).

Besides the commonly deposited genome platforms such as NCBI or EMBL genome servers, genome databases of types of species are also curated and maintained at the web server of research institutes. A few examples are the genome database maintained in the EzBioCloud server and Joint Genome Institute-Integrated Microbial Genomes and Microbiomes (JGI-IMGM) Server. In the case of the EzBioCloud database, phylogenetic inference of the users' annotated genomes can be inferred by using an integrated UBCG software suite (Na et al. 2018). This software suit identifies a core set of bacterial genes that are present in the majority of the analyzed genomes, aligned and concatenated, and the multiple-alignment files can then be used to generate a phylogenetic tree either by the inbuild tools or by using other phylogenetic inference programs.

Phylogenomic approaches have improved the classification of many poorly classified taxa over the decades (Claesson et al. 2008; Nouioui et al. 2018; Liao et al. 2020). Useful bioinformatic tools that can be considered for use in microbial classification are listed in Table 1.1.

Table 1.1 Some useful bioinformatics tools and databases in microbial taxonomic inference

Software	Function	Reference
CheckM	Check Genome contamination	Parks et al. (2015)
AMPHORA2	Phylogenomic inference tools for bacteria and archaea	Wu and Scott (2012)
MiGA	Web server-based taxonomic and gene diversity analysis using genome sequence	Rodríguez-R et al. (2018)
TYGS	Web server-based taxonomic analysis software with integration with LPSN	Meier-Kolthoff et al. (2022)
UBCG	Phylogenomic inference of bacteria based on core-gene cluster analysis	Na et al. (2018)
UACG	Phylogenomic inference using archaeal core genes	https://www.ezbiocloud.net/tools/uacg
MEGA	Phylogeny inference	Tamura et al. (2021)
ANI calculator	Calculation of ANI	Goris et al. (2007)
JSpecies	Calculation of ANI	Richter and Rosselló-Móra (2009), Richter et al. (2016)
EzAAI	Calculation of AAI	Kim et al. (2021)
GGDC	Calculation of dDDH	Meier-Kolthoff et al. (2014)
KEGG	Taxonomy-based analysis of pathways and genomes	Kanehisa et al. (2023)
RAST	Genome annotation	Overbeek et al. (2014)

1.4.3 Phenotypic Characterization

Phenotypic characterization of a microbial isolate includes defining the chemical composition of the cell and its morphological features and recording the response of the isolate against different physiological and biochemical experiments. Chemical characterization typically deals with the detection of the various structural elements of the cells including the outer cell layers and the cell membranes.

The peptidoglycan structure is commonly used in the characterization of gram-positive bacteria. For instance, the presence of LL-diamino pimelic acid is often a characteristic feature of the genus *Streptomyces* and can be used as a diagnostic characteristic (Kämpfer 2015). The mode of peptide linkage and/or the composition of the amino acid chain can all provide characteristic features of one or a certain group of bacteria. The chemical features of isolates belonging to a similar species are more or less similar in composition and features and are therefore considered a good biomarker for the identification and characterization of new microbial isolates. The earlier method of characterization of peptidoglycan was often by running a thin-layer chromatography of the whole-cell extract (Staneck and Roberts 1974). This method while still being used finds an alternative with the implementation of HPLC technique (Tang et al. 2009).

Another important chemical feature of the cell regularly used in characterization is the respiratory lipoquinones. The most common among bacterial species is the ubiquinones, which often is determined by HPLC or GLC techniques. Prokaryotes produce diverse types of polar lipids. Diagnostic polar lipids are typically determined by running a two-dimensional chromatography and staining with dyes that react with particular functional groups of the extracted polar lipids. Similarly, mycolic acids are characteristics of only a few group of high G + C gram-positives bacteria, such as the *Rhodococcus* species. Other chemical features that provide diagnostic features are fatty acids composition, isoprenoid types, etc. A detailed description of these features is given in the publication of Tindall et al. (2010). While many of these chemical features may be diagnostic in nature, it may be high time to embrace new techniques such as MALDI-TOF or even in silico characterization of chemotaxonomic features by analysis of genomic data (Vandamme and Sutcliffe 2021).

The variability in the case of the morphological, physiological, and biochemical properties should be reported wherever feasible. Examination of the morphological features provides information about the cell shape and presence of flagella and/or inclusion bodies, while the color, dimension, and colony features can be observed on culture plates. Besides the cell morphology, analyses of fermentation products and related enzymatic activities and their abilities to utilize different carbohydrate substrates are still important diagnostic tools in clinical laboratories. Determining such features often is important for understanding the ecological adaptive behavior of the isolates. In the present context, the relationship between bacterial genotype and phenotype is being predicted accurately using applied bioinformatic tools (Lees et al. 2020). With time, genome-based phenotypic analyses may be applied either in support or in place of the experimental phenotypic data obtained.

1.5 Conclusion

With the coming of age in sequencing technology, the shift of microbial taxonomy methods toward genome-based taxonomy is becoming more prominent. Nevertheless, it is important to recognize both the classical bacterial systematics through polyphasic taxonomic approaches and the genomic microbial taxonomic approaches for stable profiling of taxonomic status for all recognized microbial taxa and newly isolated microbes.

References

Auch AF, von Jan M, Klenk H-P, Göker M (2010) Digital DNA-DNA hybridization for microbial species delineation by means of genome-to-genome sequence comparison. Stand Genomic Sci 2:117–134

Bergey DH, Harrison FC, Breed RS, Hammer BW, Huntoon FM (1925) Bergey's Manual of Determinative Bacteriology, 2nd edn. The Williams & Wilkins Co, Baltimore

Bisset KA (1962) The phylogenetic concept in bacterial taxonomy. Symp Soc Gen Microbiol 12:361

Brenner DJ (1973) Deoxyribonucleic acid reassociation in the taxonomy of enteric bacteria. Int J Syst Bacteriol 23:298–307

Buchanan RE (1916) Studies in the nomenclature and classification of bacteria - The problem of bacterial nomenclature. J Bacteriol 1:591–596

Buchanan RE (1917) Studies in the nomenclature and classification of bacteria II. The primary subdivisions of the Schizomycetes. J Bacteriol 2:155–164

Buchanan RE, St. John-Brooks R, Breed RS (1948) International bacteriological code of nomenclature. J Bacteriol 55:287–306

Chun J, Oren A, Ventosa A, Christensen H, Arahal DR, da Costa MS, Rooney AP, Yi H, Xu X-W, De Meyer S, Trujillo ME (2018) Proposed minimal standards for the use of genome data for the taxonomy of prokaryotes. Int J Syst Evol Microbiol 68:461–466

Claesson MJ, van Sinderen D, O'Toole PW (2008) *Lactobacillus* phylogenomics—towards a reclassification of the genus. Int J Syst Evol Microbiol 58:2945–2954

Cohn F (1875) Die Entwicklungsgeschichte der Gattung *Volvox*. Beitr Biol Pflanz 1:93–115

Cole JR, Wang Q, Fish JA, Chai B, McGarrell DM, Sun Y, Brown CT, Porras-Alfaro A, Kuske CR, Tiedje JM (2014) Ribosomal Database Project: data and tools for high throughput rRNA analysis. Nucleic Acids Res 42:D633–D642

Colwell RR (1970) Polyphasic taxonomy of the genus *Vibrio*: numerical taxonomy of *Vibrio cholerae*, *Vibrio parahaemolyticus*, and related *Vibrio* species. J Bacteriol 104:410–433

Dong L, Han MX, Wang D, Liu F, Asem MD, Jiao JY, Xiao M, Salam N, Li WJ (2019) *Flaviflagellibacter deserti* gen. nov., sp. nov., a novel member of the order *Rhizobiales* isolated from a desert soil. Antonie Van Leeuwenhoek 112:947–954

Edgar RC (2004) MUSCLE: multiple sequence alignment with high accuracy and high throughput. Nucleic Acids Res 32:1792–1797

Felsenstein J (1981) Evolutionary trees from DNA sequences: a maximum likelihood approach. J Mol Evol 17:368–376

Fitch WM (1971) Toward defining the course of evolution: minimum change for a specified tree topology. Syst Zool 20:406–416

Fox GE, Pechman KR, Woese CR (1977) Comparative cataloguing of 16S ribosomal ribonucleic acid: molecular approach to prokaryotic systematics. Int J Syst Bacteriol 27:44–57

Goris J, Konstantinidis KT, Klappenbach JA, Coenye T, Vandamme P, Tiedje JM (2007) DNA-DNA hybridization values and their relationship to whole-genome sequence similarities. Int J Syst Evol Microbiol 57:81–91

Guo Y, Zheng W, Rong X, Huang Y (2008) A multilocus phylogeny of the *Streptomyces griseus* 16S rRNA gene clade: use of multilocus sequence analysis for streptomycete systematics. Int J Syst Evol Microbiol 58:149–159

Haeckel E (1866) Generelle Morphologie der Organismen, vol I and II. (Georg Reimer)

Kämpfer P (2015) *Streptomyces. In*: Bergey's Manual of systematics of archaea and bacteria. John Wiley & Sons. Inc. https://doi.org/10.1002/9781118960608.gbm00191

Kanehisa M, Furumichi M, Sato Y, Kawashima M, Ishiguro-Watanabe M (2023) KEGG for taxonomy-based analysis of pathways and genomes. Nucleic Acids Res 51:D587–D592

Kim D, Park S, Chun J (2021) Introducing EzAAI: a pipeline for high throughput calculation of prokaryotic average amino acid identity. J Microbiol 59:476–480

Kim M, Oh HS, Park SC, Chun J (2014) Towards a taxonomic coherence between average nucleotide identity and 16S rRNA gene sequence similarity for species demarcation of prokaryotes. Int J Syst Evol Microbiol 64:346–351

Larkin MA, Blackshields G, Brown NP, Chenna R, McGettigan PA, McWilliam H, Valentin F, Wallace IM, Wilm A, Lopez R, Thompson JD, Gibson TJ, Higgins DG (2007) Clustal W and Clustal X version 2.0. Bioinformatics 23:2947–2948

Lees JA, Mai TT, Galardini M, Wheeler NE, Horsfield ST, Parkhill J, Corander J (2020) Improved prediction of bacterial genotype-phenotype associations using interpretable pangenome-spanning regression. MBio 11:e01344-20. https://doi.org/10.1128/mBio.01344-20

Liao H, Lin X, Li Y, Qu M, Tian Y (2020) Reclassification of the Taxonomic Framework of orders *Cellvibrionales*, *Oceanospirillales*, *Pseudomonadales*, and *Alteromonadales* in Class Gammaproteobacteria through phylogenomic tree analysis. mSystems 5:e00543-20. https://doi.org/10.1128/mSystems.00543-20

Liu B-B, Salam N, Narsing Rao MP, Cheng S, Xie YG, Wang L-Y, Zhang Y-r, Yu X-Y, Guo S-x, Li W-J (2022) *Haloterrigena gelatinilytica* sp. nov., a new extremely halophilic archaeon isolated from salt-lake. Arch Microbiol 204:176. https://doi.org/10.1007/s00203-022-02783-3

Ludwig W, Strunk O, Westram R, Richter L, Meier H, Kumar Y, Buchner A, Lai T, Steppi S, Jobb G, Förster W, Brettske I, Gerber S, Ginhart AW, Gross O, Grumann S, Hermann S, Jost R, König A, Liss T, Lüßmann R, May M, Nonhoff B, Reichel B, Strehlow R, Stamatakis A, Stuckmann N, Vilbig A, Lenke M, Ludwig T, Bode A, Schleifer K-H (2004) ARB: a software environment for sequence data. Nucleic Acids Res 32:1363–1371

Meier-Kolthoff JP, Carbasse JS, Peinado-Olarte RL, Göker M (2022) TYGS and LPSN: a database tandem for fast and reliable genome-based classification and nomenclature of prokaryotes. Nucleic Acids Res 50:D801–D807

Meier-Kolthoff JP, Göker M (2019) TYGS is an automated high-throughput platform for state-of-the-art genome-based taxonomy. Nat Commun 10:2182. https://doi.org/10.1038/s41467-019-10210-3

Murray RGE, Holt JG (2015) The history of Bergey's manual. *In*: Bergeys's manual of systematics of archaea and bacteria. doi: https://doi.org/10.1002/9781118960608.bm00009

Na S-I, Kim YO, Yoon S-H, Ha S-m, Baek I, Chun J (2018) UBCG: up-to-date bacterial core gene set and pipeline for phylogenomic tree reconstruction. J Microbiol 56:280–285

Notredame C, Higgins DG, Heringa J (2000) T-coffee: a novel method for fast and accurate multiple sequence alignment. J Mol Biol 302:205–217

Nouioui I, Carro L, Garcia-Lopez M, Meier-Kolthoff JP, Woyke T, Kyrpides NC, Pukall P, Klenk H-P, Goodfellow M, Goker M (2018) Genome-based taxonomic classification of the phylum *Actinobacteria*. Front Microbiol 9:2007. https://doi.org/10.3389/fmicb.2018.02007

Orla-Jensen S (1909) Die Hauplinien des naturlichen Bakterien-systems. Zentralbl Bakteriol Parasitenkd Infektionskr Hyg Abt II 22(97-98):305–346

Overbeek R, Olson R, Pusch GD, Olsen GJ, Davis JJ, Disz T, Edwards RA, Gerdes S, Parrello B, Vonstein V, Wattam AR, Xia F, Stevens R (2014) The SEED and the rapid annotation of microbial genomes using subsystems technology (RAST). Nucleic Acids Res 42:D206–D214

Parks DH, Imelfort M, Skennerton CT, Hugenholtz P, Tyson GW (2015) CheckM: assessing the quality of microbial genomes recovered from isolates, single cells, and metagenomes. Genome Res 25:1043–1055

Parte AC, Sardà Carbasse J, Meier-Kolthoff JP, Reimer LC, Göker M (2020) List of prokaryotic names with standing in nomenclature (LPSN) moves to the DSMZ. Int J Syst Evol Microbiol 70:5607–5612

Porter JR (1976) Antony van Leeuwenhoek: tercentenary of his discovery of bacteria. Bacteriol Rev 40:260–269

Prevot AR (1958) Manuel de classification et determination des bacteries anaerobies. Masson, Paris

Quast C, Pruesse E, Yilmaz P, Gerken J, Schweer T, Yarza P, Peplies J, Glöckner FO (2013) The SILVA ribosomal RNA gene database project: improved data processing and web-based tools. Nucleic Acids Res 41:D590–D596

Richter M, Rosselló-Móra R (2009) Shifting the genomic gold standard for the prokaryotic species definition. Proc Natl Acad Sci U S A 106:19126–19131

Richter M, Rosselló-Móra R, Oliver Glöckner F, Peplies J (2016) JSpeciesWS: a web server for prokaryotic species circumscription based on pairwise genome comparison. Bioinformatics 32:929–931

Rodríguez-R LM, Gunturu S, Harvey WT, Rosselló-Mora R, Tiedje JM, Cole JR, Konstantinidis KT (2018) The microbial genome atlas (MiGA) webser: taxonomic and gene diversity analysis of Archaea and Bacteria at the whole genome level. Nucleic Acids Res 46:W282–W288

Rossi-Tamisier M, Benamar S, Raoult D, Fournier PE (2015) Cautionary tale of using 16S rRNA gene sequence similarity values in identification of human-associated bacterial species. Int J Syst Evol Microbiol 65:1929–1934

Saitou N, Nei M (1987) The neighbour-joining method: a new method for reconstructing phylogenetic trees. Mol Biol Evol 4:406–425

Smith EF (1905) Bacteria in relation to Plant Diseases - Volume 1, Methods of Work and General Literature of Bacteriology exclusive of Plant Diseases. Washington

Sneath PHA, Sokal RR (1973) Numerical taxonomy: the principles and practice of numerical classification. W.H. Freeman, San Francisco

Stackebrandt E, Goebel BM (1994) Taxonomic note: a place for DNA-DNA reassociation and 16S rRNA sequence analysis in the present species definition in bacteriology. Int J Syst Bacteriol 44:846–849

Staneck JL, Roberts GD (1974) Simplified approach to identification of aerobic actinomycetes by thin-layer chromatography. Appl Microbiol 28:226–231

Tamura K, Stecher G, Kumar S (2021) MEGA11: molecular evolutionary genetics analysis version 11. Mol Biol Evol 38:3022–3027

Tang S-K, Wang Y, Chen Y, Lou K, Cao L-L, Xu L-H, Li W-J (2009) *Zhihengliuella alba* sp. nov., and emended description of the genus *Zhihengliuella*. Int J Syst Evol Microbiol 59:2025–2032

Thompson CC, Chimetto L, Edwards RA, Swings J, Stackebrandt E, Thompson FL (2013) Microbial genomic taxonomy. BMC Genomics 14:913. https://doi.org/10.1186/1471-2164-14-913

Tindall BJ, Rosselló-Móra R, Busse H-J, Ludwig W, Kämpfer P (2010) Notes on the characterization of prokaryote strains for taxonomic purposes. Int J Syst Evol Microbiol 60:249–266

Vandamme P, Sutcliffe I (2021) Out with the old and in with the new: time to rethink twentieth century chemotaxonomic practices in bacterial taxonomy. Int J Syst Evol Microbiol 71:005127. https://doi.org/10.1099/ijsem.0.005127

Wayne LG, Brenner DJ, Colwell RR, Grimont PAD, Kandler O, Krichevsky MI, Moore LH, Moore WEC, Murray RGE, Stackebrandt E, Starr MP, Truper HG (1987) Report of the Ad Hoc Committee on reconciliation of approaches to bacterial systematics. Int J Syst Bacteriol 37:463–464

Westram R, Bader K, Prüße E, Kumar Y, Meier H, Glöckner FO, Ludwig W (2011) ARB: a software environment for sequence data. In: de Bruijn FJ (ed) Handbook of molecular microbial ecology I: metagenomics and complementary approaches. Wiley-Blackwell

Wu M, Eisen JA (2008) A simple, fast and accurate method of phylogenomic inference. Genome Biol 9:R151. https://doi.org/10.1186/gb-2008-9-10-r151

Wu M, Scott AJ (2012) Phylogenomic analysis of bacterial and archaeal sequences with AMPHORA2. Bioinformatics 28:1033–1034

Yilmaz P, Parfrey LW, Yarza P, Gerken J, Pruesse E, Quast C, Schweer T, Peplies J, Ludwig W, Glöckner FO (2014) The SILVA and "All-species Living Tree Project (LTP)" taxonomic frameworks. Nucleic Acids Res 42:D643–D648

Yoon SH, Ha SM, Kwon S, Lim J, Kim Y, Seo H, Chun J (2017) Introducing EzBioCloud: a taxonomically united database of 16S rRNA gene sequences and whole-genome assemblies. Int J Syst Evol Microbiol 67:1613–1617

Chapter 2
Naming New Taxa of Prokaryotes: Rules and Recommendations

Aharon Oren

וַיִּקְרָא הָאָדָם שֵׁמוֹת לְכָל-הַבְּהֵמָה וּלְעוֹף הַשָּׁמַיִם וּלְכֹל חַיַּת הַשָּׂדֶה;

"And Adam gave names to all cattle, and to the fowl of the air, and to every beast of the field"

—Genesis 2:20; King James translation

2.1 Introduction

From ancient times, man has given names to newly discovered organisms. Nomenclature is an essential part of biological sciences, including microbiology. In the period 2017–2021, names of around 960 new species and 175 new genera of prokaryotes were validly published annually, on the average (LPSN - List of Prokaryotic names with Standing in Nomenclature 2022). All those new taxa were given names in agreement with the rules of the *International Code of Nomenclature of Prokaryotes* (ICNP, Parker et al. 2019; Oren et al. 2023).

The ICNP follows the Linnaean binomial nomenclature system. Names of taxa are Latin or latinized words, usually taken from Latin or classical Greek. Naming new taxa of prokaryotes therefore requires a basic knowledge of the Latin lexicon and a few rules of Latin grammar. Most microbiologists who describe new taxa lack an elementary understanding of Latin and the ways Latin and Greek word elements are used to name newly discovered taxa of bacteria and archaea. However, one does not have to be an accomplished linguist to propose names of new genera and species that meet the demands of the nomenclature code. One only needs to understand the relevant rules of the ICNP and its orthography appendix (Appendix 9), know how to use a Latin dictionary, and have some understanding of the way Latin (and possibly also Greek) nouns and adjectives of different genders (masculine, feminine, neuter) and different declensions are formed. Useful information can also be found in

---

A. Oren (✉)
Department of Plant and Environmental Sciences, The Institute of Life Sciences, The Hebrew University of Jerusalem, Edmond J. Safra Campus, Jerusalem, Israel
e-mail: aharon.oren@mail.huji.ac.il

W.-J. Li et al. (eds.), *Modern Taxonomy of Bacteria and Archaea*,
https://doi.org/10.1007/978-981-99-5720-0_2

earlier published essays on prokaryotic nomenclature (MacAdoo 1993; Oren 2011; Oren 2019; Trüper 1996; Trüper 1999; Trüper 2001).

This chapter presents the most relevant rules and recommendation of the ICNP and the few rules of Latin and Greek grammar that are essential to correctly form new prokaryotic names. I hope that this chapter will assist future generations of microbiologists to name the wealth of novel taxa to be discovered during the future exploration of prokaryotic diversity on Earth.

2.2 The Rules of the International Code of Nomenclature of Prokaryotes

The nomenclature of cultivated prokaryotes is regulated by the rules of the ICNP, approved by the International Committee on Systematics of Prokaryotes (ICSP). For naming of yet-uncultivated prokaryotes, an alternative nomenclature code, the SeqCode (*Code of Nomenclature of Prokaryotes Described from Sequence Data*), was recently proposed (SeqCode Publications 2022). It contains similar Latin-based nomenclature rules as the ICNP. Chapter 9 provides further information about the SeqCode.

The most important regulations of the ICNP (2022 Revision; Oren et al. 2023) for naming new genera and species are highlighted below, with explanations and examples.

Principle 2
The nomenclature of prokaryotes is not independent of botanical and zoological nomenclature. When naming new taxa in the rank of genus or higher, due consideration is to be given to avoiding names which are regulated by the International Code of Zoological Nomenclature and the International Code of Nomenclature for algae, fungi, and plants.

Note. This principle takes effect with publication of acceptance of this change by the ICSP (from January 1, 2001) and is not retroactive.

Some older generic names of prokaryotes have homonyms in the eukaryotic world. For example, *Bacillus* is both a genus of aerobic endospore-forming bacteria and a genus of stick insects; *Proteus* is both a member of the *Gammaproteobacteria* and an amphibian living in karstic caves. The older text of Principle 2 ("The nomenclature of prokaryotes is independent of botanical nomenclature, …") was modified by the ICSP in 1999, and the change took effect in January 2001. Thus, newly proposed names of prokaryotic genera may not be homonyms of generic names in use in botany or zoology. Therefore, the names *Parvimonas* and *Quatrionicoccus* were proposed to replace the illegitimate, prokaryotic generic names *Micromonas* and *Quadricoccus*, respectively (Tindall and Euzéby 2006). Recently, more cases of such illegitimate names were found: *Gramella*, *Holzapfelia*, *Litoricola*, *Melitea*, *Membranicola*, and *Nicolia* all have homonyms in botany or zoology, and replacement names were proposed – *Christiangramia*, *Holzapfeliella*, *Litorivicinus*, *Neomelitea*, *Membranihabitans*, and *Nicoliella*, respectively (Deshmukh and Oren 2022a, b, 2023). The "Catalogue of Life" website (Catalogue of Life 2022; Hobern et al. 2021) contains a good record of botanical and zoological genus names, but not all abovementioned names were listed in the Catalogue of Life registry.

Principle 3
The names of all taxa are Latin or latinized words treated as Latin, regardless of their origin. They are usually taken from Latin or Greek.

Rule 6
The scientific names of all taxa must be treated as Latin; names of taxa above the rank of species are single words.

Similar to the botanical and zoological nomenclature, Latin and classical Greek word elements are the basis of the prokaryotic nomenclature. Grammatically all names are treated as Latin, with endings based on the declensions of Latin nouns and adjectives. Also when words derived from other languages are included, the rules of Latin grammar must be followed.

Words from languages other than Latin or classical Greek should be avoided as long as equivalents exist in Latin or classical Greek or can be constructed by combining word elements from these two languages. Exceptions: names derived from typical local items, such as food, drink or geographical localities for which no Latin or classical Greek names exist. …

Examples of names where words for local foods and drinks were treated as Latin are *Leuconostoc kimchii* ("of kimchi," a kind of Korean fermented vegetables), *Lactobacillus sakei* ("of sake," Japanese rice wine), and *Lactobacillus odoratitofui* ("of stinky tofu"), a name combining word elements from Latin (*odoratus*, "fragrant") and Chinese (tofu).

Recommendation 6
To form new prokaryotic names and epithets, authors are advised as follows:

(1) Avoid names or epithets that are long or difficult to pronounce.
(2) Make names or epithets that have an agreeable form that is easy to pronounce when latinized.

A recommendation similar to Recommendation 6(1) was already proposed by Linnaeus in his *Philosophia Botanica* of 1751 (Rule 249): *Nomina generica sesquipedalia, enuntiatu difficilia, vel nauseabunda, fugienda sunt* ("Generic names that are a foot and a half long, those that are difficult to pronounce, or are disgusting, should be avoided"). The longest genus name in use is currently *Hydrogenoanaerobacterium*. The longest species name is *Myxococcus llanfairpwllgwyngyllgogerychwyrndrobwllllantysiliogogogochensis*. Whether names have an "agreeable form" and are "easy to pronounce" depends on the native language and education of the speaker. Most microbiologists will find *Terasakiispira papahanaumokuakeensis* (referring to Papahānaumokuākea, the Marine National Monument in

Hawaii) difficult to pronounce. The shortest names are currently *Sala cibi* (eight characters) and *Baia soyae* (nine characters).

Rule 10a

The name of a genus or subgenus is a noun, or an adjective used as a noun, in the singular number in the nominative case, and written with an initial capital letter. The name may be taken from any source and may even be composed in an arbitrary manner. It is treated as a Latin noun.

Examples: single Greek stem, *Clostridium*; two Greek stems, *Haemophilus*; single Latin stem, *Spirillum*; two Latin stems, *Lactobacillus*; hybrid name, Latin-Greek stems, *Flavobacterium*; latinized personal name, *Shigella*; arbitrary name, *Afipia*, *Desemzia*, *Waddlia*, or *Cedecea*

When naming a new genus, it is important to state its gender, as adjectives used as specific epithets must agree in gender with the generic name (Rule 12c(1)). Genus names of the first/second declension (see Sect. 2.2) that end in *–us*, *–a*, and *–um* are generally masculine, feminine, and neuter, respectively. Thus, *Alkalimarinus* ("a marine bacterium living under alkaline conditions") is masculine, *Natrialba* ("sodium white," referring to its high sodium requirement and white colonies) is feminine, and *Halorubrum* ("salt (-requiring) and red") is neuter. For words of the third declension, the gender is not always obvious. Grammatically, *Methylobrevis* ("methyl-using short rod") can be either masculine or feminine; the authors who described the genus chose the masculine gender. Adjectives such as *vorax* (devouring, voracious) and *vivax* (long-lived, vivacious) have the same form in all three genders. The authors who described *Rubrivivax* ("red and long-living bacterium") and *Halovivax* ("long-living halophile") decided for the masculine gender (therefore: *Rubrivivax albus*, *Halovivax ruber*), but *Proteinivorax* ("protein devouring") was given the neuter gender. Therefore, we have *Proteinivorax tanatarense* and not *tanatarensis*. The gender of compound genus names is determined by the gender of the last component. For example, *Maribacter* ("a rod from the sea"), *Fluviispira* ("a spiral from a river"), and *Undibacterium* ("a rod from water") are, respectively, masculine, feminine, and neuter, although the Latin *mar* is neuter, *fluvius* is masculine, and *unda* is feminine.

Latin present participles (see Sect. 2.3) have the same form in all genders. Therefore, a gender must be selected when such participles are used in generic names. *Agarivorans* ("agar-devouring"), *Saccharedens* ("sugar-eating"), and *Mangrovihabitans* ("mangrove-inhabiting") were declared masculine. *Litorisediminivivens* ("living in coastal sediment") is also masculine, but *Litorivivens* ("living at the seashore") was described as feminine. The authors who described *Aestuariivivens* ("living in a tidal flat") and *Saccharofermentans* ("sugar-fermenting") chose the neuter gender.

Recommendation 10a

The following recommendations apply when forming new generic or subgeneric names.

(1) Refrain from naming genera and subgenera after persons unconnected with microbiology or at least with natural science.
(2) Give a feminine form to all personal generic and subgeneric names, whether they commemorate a man or a woman.

The recommendation to give a feminine form to all personal generic names, whether they commemorate a man or a woman, refers to names formed by adding a Latin ending to the name of the person: the names *Escherichia* (honoring Theodor Escherich), *Colwellia* (honoring Rita Colwell), *Pasteurella* (honoring Louis Pasteur), and *Gandjariella* (honoring Indrawati Gandjar) are all feminine. The recommendation does not apply to compound names where the person's name precedes another word element: the gender of the last word always determines the gender of the genus. For example, *Youngiibacter* (honoring Lily Young) is masculine, *Terasakiispira* (honoring Yasuke Teraraski) is feminine, and *Geitlerinema* (honoring Lothar Geitler) is neuter.

Rule 12c

A specific epithet may be taken from any source and may even be composed arbitrarily.

Example: *thetaiotaomicron* in *Bacteroides thetaiotaomicron* derived from a combination of the Greek letters *theta*, *iota*, and *omicron*

Words from languages other than Latin or Greek should be avoided as parts of a specific epithet as long as equivalents exist in Latin or Greek or can be constructed by combining word elements from these two languages. Exceptions can be made for names derived from typical local items, such as food, drink, or geographical localities for which no Latin or Greek names exist or for names based on acronyms.

Example: *safensis* in *Bacillus safensis*, arbitrarily derived from SAF (the spacecraft-assembly facility at the Jet Propulsion Laboratory, Pasadena, CA, USA)

A specific epithet must be treated in one of the three following ways:

(1) As an adjective in the singular number in the nominative case that must agree in gender with the generic name
    Example: *aureus* in *Staphylococcus aureus*
(2) As a substantive (noun) in apposition in the nominative case
    Examples: *Blautia obeum*
(3) As a substantive (noun) in the genitive case
    Example: *coli* in *Escherichia coli*

In grammar, apposition means the placing of a word beside another so that the second explains and has the same grammatical construction as the first. Thus, *Desulfovibrio gigas* may be understood as "*Desulfovibrio*, called the giant." In contrast to adjectives that that must agree in gender with the generic name, nouns in apposition used as specific epithets may have a different gender. In *Blautia obeum*, *Blautia* is feminine, while *obeum* ("the egg") is neuter (note: the word ὤβεον, latinized *obeum*, is a rare word in Greek, mentioned by the grammarian Hesychius of Alexandria; the common Greek word for egg was ᾠον). Specific epithets ending in the Latin suffixes –*cola*, derived from *incola*, "inhabitant," "dweller" (e.g., *Blautia faecicola,* "inhabiting feces"), and *–cida*, "killer" (e.g., *Aeromonas salmonicida*, "salmon killer"), are also examples of nouns in apposition used as specific epithets. As *incola* and *–cola* can be masculine or feminine, authors who use these endings in generic names should indicate the gender of the genus. *Aquincola* ("water dweller") and *Fervidicola* ("inhabitant of thermal waters") were published as masculine and *Thermincola* ("inhabitant of a hot spring") and *Tabrizicola* ("inhabitant of Tabriz") as feminine.

Recommendation 12c

(1) Choose a specific epithet that gives some indication of a property or of the source of the species.
(2) Avoid those that express a character common to all, or nearly all, the species of a genus.
(3) Specific epithets should not honor the author or coauthors of the proposed species or subspecies, or any persons not connected with microbiology or at least with natural science.
(4) Avoid the same genus epithets which are very much alike, especially those that differ only in their last letters.
(5) Avoid the use of the genitive and the adjectival forms of the same specific epithet to refer to two different species of the same genus.
(6) If an ordinal adjective used for enumeration is chosen, then they may include numbers up to ten.

Example: *primus*, *secundus*

The recommendation to avoid epithets that "express a character common to all, or nearly all, the species of a genus" was not always followed, as shown by *Maribacter maritimus* ("marine"), *Halorubrum halophilum* ("salt-loving"), and *Thiobacillus* (*Acidithiobacillus*) *thiooxidans* ("sulfur oxidizing"). Examples of epithets used in the same genus which are very much alike are *Vibrio marinus*-*Vibrio maritimus* ("marine") and *Flavobacterium aquaticum*-*Flavobacterium aquatile* ("aquatic"). An example where the recommendation to "avoid the use of the genitive and the adjectival forms of the same specific epithet to refer to two different species of the same genus" was not followed is *Flavobacterium fluvii* (genitive noun) and *Flavobacterium fluvatile* (neuter adjective), both meaning "of a river."

Rule 24a

… Priority of publication dates from January 1, 1980. On that date, all names published prior to January 1, 1980, and included in the Approved Lists of Bacterial Names are treated, for nomenclatural purposes, as though they had been validly published for the first time on that date, the existing types being retained.

Note 1. … Lists of names were prepared together with the names of the authors who originally proposed the names. These Approved Lists of Bacterial Names were approved by the ICSB and published in the IJSB on January 1, 1980.

With the publication of the *Approved Lists of Bacterial Names* in 1980 (Skerman et al. 1980), all older names that were not included in those lists lost their standing in the nomenclature. It remains possible to revive such names as *nomina revicta*. Examples are *Actinobacillus seminis* and *Leptothrix discophora*, names revived in 1990 and 1997, respectively. Rule 33c gives further guidelines. The *Index Bergeyana* (Buchanan et al. 1966) is an excellent source of information about old names that lost standing in the nomenclature with the publication of the 1980 Approved Lists.

Rule 64
Diacritic signs are not used in the nomenclature of prokaryotes.

In names or epithets derived from words with such signs, the signs must be suppressed and the letters transcribed as follows: (1) *ä*, *ö*, and *ü* become *ae*, *oe*, and *ue*; (2) *é*, *è*, and *ê* become *e*; (3) *ø*, *æ*, and *å* become *oe*, *ae*, and *aa*, respectively.

Examples are *Streptomyces kaempferi*, *Azospirillum doebereinerae*, and *Truepera*, honoring, respectively, Peter Kämpfer, Johanna Döbereiner, and Hans Georg Trüper. *Prevotella* was named after André Romain Prévot, and *Oerskovia* honors Jeppe Ørskov.

Rule 65
Note. As of January 1, 2023, generic names ending in *–oides* (from Gr. suff. *–eides* derived from Gr. neut. n. *eidos* that which is seen, form, shape, figure) will have the neuter gender, irrespective of the gender of the word or word element that precedes the *–oides* ending, and names ending in *–opsis* (from Gr. fem. n. *opsis* aspect, appearance) must be treated as feminine.

In the past, different genders were assigned to generic names ending in *–oides*: *Bacteroides*, *Dehalococcoides*, *Myroides*, and *Thermococcoides* are masculine, the cyanobacterial genus *Planktothricoides* is feminine, and *Methanococcoides* was given the neuter gender. The ending *–oides* is based on the Greek suffix *–eides*, derived from the Greek neuter noun *eidos* ("that which is seen," "form," "shape"). Therefore, new generic names ending in *–oides* should be given the neuter gender (Oren and Schink 2016).

Arbitrarily formed generic names or vernacular names used as generic names take the gender assigned to them by their authors but must be based on the usage of comparable words in Latin where appropriate. If the original author(s) failed to indicate the gender, subsequent author(s) have the right of choice.

Examples: *Desemzia* Stackebrandt et al. 1999 was assigned the feminine gender; *Bergeyella* Vandamme et al. 1994 was assigned the feminine gender; *Aestuariivivens* Park et al. 2015 was given the neuter gender; no gender was assigned yet to *Pontivivens* Park et al. 2015; *Marivivens* Park et al. 2016 was given as masculine or feminine; a subsequent author may choose.

For several genus names whose gender was not assigned by the authors, the gender cannot be deducted from the Latin ending. For *Pontivivens*, the authors gave the name as N.L. part. adj. used here as a noun, which may have any gender. The epithets of the two species classified in the genus, *Pontivivens insulae* and *Pontivivens yutae*, are nouns in the genitive case (Rule 12c(3)). When in the future a new representative of the genus will be described with an adjective as specific epithet (Rule 12c(1)), the authors should choose the gender of the genus and propose an adjectival epithet that agrees in gender with the generic name. A similar case is *Marivivens*. Its authors did not state the gender. As the type species was named *Marivivens donghaensis*, the gender can be masculine or feminine (for neuter, *donghaense* would have been correct). The epithet of the second species of the genus, *Maribacter niveibacter*, is a noun in apposition in the nominative case (Rule 12c(2)), which does not need to agree in gender with the generic name.

The orthography appendix of the ICNP (Appendix 9; Trüper and Euzéby 2009) provides useful guidelines for the naming of new prokaryotes. Section A shows how compound names, generic names as well as specific epithets, are formed by combining two or more words or word elements of Latin and/or Greek origin. Examples for generic names (two word elements) are *Staphylococcus* (Greek *staphyle*, "bunch of grapes"; Greek *kokkos*, "grain"), (three word elements) *Flavihalobacter* (Latin *flavus*, "yellow"; Greek *hals*, genitive *halos*, "salt"; Neo-Latin *bacter*, "rod"), and (four word elements) *Ectothiorhodosinus* (Greek *ektos*, "outside"; Greek *theion*, "sulfur"; Greek *rhodon*, "rose"; Latin *sinus*, "bend"); examples for specific epithets are (two word elements) *carniphilus* (Latin *caro*, genitive *carnis*, "meat"; Gr. masc. adj. *philos*, "loving") and (three word elements) *psychromarinicola* (Greek *psychros*, "cold"; Latin *marinus*, "marine"; Latin *–cola,* "inhabitant"). The rules to form such compound names are as follows:

- Except for the last word element, the word stems are used; the stems of Greek and Latin words are generally found in the genitive case. Therefore, we have *Flavihalobacter* and not *Flavihalsobacter*. Salt in Greek is *hals*, the genitive is *halos*, and therefore the stem is *halo–*. "Meat" in Latin is *caro*, genitive *carnis*, and therefore we have *Carnimonas* and *Vagococcus carniphilus* and not *Carimonas* and *cariphilus*. In the past this guideline was not always implemented. "Blood" in Greek is αἷμα, genitive αἵματος, transliterated as *haema*, *haematos*, stem *haemato–*. Therefore, we have the genus *Haematobacter* and the species *Phenylobacterium haematophilum*, but we also have *Haemobartonella* ("the blood-inhabiting *Bartonella*") and *Mycoplasma haemocanis* ("of dog blood"), the preferred forms being *Haematobartonella* and *haematocanis*. "Blood" in Latin is *sanguis*, genitive *sanguinis*, stem: *sanguini–*. Therefore, we have *Photobacterium sanguinicancri* ("of crab blood"). Instead of *Sanguibacter*, *Sanguinibacter* would be the preferred form. Sometimes, the stem of the word as found in the genitive case is very different from the nominative case. Therefore, we have *Mycoplasmopsis bovigenitalium* ("of bovine genitalia"), from Latin *bos* ("cow," "cattle"), genitive *bovis*, stem: *bovi–*.
- For compound names that contain a noun or adjective in a nonfinal position, the connecting vowel is *-i-* if the preceding word element is of Latin origin and *-o-* if the preceding word element is Greek. Therefore, we have in the above-cited examples *staphylo–*, *halo–*, *ecto–*, and *psychro–* derived from Greek and *flavi–*, *carni–*, and *bovi–* from Latin words. In the past, this rule was not always applied consistently, as shown by the abovementioned *Vagococcus* (from the Latin adjective *vagus*, "wandering"), *Flavobacterium* (from the Latin adjective *flavus*, "yellow"), and *Lactobacillus* (from the Latin noun *lac*, genitive *lactis*, "milk").

Lactivibrio and *Lactiplantibacillus* are examples of genera having *-i-* as the connecting vowel. The Greek equivalent of the Latin *lac*, genitive *lactis*, is *gala*, genitive *galaktos*, used in the genus name *Galactobacter*. In Sect. 2.11, an exception to this rule is discussed: in inorganic chemistry, the vowels "i" and "o" are used to indicate oxidation levels of cations, e.g., ferro and ferri for divalent and trivalent iron; both forms can be used in compound names where appropriate.

- Greek is more flexible than Latin about the connecting vowel, and other connecting vowels than *-o-* may be used if a precedent is found in Greek. An example is the genus *Corynebacterium*, from the Greek noun *koryne* ('club'); the word *korynephoros* ("club-bearing") existed in classical Greek.
- No connecting vowel is added when the following word element starts with a vowel (a, e, i, o, u). Thus, we have *Salinarchaeum* ("archaeon from salt works"; from the Latin *salinae*, "salterns"; and Neo-Latin *archaeum*, "archaeon") and *Lactovum* ("egg from milk," from the Latin *lac*, genitive *lactis*, "milk," and the Latin noun *ovum*, "egg"). Examples of specific epithets are *Methanothermobacter thermautotrophicus* ("thermophilic and autotrophic," from Greek *thermos*, "hot"; Greek *autos*, "self"; and Greek *trophikos*, "nursing," "tending") and *Streptococcus ursoris* ("of the mouth of a bear," from Latin *ursus,* "bear," and Latin *os*, genitive *oris*, "mouth"). Exceptions to this guideline can be made for well-established word elements from chemistry or physics such as *thio–* (from Greek *theion*, "sulfur," and *radio–*). In word components such as *allo–*, *bio–*, *geo–*, *halo–*, *hetero–*, *iso–*, *meso–*, *neo–*, *macro–*, *micro–*, etc., the connecting vowel *-o-* may be retained for reasons of clarity or of previous usage when the following component begins with a vowel. In the case of *halo–* (from Greek *hals*, genitive *halos*, "salt"), use of the connecting vowel *-o-* was inconsistent. No connecting vowel was used in names such as *Haladaptatus* ("adapted to salt"), *Halalkalicoccus* ("a coccus living in salty alkaline environments"), and *Isoptericola halalbus* ("salty and white"). The connecting vowel was used in *Haloarcula* ("salt small box"), *Haloactinomyces* ("a salt-loving actinomycete"), and *Longimonas haloalkaliphila* ("salt and alkali loving"). This is also the case for *Sphingomonas haloaromaticamans* ("loving haloaromatic compounds"), following the use in chemistry.
- Greek prepositions and prefixes such as *meta* ("besides") and *para* ("besides," "near," "like") are not followed by a connecting vowel in names such as *Metamycoplasma* and *Paracoccus*. When Greek prepositions and prefixes that end in a vowel such as *epi* ("on"), *meta*, or *para* are attached to word elements that begin with a vowel, the final vowel is elided, for example, *Eperythrozoon*, *Paralcaligenes*, *Parendozoicomonas*, and *Vibrio metoecus* (Oren et al. 2019). In the case of *para*, this was not always implemented as shown for *Paraeggerthella*, *Paraoerskovia*, *Keratinibaculum paraultunense*, and *Streptococcus parauberis*. However, for the prepositions *peri* ("around," "nearby") and *pro* ("before"), the final vowel is not elided, for example, *Fusobacterium periodonticum*. This is also the case for *amphi* ("around"), *ana* ("up," "upon"), *anti* ("opposite"), *apo* ("from"), *dia* ("through"), *epi* ("on"), and *hypo* ("under"). Prepositions formed from Greek adjectives (e.g., *poly–*, "many," and *mega–*, "big") and adverbs (e.g., *exo–*, "outside," and *eu–*, "good") also do not elide, for example, *Polyangium* and

Clostridium polyendosporum. Latin prepositions and prefixes are not followed by a connecting vowel. When Latin prepositions and prefixes that end in a vowel (e.g., *intra*, "inside," *extra*, "outside") are attached to word elements that begin with a vowel, the final vowel is not elided, conforming to the usage in classical Latin. I did not find examples in the prokaryotic nomenclature; an example from the zoological nomenclature is *Amblyomma extraoculatum*.

- Adverbs are rarely used in compound names. For Latin adverbs such as *paene*, "almost," the connecting vowel *-i-* may be used, for example, *Paenibacillus*. The connecting vowel is dropped if the following word element starts with a vowel, for example, *Paenalcaligenes*, *Paenochrobactrum*. In *Asticcacaulis benevestitus*, the adverb *bene* ("well") was used without a connecting vowel.

2.3 Basic Latin Grammar as Relevant to Nomenclature

As shown in Principle 3 and Rule 6 of the ICNP, prokaryotic nomenclature is based on Latin. A basic understanding of Latin vocabulary and grammar is therefore essential for naming new taxa of prokaryotes. The rules of the ICNP mention numerous grammatical terms. Examples (printed in boldface) are as follows: (Rule 7) the name of a taxon above the rank of genus up to and including order is a **noun** or an **adjective** used as a **noun** of Latin or Greek origin, or a latinized word. It is in the **feminine gender**, the **plural** number …. (Rule 8) The name of a class is in the **neuter gender** and is formed by the addition of the suffix *–ia* to the **stem** of the name of the type genus of the type order of the class. (Rule 10a) The name of a genus or subgenus is a **noun**, or an **adjective** used as a **substantive**, in the **singular number** in the **nominative case**. (Rule 12c) … A specific epithet must be treated in one of the three following ways: (1) as an **adjective** that must agree in **gender** with the generic name, (2) as a **noun in apposition** in the **nominative case**, and (3) as a **noun** in the **genitive case**. Therefore, familiarity with the meaning of terms such as "substantive," "adjective," "gender" (masculine, feminine, neuter), "case" (nominative, genitive), and "number" (singular, plural) is essential. One does not need to master all the intricacies of Latin grammar to name new taxa. Scholarly textbooks of Latin grammar such as Greenough et al. (1903) can be helpful, but most information found there is of little relevance to prokaryotic nomenclature. The delightful essay by Thomas MacAdoo (1993) on *Nomenclatural Literacy* is highly recommended. It was written by a classics scholar who was not a microbiologist but had a thorough understanding of the way prokaryotic nomenclature works. It was he who proposed the name *Blautia obeum* discussed above.

Latin is an inflected language, i.e., a language that changes the form or ending of many words when the way in which they are used in sentences changes. The inflection of nouns, adjectives, pronouns, and participles to denote gender, number, and case is called declension. For the purpose of naming microorganisms, it is not necessary to master the complete system of declensions for nouns and adjectives. For nouns, only three forms are relevant as shown in Rules 10a and 12c of the ICNP – nominative singular, genitive singular, and genitive plural – for each of the three genders: masculine, feminine, and neuter. Nominative plural forms are only used for naming the higher taxa above the rank of genus, as shown in Rules 7 and 8. The other cases in Latin (dative, accusative, ablative, vocative, locative) are not relevant to biological nomenclature. For adjectives, one only needs the nominative singular form. The complicated conjugation of Latin verbs has little impact on the naming of taxa, but a basic understanding is needed for the formation of present and past participles of the verb that are often used in generic names and in specific epithets.

2.3.1 The Five Declensions of the Latin Noun

First Declension

Latin nouns of the first declension end in *–a*. The genitive singular ends in *–ae* and the genitive plural in *–arum*. Most Latin nouns that end in *–a* have the feminine gender, but there are exceptions such as *nauta* ("sailor") and names of male persons such as *Seneca*. The Latin noun *incola* ("inhabitant"), used in many compound generic names and specific epithets, is another example.

Examples of generic names that end in *–a* and are based on Latin nouns of the first declension are *Aquicella* ("a cell from water"; *cella*, "chamber," "cell"), *Oceanivirga* ("a rod from the ocean"; *virga*, "twig," "rod"), and *Muricauda* (*Allomuricauda*) ("mouse tail"; *cauda*, "tail"). Genera named to honor famous microbiologists are formed as feminine nouns of the first declension: *Escherichia* for Theodor Escherich, *Nocardia* for Edmond Nocard, etc. Examples of specific epithets formed as nouns of the first declension in the genitive case (Rule 12c(3)) are *Lactobacillus oryzae* ("of rice"; *oryza*, "rice"), *Mycobacterium vaccae* ("of a cow"; *vacca*, "cow"), and *Roseibium aquae* ("of water"; *aqua*, "water").

Not all generic names of prokaryotes that end in *–a* are based on Latin feminine words of the first declension. Some are derived from Greek, have the neuter gender, and belong to the Greek third declension. Examples of Greek neuter nouns ending in *–a* used in prokaryotic names are *derma* ("skin"), *nema* ("thread"), *plasma* ("something formed or molded," "a form"), and *soma* ("body"). Their genitives in Greek end in *–atos*, latinized to *–atis*. Therefore, we have *Pasteurella stomatis* (*stoma*, "mouth"), *Neisseria zoodegmatis* ("of an animal's bite"; *degma*, "bite"), *Mycobacterium smegmatis* (*smegma*, "cleansing medicine," "detergent," a word also used in classical Latin), and *Chlamydia trachomatis* (from *trachysma*, "roughness," from which the Neo-Latin word *trachoma* was formed).

As the stems of Greek and Latin words are generally found in the genitive case, the names of the genus *Dermatophilus* and the orders *Brevinematales* and *Mycoplasmatales* were correctly formed, based on stems ending in -at-. However, the name *Spirosomaceae*, derived from the generic name *Spirosoma*, is malformed. The corrected version *Spirosomataceae* is given in the List of Prokaryote names with Standing in Nomenclature (LPSN 2022), but correction of the name was never formally proposed. There are more such malformed names in the 1980 Approved Lists (Skerman et al. 1980) where authors may have confused Greek nouns of the third declension ending in *–a* with Latin nouns of the first declension: *Bactoderma alba* should have been *Bactoderma album*, and *Bactoderma rosea* should be *Bactoderma roseum*. A similar error was made in the description of the cyanobacterium *Phyllonema ansata* under the provisions of the International Code of Nomenclature for algae, fungi, and plants. Instead of the feminine adjective *ansata*, the neuter form *ansatum* ("with loops," "having a handle") would have been correct.

Second Declension

Most nouns of the second declension end in *–us* and are masculine or end in *–um* and have the neuter gender. The genitive singular ends in *–i* and the genitive plural in *–orum*. Examples of masculine generic names are *Bacillus* ("small staff"), *Streptococcus* ("a pliant coccus," the Neo-Latin *coccus* being derived from the Greek masculine noun *kokkos*, "grain," "seed"), and *Ferroglobus* ("iron ball," from the Latin masculine noun *globus*, "ball"). Neuter generic names that end in *–um* are, e.g., *Desulfobaculum* ("sulfate-reducing rod," from the Latin neuter noun *baculum*, "rod"), *Mycobacterium* ("fungus rodlet," from the Neo-Latin neuter noun *bacterium*, derived from the Greek neuter diminutive noun *bakterion*, "small staff"), and *Geofilum* ("a thread from the Earth," from the Latin neuter noun *filum*, "thread"). Examples of specific epithets formed as masculine nouns of the second declension in the genitive case are *Acinetobacter equi* ("of a horse," from Latin *equus*, "horse"), *Alteromonas oceani* ("of the ocean," from Latin *oceanus*, "ocean"), and *Halobellus limi* ("of mud," from Latin *limus*, "mud"). Examples of specific epithets formed as neuter nouns of the second declension in the genitive case are *Kocuria soli* ("of soil," from Latin *solum*, "soil"), *Bacteroides luti* ("of mud," from Latin *lutum*, "mud"), and *Arenimonas metalli* ("of a mine," "of a quarry," from Latin *metallum*, "mine").

Not all nouns of the second declension end in *–us* or *–um*. Examples are *ager* ("field"), *puer* ("boy"), and *vir* ("man"). The genitive forms are *agri*, *pueri*, and *viri*, respectively. The specific epithet *agri* ("of a field") is used in many genera. Note that the epithet *pueri* used for species of *Paenibacillus* and *Pullulanibacillus* has nothing to do with the Latin noun but was derived from "puer," a kind of tea from China. Transliterated as "puerh," it was also used in the names *Puerhibacterium*, *Massilia puerhi*, and *Paenibacillus puerhi*. Not all nouns of the second declension that end in *–us* are masculine. The word *humus*, genitive *humi* ("earth," "soil"), is feminine, and so is *abyssus* ("abyss," "the sea," derived from the feminine Greek noun *abyssos*); *alvus* ("belly," "the bowels") was generally used as feminine. The Neo-Latin noun *rhabdus* ("rod") is also feminine as it is derived from the Greek feminine noun *rhabdos* (Euzéby and Boemare 2000).

The Latin names of many trees that end in *–us* are feminine: *Alnus* ("alder"), *Corylus* ("hazel tree"), *Malus* ("apple"), *Pinus* ("fir"), *Populus* ("poplar"), *Prunus* ("plum tree"), *Sambucus* ("elder tree"), *Ulmus* ("elm"), and many more. We find these words as nouns in the genitive case: *Frankia alni*, *Streptomyces coryli*, *Sphingomonas mali*, *Paenibacillus pini*, *Kurthia populi*, *Sphingomonas pruni*, and *Agromyces ulmi*. The name *Glycomyces sambucus* is malformed; correct would be *sambuci*. A few words of the second declension that end in *–us* have the neuter gender. Examples are *pelagus* ("the sea," from the Greek neuter noun *pelagos*) and *virus* ('poison').

Not all Latin nouns that end in *–us* belong to the second declension. All masculine and feminine nouns of the fourth declension end in *–us*, and so do a few nouns of the third declension of the neuter gender. Examples relevant to the prokaryotic nomenclature are *glomus*, genitive *glomeris* ("ball"), and *frigus*, genitive *frigoris* ("coldness"). Thus, we have the genera *Dictyoglomus* ("net ball") with species *Dictyoglomus thermophilum* and *Dictyoglomus turgidum* and *Desulfofrigus* ("sulfate reducer living in the cold") with species *Desulfofrigus fragile* and *Desulfofrigus oceanense*. Note the neuter endings *–um* and *–e* of the specific epithets.

Third Declension

Nouns of the third declension end in *–a*, *–e*, *–i*, *–o*, *–y*, *–c*, *–l*, *–n*, *–r*, *–s*, *–t*, or *–x*. The genitive singular ends in *–is* and the genitive plural in *–um* or *–ium*. Nouns of the third declension can have any gender, and words are declined differently, depending on the stem and the gender. Textbooks of Latin grammar provide declension tables for nouns with stems with different endings. There also are irregular nouns such as *nix*, genitive *nivis* ("snow"); *caro*, genitive *carnis* ("meat"); and *bos*, genitive *bovis*, genitive plural *boum* ("cow," "cattle"). The following examples show how nouns of the third declension and their stems are used in the prokaryotic nomenclature.

Masculine

aries, genitive *arietis* ("ram"): *Seminibacterium arietis* ("of a ram")
caput, genitive *capitis* ("head"): *Staphylococcus capitis* ("of the head")
ignis, genitive *ignis* ("fire"): *Ignisphaera* ("fireball")
lapis, genitive *lapidis* ("stone"): *Agromyces lapidis* ("of a stone"), *Tessaracoccus lapidicaptus* ("stone-captive")
leo, genitive *leonis* ("lion"): *Pseudophaeobacter leonis* ("of a lion"), *Mycoplasma leonicaptivi* ("of a captive lion")
rex, genitive *regis* ("king"): *Amycolatopsis regifaucium* ("of Kings Canyon, Australia")

Feminine

caro, genitive *carnis* ("meat"): *Carnimonas* ("a monad of meat"), *Clostridium carnis* ("of meat")
feles, genitive *felis* ("cat"): *Chlamydia felis* ("of a cat")
nix, genitive *nivis* ("snow"): *Herbinix* ("snow-white of plants"), *Hymenobacter nivis* ("of snow")

radix, genitive *radicis*, genitive plural *radicium* ("root"): *Paenibacillus radicis* ("of a root"), *Actinoallomurus radicium* ("of roots")
urbs, genitive *urbis* ("city"): *Bacillus urbisdiaboli* ("of Devil City"), *Oxalicibacterium solurbis* ("of city soil")

Masculine or Feminine

bos, genitive *bovis*, genitive plural *boum* ("cattle"): *Corynebacterium bovis* ("of a cow"), *Bifidobacterium boum* ("of cattle")
canis, genitive *canis* ("dog"): *Enterococcus canis* ("of a dog"), *Capnocytophaga canimorsus* ("of a dog bite")
sus, genitive *suis*, genitive plural *suum* ("pig"): *Suicoccus* ("a coccus from a hog"), *Mycoplasma suis* ("of a pig"), *Paracoccus suum* ("of pigs")

Neuter

aequor, genitive *aequoris* ("the calm sea"): *Aequorivita* ("living at the sea surface"), *Litorivivens aequoris* ("of the sea")
animal, genitive *animalis*, genitive plural *animalium* ("animal"): *Neisseria animalis* ("of an animal"), *Moraxella pluranimalium* ("of many animals")
corpus, genitive *corporis* ("body"): *Prevotella corporis* ("of the body")
mare, genitive *maris* ("sea"): *Dietzia maris* ("of the sea"), *Bermanella marisrubri* ("of the Red Sea")
os, genitive *ossis* ("bone"): *Psychromonas ossibalenae* ("of a bone of a whale")

Fourth Declension

Nouns of the fourth declension end in *–us* (masculine, feminine) or *–u* (neuter). There only are four neuter nouns of the fourth declension ending in *–u* in the Latin lexicon, and these have never been used in names of prokaryotic taxa. The stems end in *–u*, the genitive singular ending is *–us*, and the genitive plural ending is *–uum*. Most words ending in *–us* are masculine; among the few exceptions are names of trees: *quercus* ("oak tree"), *ficus* ("fig tree"), and *pinus* ("pine tree"). Here are examples of names of prokaryotes that include nouns of the fourth declension:

abortus, genitive *abortus* ("miscarriage"): *Chlamydia abortus* ("of a miscarriage")
abscessus, genitive *abscessus* ("abscess"): *Mycobacteroides abscessus* ("of an abscess")
aestus, genitive *aestus* ("the tide of the sea"): *Alcaligenes aestus* ("of the tide"). When in 1983 this taxon was transferred to the genus *Deleya*, the authors apparently misunderstood the grammatical context of the specific epithet, considering it to be an adjective of the first/second declension. They therefore described *Deleya aesta*; correct would have been *Deleya aestus*.
arcus, genitive *arcus* ("arc"): *Nitratiruptor tergarcus* ("of a back arc basin," a geological term for submarine features associated with island arcs and subduction zones)
currus, genitive *currus* ("chariot"): *Nocardioides currus* ("of a vehicle")

fructus, genitive *fructus* ("fruit"): *Fructilactobacillus* ("fruit-loving lactobacillus"), *Alicyclobacillus fructus* ("of fruit")
lacus, genitive *lacus* (but the genitive *laci* is also found in classical Latin) ("lake"): *Lacimicrobium* ("a lake microbe"), *Rhodobacter lacus* ("of a lake"), *Polynucleobacter campilacus* ("of field lake," referring to Lake Feldsee in Germany), *Pelagibacterium montanilacus* ("of a mountain lake"), *Flavobacterium lacicola* ("dweller of lakes"), *Halobiforma lacisalsi* ("of a salt lake"), *Alkalitalea saponilacus* ("of Soap Lake, Washington, USA"), *Amantichitinum ursilacus* ("of Bear Lake" (Bärensee) nearby Stuttgart, Germany)
morsus, genitive *morsus* ("a bite"): *Capnocytophaga canimorsus* ("of a dog bite"), *Vandammella animalimorsus* ("of an animal bite")
portus, genitive *portus* ("harbor," "port"): *Portibacter* ("a rod from a harbor"), *Alteromonas portus* ("of a harbor"), *Phaeobacter porticola* ("inhabitant of a harbor")
quercus, genitive *quercus* ("an oak tree"): *Paenibacillus quercus* ("of an oak tree")
sinus, genitive *sinus* ("bent surface," "bay"): *Oribacterium sinus* ("of the sinus"), *Nocardioides sinuspersici* ("of the Persian Gulf'), *Flexistipes sinusarabici* ("of the Arabian Gulf')

The feminine nouns *ficus* ("fig tree") and *pinus* ("pine tree") can be declined using the second declension (genitive *fici*, *pini*) or the fourth declension (genitive *ficus*, *pinus*). The fourth declension was used in *Deinococcus ficus* ('of a fig tree') and the second declension in *Paenibacillus pini* and *Streptomyces pini* ("of a pine tree").

Fifth Declension
Nouns of the fifth declension end in *–es*, the genitive in *–ei*, and the genitive plural in *–erum*. The Latin dictionary has only few nouns of the fifth declension. Three, both of the feminine gender, were thus far used in the prokaryotic nomenclature:

egeries, genitive *egeriei* ("excrement," "dung"): *Candidatus* Egerieimonas ("a form from excrement"), *Candidatus* Egerieicola ("an inhabitant of excrement")
facies, genitive *faciei* ("shape," "face"): *Candidatus* Pellibacterium faciei ("of a face")
glacies, genitive *glaciei* ("ice"): *Glacieibacterium* ("a bacterium from ice"), *Planococcus glaciei* ("of ice"), *Hymenobacter glacieicola* ("inhabitant of ice"), *Nocardioides glacieisoli* ("of glacier soil"). Note that the genus names *Glaciibacter*, *Glaciihabitans*, *Glaciimonas*, and *Glaciecola* were incorrectly formed: correct would be *Glacieibacter*, *Glacieihabitans*, *Glacieimonas*, and *Glacieicola*, respectively.

Nouns in the nominative case in the plural number are used in the prokaryotic nomenclature only for names of taxa above the rank of genus. Names of phyla and classes end in *–ota* and *–ia*, respectively, and have the neuter gender; names of subclasses, orders, suborders, families, and tribes end in *–idea*, *–ales*, *–ineae*, *–aceae*, and *–eae*, respectively, and have the feminine gender (Rules 7 and 8 of the ICNP), for example, *Pseudomonadota*, *Flavobacteriia*, *Sphaerobacteridae*, *Clostridiales*, *Caulobacterineae*, *Halobacteriaceae*, and *Bacteroidaceae*.

Nouns in the genitive case in the plural number used as specific epithets end in *–arum* (first declension), *–orum* (second declension), or *–um* or *–ium* (third declension). Below are a number of examples. No examples are yet found in the nomenclature of words of the fourth declension (genitive plural ending *–uum*) or the fifth declension (genitive plural ending *–erum*).

First Declension

Cohnella algarum ("of algae"; from *alga*, "seaweed")
Aeromonas bestiarum ("of beasts"; from *bestia*, "beast")
Agrobacterium fabacearum ("of the botanical family *Fabaceae*")
Pedobacter quisquiliarum ("of refuse"; from *quisquiliae*, "refuse")
Rhodobacter thermarum ("of hot springs"; from *thermae*, "hot springs")
Ralstonia solanacearum ("of the botanical family *Solanaceae*")

Second Declension

Helicobacter equorum ("of horses"; from *equus*, "horse")
Microbacterium foliorum ("of leaves"; from *folium*, "leaf"). Note: the name *Marinomonas foliarum* ("of leaves") was probably a typographic error.
Bacillus fungorum ("of mushrooms"; from *fungus*, "mushroom")
Aeromonas molluscorum ("of the zoological phylum Mollusca")
Bergeyella porcorum ("of pigs"; from *porcus*, "pig")
Tamlana sedimentorum ("of sediments"; from *sedimentum*, "sediment")

Third Declension

Actinobacillus anseriformium ("of the zoological order Anseriformes")
Brevibacterium avium ("of birds"; from *avis*, "bird")
Bifidobacterium boum ("of cattle"; from *bos*, "cow")
Bifidobacterium dentium ("of teeth"; from *dens*, genitive *dentis*, "tooth")
Enterococcus faecium ("of feces"; from *faex*, genitive *faecis*, "feces")
Lactobacillus florum ("of flowers"; from *flos*, genitive *floris*, "flower")
Mycoplasmoides genitalium ("of genitals"; from the Latin adjective *genitalis*, "genital")
Lactococcus piscium ("of fishes"; from *piscis*, "fish")
Streptococcus pluranimalium ("of many animals"; from *animal*, genitive *animalis*, "animal")
Bifidobacterium primatium ("of the zoological order Primates")
Actinoallomurus radicium ("of roots"; from *radix*, genitive *radicis*, "root")
Necropsobacter rosorum ("of rodents"; *rosor*, genitive *rosoris*, "rodent," from the verb *rodo*, "to gnaw")
Oscillibacter ruminantium ("of the zoological suborder Ruminantia")
Paracoccus suum ("of pigs"; from *sus*, genitive *suis*, "pig")
Chryseobacterium viscerum ("of the viscera"; from *viscera*, "the internal organs")

2.3.2 Adjectives

Rule 12c(1) regulates the way in which adjectives are used as specific epithets: in the singular number in the nominative case, the gender agreeing with the gender of the generic name.

Latin adjectives either follow a combination of the first and the second declension or the third declension. Some adjectives of the third declension have three different terminations for the masculine, feminine, and neuter gender, others have two, and still others have the same form for all genders.

First and Second Declension

longus ("long"): *Erythrobacter longus* (masculine), *Salisaeta longa* (feminine), *Acetonema longum* (neuter)

ruber ("red"): *Salinibacter ruber* (masculine), *Halovenus rubra* (feminine), *Rhodospirillum rubrum* (neuter)

Ordinal adjectives used for enumeration also belong to this category; they may include numbers up to ten (Recommendation 12c(6)):

primus ("the first"): *Cephaloticoccus primus* (not validly published; masculine), *Mesotoga prima* (feminine)

secundus ("the second"): *Coprobacter secundus* (masculine), *Parasutterella secunda* (feminine)

tertius ("the third"): *Clostridium tertium* (neuter)

sextus ("the sixth"): *Candidatus* Euplotella sexta (feminine)

Third Declension, Three Terminations

celer ("quick"): *Thermococcus celer* (masculine), *Aliidiomarina celeris* (feminine), *Brevibacterium celere* (neuter)

Third Declension, Two Terminations

agilis ("quick"): *Achromobacter agilis* (masculine), *Massilia agilis* (feminine), *Anaerofilum agile* (neuter)

Third Declension, One Termination

fallax ("deceptive"): *Pyxidicoccus fallax* (masculine), *Leptolyngbya fallax* (feminine), *Clostridium fallax* (neuter)

Adjectives can be used in the positive, the comparative, and the superlative degree. The positive is the base form, comparative is a higher degree ("more"), superlative is the highest ("most"). The endings of the comparative form are *–ior* (masculine, feminine) and *–ius* (neuter):

crassus ("thick"), comparative *crassior* ("thicker"): *Leptolyngbya crassior* (feminine)

A few adjectives have irregular comparative and superlative forms, e.g., *magnus* ("big"), *maior* ("bigger"), *maximus* ("biggest"), *parvus* ("small"), *minor* ("smaller"), *minimus* ("smallest"), *bonus* ("good"), *melior* ("better"), and *optimus* ("best"). Therefore, we have

maior ("bigger"): *Flectobacillus major* (masculine); instead of the classical Latin spelling *maior*, the authors used *major*.
minor ("smaller"): *Weissella minor* (feminine)
melior ("better"): used in the compound generic name *Melioribacter* ("a superior rod")

The superlative form generally ends in *–issimus* (masculine), *–issima* (feminine), –and *issimum* (neuter):

brevis ("short"), superlative *brevissimus* ("the shortest"): *Calothrix brevissima* (feminine)
caldus ("hot"), superlative *caldissimus* ("the hottest"): *Saccharolobus caldissimus* (masculine)
fulvus ("yellow"), superlative *fulvissimus* ("the yellowest"): *Streptomyces fulvissimus* (masculine)
minutus ("small"), superlative *minutissimus* ("the smallest"): *Allochromatium minutissimum* (neuter)
musocus ("slimy"), superlative *mucosissimus* ("the slimiest"): *Sphingomonas musosissima* (feminine)
subtilis ("thin"), superlative *subilissimus* ("the thinnest"): *Leptolyngbya subtilissima* (feminine)

For adjectives that end in *–er*, the ending *–rimus* (*–rima*, *–rimum*) is added to form the superlative:

tener ("delicate"), superlative *tenerrimus* ("the most delicate"): *Trichocoleus tenerrimus* (masculine)

Irregular forms:

maximus ("the biggest"): *Sarcina maxima* (feminine)
minimus ("the smallest"): *Streptococcus acidominimus* (masculine), probably intended to mean that this organism produces the least amount of acid

According to Rule 10a of the ICNP, adjectives can be used as nouns to name genera. Rare examples where Latin or Greek adjectives were thus used as a genus names are *Formosa* ("beautiful") and *Amorphus* (from Greek *amorphos*, "without shape"). However, there are many examples of compound generic names whose last word element is an adjective:

First and Second Declension
Masculine
Cupriavidus (*avidus*, "eager for," "loving"), *Halobellus* (*bellus*, "beautiful"), *Methanoplanus* (*planus*, "flat"), *Halopelagius* (*pelagius*, "pertaining to the sea"), *Roseovarius* (*varius*, "diverse"), *Salipiger* (*piger*, "lazy")

Feminine
Natrialba (*albus*, "white")

Neuter
Methylorubrum (*ruber*, "red"), *Methanosalsum* (*salsus*, "salted")

Third Declension
Masculine
Methylobrevis (*brevis*, "short")

Feminine
Halorientalis (*orientalis*, "from the east")

Neuter
Mariniflexile (*flexilis*, "flexible," "pliant")

When forming generic names from adjectives, authors may choose the gender. Several Latin adjectives have been used in different genders in generic names:

(*caldus*, "hot"): *Acidicaldus* (masculine), *Methylocaldum* (neuter)
(*marinus*, "marine"), *Roseimarinus* (masculine), *Flavimarina* (feminine), *Methylomarinum* (neuter)
(*profundus*, "deep"): *Mariprofundus* (masculine), *Thioprofundum* (neuter)
(*curvus*, "bent"): *Desulfocurvus* (masculine), *Candidatus* Lachnocurva (feminine), *Sulfuricurvum* (neuter)

For adjectives of the third declension with one termination, authors must indicate the gender. The same adjective can be used in different genders:

ferax ("fertile"): *Rhodoferax* (masculine), *Propioniferax* (feminine), *Haloferax* (neuter)
vorax (voracious): *Variovorax* (masculine), *Herbivorax* (feminine), *Proteinivorax* (neuter)

2.3.3 Verbs and Participles

Latin verbs do not qualify as generic names and specific epithets based on Rules 10a and 12c of the ICNP. However, present participles and past participles, used as adjectives, are used in many names. Participles were also used as nouns in a number of generic names. Therefore, basic knowledge of the Latin verb is needed to correctly form participles.

The inflection of verbs is called conjugation. Latin has four conjugations. A full understanding of the complicated conjugation systems found in textbooks of Latin grammar is not required for the correct formation of names. The following table shows a few examples of verbs used as participles in prokaryotic names.

	Infinitive	First person singular present indicative	Present participle	Past participle
First conjugation	*vorare* ("to devour")	*voro*	*vorans*	*voratus* (*–a*, *–um*)
	aggregare ("to add to something")	*aggrego*	*aggregans*	*aggregatus* (*–a*, *–um*)
Second conjugation	*miscere* ("to mix")	*misceo*	*miscens*	*mixtus* (*–a*, *–um*)
	inhibere ("to inhibit")	*inhibeo*	*inhibens*	*inhibitus* (*–a*, *–um*)
Third conjugation	*flectere* ("to bend")	*flecto*	*flectens*	*flectus* (*–a*, *–um*)
	facere ("to make")	*facio*	*faciens*	*factus* (*–a*, *–um*)
Fourth conjugation	*bullire* ("to boil," "to bubble")	*bullio*	*bulliens*	*bullitus* (*–a*, *–um*)

Latin present participles end in *–ns* (compare the *–ing* ending for such words in English): *–ans* for verbs of the first conjugation, *–ens* for the second conjugation, *–ens or –iens* for the third conjugation, and *–iens* for the fourth conjugation. Present participles have the same ending in the three genders. They are declined (e.g., *necans* ("killing"), genitive *necantis*). This is relevant for naming of higher taxa by addition of the appropriate suffix to the stem of the type genus: the family name *Calderihabitantaceae* was formed from the generic name *Calderihabitans* ("inhabiting a caldera"), and the *Candidatus* Methanoperdentaceae family was formed from the generic name *Candidatus* Methanoperedens ("methane consuming").

Here are examples of present participles used as adjectives in specific epithets:

First Conjugation

Bacillus coagulans (from the verb *coagulare*, "to cause a fluid to curdle")

Second Conjugation

Lactovum miscens (from the verb *miscere*, "to mix"; the epithet was chosen to indicate a mixed fermentative metabolism)

Third Conjugation

Phaseolibacter flectens (from the verb *flectere*, "to bend")
Serratia liquefaciens (from the verb *liquefacere,* "to make liquid")
Frateuria defendens (from the verb *defendere*, "to protect," as it protects plants from pathogens)

Fourth Conjugation

Enterococcus bulliens (from the verb *bullire*, "to bubble")

Present participles used as adjectives are also found in generic names as shown by the following examples:

First Conjugation

Agarivorans ("agar eater," from the verb *vorare*, "to devour")
Aquihabitans ("an inhabitant of freshwater," from the verb *habitare*, "to inhabit")

Third Conjugation

Marivivens ("an organism living in the sea," from the verb *vivere*, "to live")

Nonlabens ("a non-gliding organism," from the verb *labori* (a deponent verb that only exists in the passive form), "to glide")

Saccharedens ("a sugar-eating organism"; from the verb *edere*, "to eat")

Latin perfect participles end in *–tus* (masculine), *–ta* (feminine), *–tum* (neuter), or *–sus* (*–sa*, *–sum*) (compare the *–ed* ending for such words in English). They are declined as adjectives of the first and second declension:

First Conjugation

Desulfobacter curvatus ("curved"; from the verb *curvare*, "to bend")

Geodermatophilus siccatus ("dried"; from the verb *siccare*, "to dry")

Second Conjugation

Telluria mixta ("mixed"; from the verb *miscere*, to mix; referring to mixed flagellation)

Achromobacter insolitus ("uncommon"; from the verb *solere*, "to be accustomed")

Third Conjugation

Halorubrum depositum ("deposited"; from the verb *deponere*, "to deposit")

Methyloceanibacter superfactus (referring to maximal cell densities; from the verb *facere*, "to make," "to produce")

Hymenobacter perfusus ("soaked"; from the verb *perfundere*, "to moisten")

Fourth Conjugation

Asticcacaulis benevestitus ("well-clad"; from the verb *vestire*, "to dress," "to clothe")

Past participles used as adjectives are sometimes used as generic names:

First Conjugation

Pectinatus ("a combed bacterium"; from the verb *pectinare*, "to comb")

Haladaptatus ("a bacterium adapted to salt"; from the verb *adaptare*, "to fit," "to adjust")

Luteipulveratus ("a bacterium forming white powdery aerial mycelium on yellow colonies"; from the verb *pulverare*, "to scatter with dust")

Second Conjugation

Mixta ("the mixed one," referring to the mixed lifestyles of species in the genus, from the verb *miscere*, "to mix")

Opitutus ("protected by Ops," from the deponent verb *tueor*, *tueri*, "to protect")

Third Conjugation

Candidatus Sabulitectum ("covered by sand," from the verb *tegere*, "to cover")

A related grammatical form is the gerundive: an adjective implying obligation or necessity ("ought" or "must" in English). Gerundive adjectives end in *–ndus* (masculine), *–nda* (feminine), and *–ndum* (neuter). Compare words such as "corrigendum" (from the verb *corrigere*; "to set straight": "a thing that must be corrected") and "agenda" (from the verb *agere*; "to do": "things that must be done"; *–nda* being here the neuter plural ending). Gerundive forms of verbs of the first conjugation are found in epithets: *Thermoactinomyces mirandus* ("to be admired"; from *mirare*, "to wonder") and *Micavibrio admirandus* ("to be admired"; from the deponent verb *admirari*, "to admire").

2.4 Use of Greek in Prokaryotic Nomenclature

Since the days of Linnaeus, Greek holds a special position in biological nomenclature: *Nomina generica, quæ ex græca, vel latina lingua radicem non habent, rejicienda sunt* ("Generic names had do not have a root from the Greek or Latin language must be rejected.") (Rule 229, Linnaeus 1751). Principle 3 of the ICNP states that the scientific names of all taxa are Latin or latinized words … that are usually taken from Latin or Greek, and Recommendation 6(3) reads: "Words from languages other than Latin or Greek should be avoided as long as equivalents exist in Latin or Greek or can be constructed by combining word elements from these two languages." Many names of genera are composed only of Greek word elements, for example, *Staphylococcus*, *Ectothiorhodospira*, *Geodermatophilus*, and *Treponema*. In the context of the ICNP, "Greek" means classical Greek as spoken until the end of the Roman Empire (Oren and Schink 2020a).

Greek letters and letter combinations are transliterated in the Latin alphabet as shown in the following table, derived in part from MacAdoo (1993):

Greek	Transliteration	Greek	Transliteration	Greek	Transliteration
α	a	ηι, ᾐ	i	π	p
αι, ᾳ	ae	ϑ	th	ρ	r
ἁ	ha	ι	i, j	ῥ	rh
αυ	au	ἱ	hi	σ, ς	s
β	b	κ	c or k	τ	t
γ	g	λ	l	υ	y
δ	d	μ	m	ὑ	hy
ε	e	ν	n	φ	ph
ἑ	he	ξ	x	χ	ch
ει	ei or i	ο	o	ψ	ps
ευ	eu (ev before vowels)	ὁ	ho	ω	o
ζ	z	οι	oe	ωι, ᾠ	oe (rare)
η	e	ου	u, rarely ou		

The Greek κ (kappa) was generally transliterated to "c" in prokaryotic names, e.g., in *Caryophanon* from καρυον (*karyon*, "nut"); *Deinococcus actinosclerus* from κοκκος (*kokkos*, "grain," "seed"), ἀκτις (*actis*, "ray," "beam"), and σκληρος (*skleros*, "hard"); and *Microcystis* from μικρος (*mikros*, "small") and κυστις ("bladder"). The "k" spelling was rarely used. Examples are *Dokdonia eikasta* from εἰκαστος (*eikastos*, "comparable"), *Dokdonia genika* from γενικος (*genikos*, "principal," "typical"), *Acholeplasma hippikon* from ἱππικος (*hippikos*, "pertaining to the horse"), and *Planococcus okeanokoites* from ὠκεανος (*okeanos*, "ocean") and κοιτη (*koite*, "bed").

There are three declensions of Greek nouns, and similar to Latin, Greek has three genders. MacAdoo (1993) gave guidelines how nominative, genitive, and nominative plural endings of Greek nouns of the different declensions can be latinized. Words correctly formed in Greek do not necessarily make correct generic names or acceptable specific epithets as they must be treated as Latin (Rule 6 of the ICNP). More detailed explanations are beyond the scope of this chapter. Further information is found in a treatise on the use of Greek in prokaryotic nomenclature (Oren et al. 2016).

Below is a list of Greek word elements that were often used in names of prokaryotic genera and species.

Greek word (transliterated)	Examples
aktis, gen. *aktinos* ("ray," "beam")	*Actinomyces*, *Hymenobacter actinosclerus*
allos ("other")	*Allorhizobium*, *Paenibacillus allorhizosphaerae*
anthropos ("man")	*Anthropogastromicrobium*, *Chryseobacterium anthropi*
archaios ("ancient")	*Archaeoglobus*, *Candidatus* Velamenicoccus archaeovorus
arthron ("joint")	*Arthrobacter*, *Microbacterium arthrosphaerae*
baktron ("staff," "stick")	*Ochrobactrum*, *Myxobaktron*
bios ("life")	*Symbiobacterium*, *Frankia asymbiotia*
blaste ("bud," "sprout")	*Blastochloris*, *Fuscovulum blasticum*
brachys ("short")	*Brachybacterium*, *Eubacterium brachy* (correct Greek, but not treated as Latin)
bradys ("slow")	*Bradyrhizobium*, *Edaphobacter bradus*
chaite ("long hair")	*Spirochaeta*, *Halodesulfovibrio spirochaetisodalis*
chlamys ("mantle")	*Chlamydia*, *Coelosphaeropsis chlamydocystis*
chloros ("green")	*Chlorobium*, *Heliobacterium chlorum*
chroma, gen. *chromatos* ("color")	*Chromatocurvus*, *Nocardiopsis chromatogenes*
chros ("skin," "flesh")	*Chroococcidiopsis*, *Azotobacter chroococcum*
chryseos ("golden")	*Chryseobacterium*, *Flavobacterium chryseum*
derma, gen. *dermatos* ("hide," "skin")	*Dermatophilus*, *Arsenicoccus dermatophilus*
diction ("net")	*Pelodictyon*, *Geodermatophilus dictyosporus*
edaphos ("ground," "soil")	*Edaphobacter*, *Sphingomonas edaphi*
endon ("inside")	*Endomicrobium*, *Flavobacterium endophyticum*
enteron ("gut")	*Enterococcus*, *Yersinia enterocolitica*
erythros ("red")	*Erythrobacter*, *Saccharopolyspora erythraea*
ge ("earth")	*Geobacter*, *Campylobacter geochelonis*
genos ("birth," "descent")	*Alcaligenes*, *Streptococcus pyogenes*
gloios ("gum," "resin")	*Zoogloea*, *Leptolyngbya gloeophila*
glycys ("sweet")	*Glycomyces*, *Streptomyces glycovorans*

Greek word (transliterated)	Examples
haima (*haema*), gen. *haimatos* (*haematos*) ("blood")	*Haematospirillum*, *Paracoccus haematequi*
hals, gen. *halos* ("salt")	*Halobacterium*, *Paracooccus halotolerans*
helix, gen. *helikos* ("coil," "curl")	*Helicobacter*, *Spiroplasma platyhelix*
hydor, gen. *hydatos* ("water") (*hydro–* in compound words in Greek)	*Hydrobacter*, *Aeromonas hydrophila*
hyphe ("web")	*Hyphomicrobium*
ichthys ("fish")	*Ichthyobacterium*, *Oceaniglobus ichthyenteri*
karyon ("nut")	*Caryophanon*, *Caedibacter caryophilus*
kloster ("spindle")	*Clostridium*, *Enterocloster clostridioformis*
kokkos ("grain," "seed")	*Streptococcus*, *Psychrobacter coccoides*
kopros ("dung")	*Coprobacter*, *Actinoallomurus coprocola*
krene ("spring," "fountain")	*Crenothrix*, *Aeromonas eucrenophila*
kyaneos ("dark blue")	*Cyanothece*, *Streptomyces cyaneus*
kystis ("bladder")	*Cystobacter*, *Roseococcus microcysteis*
leptos ("fine," "small")	*Leptospira*, *Oryzihumus leptocrescens*
leukos ("white")	*Leuconostoc*
limne ("lake," "pool")	*Limnoglobus*, *Mucilaginibacter limnophilus*
lithos ("stone")	*Thermolithobacter*, *Pedobacter lithocola*
lytikos ("lysing")	*Chitinilyticum*, *Bacillus cellulosilyticus*
makros ("large")	*Macromonas*, *Streptomyces macrosporus*
megas, gen. *megalos* ("big")	*Megalodesulfovibrio*, *Sutterella megalosphaeroides*
mikros ("small")	*Thiomicrospira*, *Streptococcus microflavus*
morphe ("shape")	*Pleomorphovibrio*, *Faecalicoccus polymorphus*
mykes ("fungus")	*Mycobacterium*, *Bacillus mycoides*
myxa ("slime")	*Myxococcus*, *Casenzaea myxofaciens*
nema, gen. *nematos* ("thread")	*Treponema*, *Desmospora profundinema*
oligos ("few," "little")	*Oligoflexus*, *Nitrosomonas oligotropha*
ornis, gen. *ornithos* ("bird")	*Ornithobacterium*, *Campylobacter ornithocola*
pedion ("plain")	*Pediococcus*, *Thiolamprovum pedioforme*
pedon ("ground," "soil")	*Pedobacter*, *Flavobacterium pedocola*
pelos ("mud")	*Pelobacter*, *Thiomicrospira pelophila*
peptos ("cooked")	*Peptococcus*, *Isachenkonia alkalopeptolytica*
petra ("rock")	*Petrotoga*, *Pedobacter petrophilus*
phagein ("to eat")	*Cytophaga*, *Alcanivorax gelatiniphagus*
phaios ("gray")	*Phaeobacter*, *Rhodovulum phaeolacus*
philos ("loving")	*Xylophilus*, *Leptospirillum ferriphilum*
phos, gen. *photos* ("light")	*Photobacterium*, *Hoeflea phototrophica*
phykos ("alga")	*Phycocomes*, *Neptunomonas phycophila*
planktos ("wandering")	*Planctomyces*, *Pseudoalteromonas haloplanktis*
plasma ("image," "figure")	*Mycoplasma*
polys ("much," "many")	*Actinopolyspora*, *Paenibacillus polymyxa*
pseudes ("false")	*Pseudomonas*, *Burkholderia pseudomallei*

Greek word (transliterated)	Examples
psychros ("cold")	*Psychrobacter*, *Lysobacter psychrotolerans*
pyr ("fire")	*Pyrodictyum*, *Aquifex pyrophilus*
rhabdos ("stick")	*Halorhabdus*, *Amycolatopsis rhabdoformis*
rhiza ("root")	*Rhizobium*, *Paenibacillus rhizoryzae*
rhodon ("rose")	*Ectothiorhodospira*, *Geofilum rhodophaeum*
soma, gen. *somatos* ("body")	*Spirosoma*, *Caballeronia megalochromosomata*
staphyle ("bunch of grapes")	*Staphylococcus*
stenos ("narrow")	*Stenoxybacter*, *Flavipsychrobacter stenotrophus*
stoma, gen. *stomatos* ("mouth")	*Stomatobaculum*, *Peptoanaerobacter stomatis*
streptos ("twisted," "flexible")	*Streptomyces*, *Treponema stenostreptum*
thalassa ("sea")	*Thalassomonas*, *Vibrio thalassae*
theion ("sulfur")	*Thiobacillus*, *Geovibrio thiophilus*
thermos ("heat")	*Thermus*, *Thermotoga*, *Bacillus thermotolerans*
thrix, gen. *trichos* ("hair")	*Leptothrix*, *Oscillochloris trichoides*
trophe ("food")	*Oligotropha*, *Paracoccus pantotrophus*
xanthos ("yellow")	*Xanthomonas*, *Myxococcus xanthus*
xenos ("guest," "stranger")	*Xenococcus*, *Sphingobium xenophagum*
zyme ("leaven")	*Zymomonas*, *Desulfocapsa thiozymogenes*

Classical Latin adopted many words from Greek. Following such words in compound names, either *-i-* (after a Latin word element) or *-o-* (after a Greek word element) may be used, as shown in the examples below.

Latin, Greek	Examples
aer, ἀηρ ("air")	*Aeromonas*, *Aeribacillus*, *Deinococcus aerolatus*, *Roseomonas aeriglobus*
ager, ἀγρος ("field")	*Agrobacterium*, *Agriterribacter*, *Nocardioides agrisoli*
glaucus, γλαυκος ("bright," "gleaming")	*Glacucospira*, *Streptomyces glaucosporus*, *Actinomadura glauciflava*
oceanus, ὠκεανος ("ocean")	*Oceanospirillum*, *Oceanicaulis*, *Flavobacterium oceanosedimentum*, *Paenibacillus oceanisediminis*
oryza, ὀρυζα ("rice")	*Oryzobacter*, *Oryzomicrobium*, *Geomonas oryzisoli*
saccharon, σακχαρ, or σακχαρον ("sugar")	*Saccharococcus*, *Saccharibacter*, *Citroniella saccharovorans*, *Terrimicrobium sacchariphilum*
sphaera, σφαιρα ("ball," "sphere")	*Sphaerobacter*, *Sphaerisporangium*, *Nocardia rhizosphaerihabitans*

Two Greek word elements found in many prokaryotic names are the endings *–oides* and *–opsis* that can be translated as '-like'. Examples of *–oides* are found in the genera *Bacteroides* ("rodlike") and *Clostridioides* ("*Clostridium*-like") and in the specific epithets of *Bacillus mycoides* ("fungus-like") and *Rhodobacter sphaeroides* ("sphere-like"). Examples for *–opsis* are the genera *Nocardiopsis* ("*Nocardia*-like") and *Amycolatopsis* (*Amycolata*-like) and the species *Streptomyces netropsis* ("spindle-like"). The ending *–oides* is based on the Greek suffix *–eides*, from the Greek neuter noun εἶδος (*eidos*, "that which is seen," "form," "shape"). In the 2022 revision of the ICNP, a note was added to Rule 65 and to Appendix 9 stating that generic names ending in *–oides* have the neuter gender, irrespective of the gender of

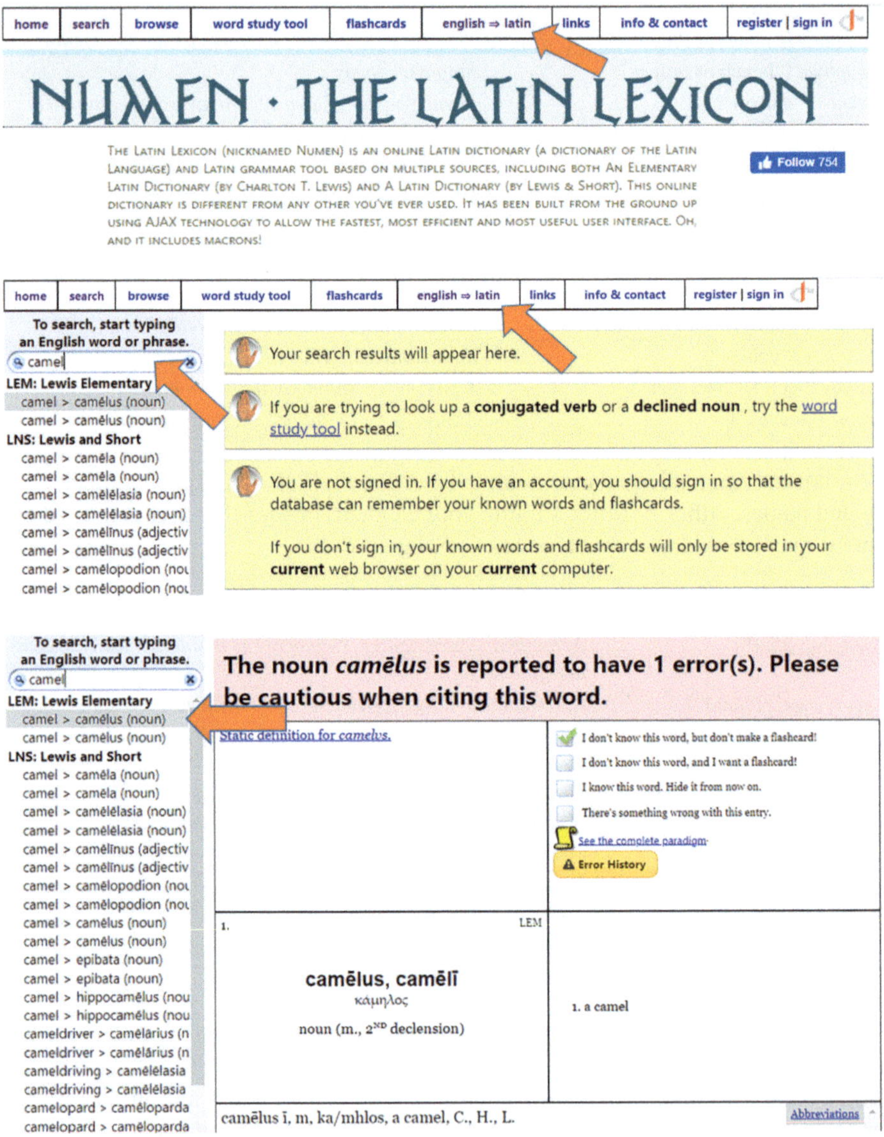

Fig. 2.1 Screenshots of *Numen: The Latin Lexicon* (https://latinlexicon.org/index.php. Accessed 20 Nov 2022), showing translation of "camel" to *camelus* and translation of "lake" to *lacus* or *stagnum*

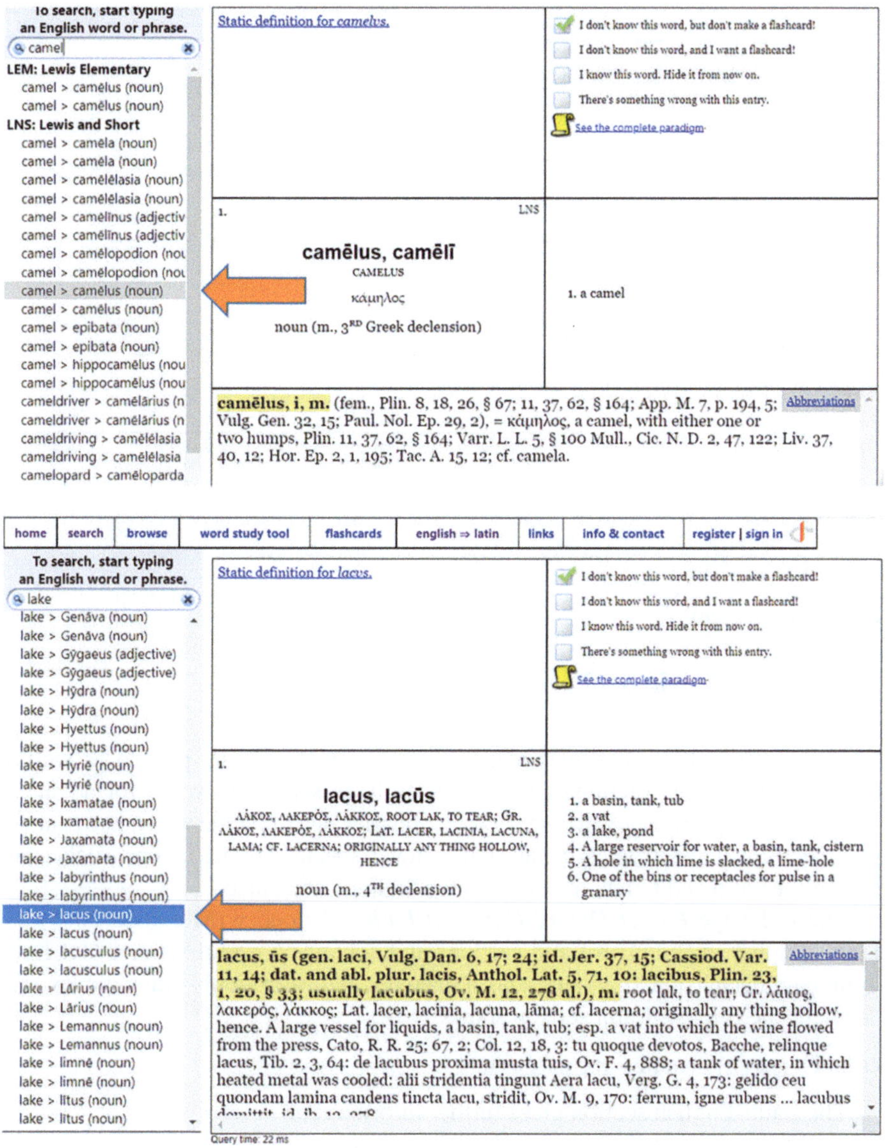

Fig. 2.1 (continued)

the word that precedes *–oides*, and names ending in *–opsis* (from the Greek feminine noun ὄψις (*opsis*, "aspect," "appearance") must be treated as feminine (Oren and Schink 2016). This emendation of the ICNP rules is not retroactive: the names *Bacteroides* and *Nocardioides* remain masculine, while *Planktothricoides* remains feminine.

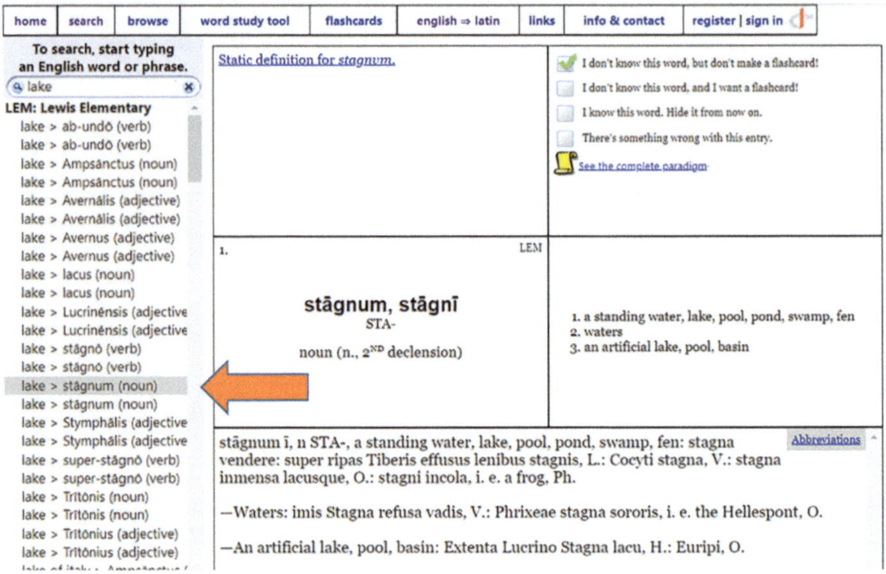

Fig. 2.1 (continued)

2.5 Use of Latin and Greek Dictionaries

There are many Latin and classical Greek lexicons available, printed as well as online, simple basic dictionaries, as well as large scholarly works for advanced users. To demonstrate how to use such dictionaries, let's assume we want to name a new genus isolated from a camel and a novel species isolated from a lake, and search for Latin and Greek translations for "camel" and "lake."

The 481-page *Pocket Oxford Classical Latin Dictionary* (Morwood 2005) has Latin-English and English-Latin sections and contains a useful summary of Latin grammar: declension of nouns and adjectives and verb conjugation tables. For "camel," the English-Latin section shows

camel *n* [noun] camelus *m* [masculine]

The Latin-English section gives further information:

camēlus, ī *m* camel, dromedary

(the stressed syllable is indicated; second declension, genitive *cameli*)

For "lake," the English-Latin section shows

lake *n* lacus *m*; stagnum *nt* [neuter]

The Latin-English section gives

lacus, **ūs** *m* lake; pond; tank, reservoir, trough

(genitive *lacus*, with the second syllable stressed; fourth declension)

stāgnum, **i** *nt* pool, lagoon, water

(second declension, genitive *stagni*)

A more advanced and highly recommended Latin-English/English-Latin dictionary is *Cassell's Latin Dictionary* (Simpson 1968). The English-Latin section (233 pages) gives less information than the 650-page Latin-English section. For "camel," the English-Latin section gives

camel, *camelus*

For further information, the Latin-English section may be consulted:

cămēlus –i, m., and f. (κάμηλος), *a camel or dromedary*: Cic., Hor., Tac.

(the stressed syllable is indicated; the genitive is *cameli* (second declension); the word can be masculine or feminine, is derived from Greek *kamelos*, and can be found in the writings of Cicero, Horatius, and Tacitus)

Surprisingly, the English-Latin section does not have an entry for "lake" but shows translations for "pond":

pond, *stagnum*, *piscina*, *lacus* (*–ūs*), *lacuna* (poet.)

(genitive is *lacus* with the second syllable stressed, fourth declension; poet. = poetically)

Further information is found in the Latin-English section:

stagnum –i, n. (connected with *sto*). LIT: *standing water*, *a pool*, *pond*, *marsh*, *swamp*; Liv., Verg., etc. TRANSF., any narrow or enclosed stretch of water: …

(genitive is *stagni*, second declension; neuter; etymologically the word is connected with the verb *sto*, *stare*, to stand; Literal meanings are …; used by Livius, Vergilius, and others; transferred (i.e., used in an altered or metaphorical sense) …

lācus –ūs, m, *a hollow*; hence: *a lake*: Lucr., Caes., Cic.: used poet. for *any large body of water*; …

(first syllable is stressed; genitive is *lacus* with the second syllable stressed, fourth declension; masculine; used by Lucretius, Caesar, and Cicero)

Far more detailed information is found in the Latin-English dictionary by Lewis and Short 1891, also known as *Harper's Latin Dictionary*. The 1891 edition (2019 pages) can be downloaded in pdf format (see the References list). All information given by Lewis and Short is also found online at https://latinlexicon.org/index.php (see below; Fig. 2.1).

The online resource Google Translate (2022) can translate English and other languages into Latin, but it should be used cautiously. The program incorrectly translates ‘camel’ as *camelum* (a neuter noun not found in the dictionaries) instead of *camelus* or (for a female camel) *camela.* “Lake” is correctly translated *lacus*, but no grammatical context is shown. When translating *lacus* back into English, Google Translate gives “lakes,” which is correct for a Latin noun of the fourth declension, but highly confusing.

An excellent online search option for translations from English into Latin is “Numen – The Latin Lexicon” (2022) (https://latinlexicon.org/index.php). It contains the complete information found in the Lewis and Short dictionary. Screenshots presented in Fig. 2.1 show translations of “camel” to *camelus* and of “lake” to *lacus* or *stagnum.* A possible problem for beginners is that the site offers many translations, and not all may be appropriate within the context needed.

For translation of English words into classical Greek, the 445-page *Pocket Oxford Classical Greek Dictionary* (Morwood and Taylor 2002) is a very basic resource. It has a short (65 pages) English-Greek section. For “camel” and “lake,” it gives

camel κάμηλος ου ὁ/ἡ

(the genitive of κάμηλος (transliterated *kamelos*) is κάμηλου (*kamelou*); the word can be masculine or feminine)

lake λιμνη ης ἡ

(the genitive of λίμνη (transliterated *limne*) is λίμνης (*limnes*); the word is feminine)

Far more detailed information is found in the Greek-English dictionary by Liddell and Scott (1996). The 1882 edition (1,774 pages + addenda) can be downloaded in pdf format (see the References list). It is recommended only for scholars familiar with the language. The new, 2-volume, 1,529-page *Cambridge Greek Lexicon* (Greek-English only) was recently published (Diggle et al. 2021).

Classical Greek is not included on the Google Translate website. There, Greek refers to the modern language. The 2022 revision of the ICNP clarifies that for the purpose of Principle 3, Rules 7, 10a, and 65, and Recommendation 6, “Greek” means classical Greek (Oren and Schink 2020a). Modern Greek words are often identical to the classical language. This is the case for λίμνη (*limne*, “lake”) but not for “camel” (καμήλα, *kamela* in modern Greek). Some authors apparently searched Google Translate for Greek words to be used in names of prokaryotic taxa, resulting in names such as *Candidatus* Thermonerobacter and *Candidatus* Tyrphobacter. The word νερό (*nero*, “water”) is not found in the lexicon of classical Greek in addition to ὕδωρ (*hydor*). The word τυρφη (*tyrphe*), “peat,” is also not found in the dictionaries of the classical language.

The online version of the Woodhouse (1910) English-Greek dictionary (https://dcthree.github.io/woodhouse/# or https://perseids-project.github.io/woodhouse-js) gives the correct translations with only little additional information:

camel P [Prose] and V. [Verse] κάμηλος, ὁ or ἡ [masculine or feminine] (Dem. [Demosthenes], also Ar. [Aristophanes])

lake P. and V. λίμνη, ἡ [feminine]

2.6 How to Present New Names and Their Etymologies

For the valid publication of names of new taxa of prokaryotes, the new name or new combination should be clearly stated and indicated as such (i.e., fam. nov., gen. nov., sp. nov., comb. nov., etc.), and the derivation (etymology) of the name must be given as part of the formal description ("protologue") of the new taxon to be included in the publication in the IJSEM or in the effectively published description of the taxon published elsewhere (ICNP, Rule 27).

The etymology lines in protologues are presented in a standardized format and typically include abbreviations. Below is a list of commonly used abbreviations (modified from Appendix 7 of the 2022 revision of the ICNP; Oren et al. 2023).

	Abbreviation	Full spelling	Explanation
Taxonomic ranks			
	class. nov.	*classis nova*	New class
	comb. nov.	*combinatio nova*	New combination, when an established epithet (taken from the basonym) is combined with another genus name to form a species name or with another genus name and another epithet to form a subspecies name
	fam. nov.	*familia nova*	New family
	gen. nov.[a]	*genus novum*	New genus
	nom. nov.[a]	*nomen novum*	New name, to be established when the establishment of a comb. nov. would result in a homonym
	nom. rev.[a]	*nomen revictum*	Revived name that existed before 1980 but not included in the Approved Lists of 1980
	ord. nov.	*ordo novus*	New order
	phyl. nov.	*phylum novum*	New phylum
	sp. nov.	*species nova*	New species
	subsp. nov.	*subspecies nova*	New subspecies
Grammatical terms			
	adj.	adjective	
	dim.	diminutive	
	fem.	feminine	
	gen.[a]	genitive	
	masc.	masculine	
	n.	noun	
	neut.	neuter	
	nom.[a]	nominative	
	part.	participle	

	Abbreviation	Full spelling	Explanation
	part. adj.	participle used as adjective	To comply with Rule 12c(1) so that a participle can be used as a specific or subspecific epithet
	pl.	plural	
	pref.	prefix	
	prep.	preposition	
	pres. part.	present participle	
	sing.	singular	
	suff.	suffix	
	v.	verb	
Source of words			
	Gr.	Greek	
	L.	Latin	Words used in classical Latin
	M.L.	Medieval Latin	In the past, M.L. was often used for Modern Latin, now to be replaced with N.L.
	N.L.	Neo-Latin	Words newly coined, based on classical Latin elements and/or latinized modern words
Other abbreviations			
	corrig.	*corrigendum*	A corrected typographical or orthographic error
	emend.	*emendavit*	Alteration of the diagnostic characters or of the circumscription of a taxon

[a]The abbreviations "nom." and "gen." mean "nomen" or "nominative" and "genus" or "genitive," respectively, depending on the context

Below are examples published in 2020 in the IJSEM, showing how etymology lines are presented, with explanations of the abbreviations used.

***Patiriisocius* gen. nov.**

Patiriisocius (Pa.ti.ri.i.so'ci.us. N.L. fem. n. *Patiria*, a starfish genus; L. masc. n. *socius*, companion; N.L. masc. n. *Patiriisocius*, a companion of *Patiria* starfish)

Explanations

Patiriisocius gen. nov. [genus novum, a new genus] (Pa.ti.ri.i.so'ci.us. [syllabification, the antepenultimate syllable stressed] N.L. fem. n. [Neo-Latin feminine noun] *Patiria*, a starfish genus; L. masc. n. [classical Latin masculine noun] *socius*, companion; N.L. masc. n. [Neo-Latin masculine noun] *Patiriisocius*, a companion of *Patiria* starfish [the second -i- of *Patiriisocius* is the connecting vowel])

***Polaribacter septentrionalilitoris* sp. nov.**

Polaribacter septentrionalilitoris (sep.ten.tri.o.na.li.li´to.ris. L. masc. adj. *septentrionalis*, northern; L. neut. n. *litus*, beach; N.L. gen. n. *septentrionalilitoris*, of Nordstrand – North Beach, where the type strain was isolated)

Explanations

Polaribacter septentrionalilitoris sp. nov. [species nova, a new species] (sep.ten.tri.o.na.li.li´to.ris. [syllabification, the penultimate syllable stressed] L. masc. adj. [classical Latin, masculine adjective] *septentrionalis*, northern; L. neut. n. [classical Latin, neuter noun] *litus*, beach; N.L. gen. n. [Neo-Latin, genitive noun] *septentrionalilitoris*, of Nordstrand – North Beach, where the type strain was isolated [the two parts of the German word "Nordstrand" were translated into Latin, and the genitive case *litoris* (of a beach) was used to indicate "of Nordstrand"])

***Natrarchaeobaculum aegyptiacum* comb. nov.**

Natrarchaeobaculum aegyptiacum (ae.gyp.ti′a.cum. L. neut. adj. *aegyptiacum*, Egyptian). Basonym: *Natronolimnobius aegyptiacus* Zhao et al. 2018

Explanations

Natrarchaeobaculum aegyptiacum comb. nov. [combinatio nova, a new combination: the species *Natronolimnobius aegyptiacus* (the basonym) was transferred to a different genus and the specific epithet was conserved] (ae.gyp.ti′a.cum. [syllabification, the antepenultimate syllable stressed] L. neut. adj. [classical Latin, neuter adjective] *aegyptiacum*, Egyptian [note that the gender of the adjective was changed from masculine in the basonym to neuter in the new combination, to comply with Rule 12c(1) of the ICNP])

***Microlunatus kandeliicorticis* nom. nov.**

Microlunatus kandeliicorticis (kan.de.li.i.cor′ti.cis. N.L. fem. n. *Kandelia*, a plant genus; L. masc. n. *cortex –icis*, bark; N.L. gen. n. *kandeliicorticis*, of the bark of *Kandelia candel*, the source of isolation of the type strain)

Basonym: *Friedmanniella endophytica* Tuo *et al.* 2016

Explanations

Microlunatus kandeliicorticis nom. nov. [nomen novum: a species (here *Friedmanniella endophytica*, the basonym) was transferred to a different genus, but the original specific epithet could not be used to form a combinatio nova as the name *Microlunatus endophyticus* already existed; a nomen novum was therefore created based on Rule 34a to avoid formation of an illegitimate homonym] (kan.de.li.i.cor′ti.cis. [syllabification, the antepenultimate syllable stressed] N.L. fem. n. [Neo-Latin feminine noun] *Kandelia*, a plant genus; L. masc. n. [classical Latin masculine noun] *cortex –icis*, [*cortex* is the nominative case, *corticis* the genitive, used for the formation of the specific epithet] bark; N.L. gen. n. [Neo-Latin genitive noun] *kandeliicorticis*, of the bark of *Kandelia candel*, the source of isolation of the type strain [the second *-i-* of *kandeliicortics* is the connecting vowel])

***Calderihabitantaceae* fam. nov.**

Calderihabitantaceae (Cal.de.ri.ha.bi.tan.ta.ce'ae. N.L. masc. n. *Calderihabitans,* a bacterial genus; *–aceae,* ending to denote a family; N.L. fem. pl. n. *Calderihabitantaceae,* the *Calderihabitans* family)

Explanations
Calderihabitantaceae fam. nov. [familia nova, a new family]. *Calderihabitantaceae* (Cal.de.ri.ha.bi.tan.ta.ce'ae. [syllabification, the penultimate syllable stressed] N.L. masc. n. [Neo-Latin masculine noun] *Calderihabitans*, a bacterial genus [the designated type genus of the family]; *–aceae*, ending to denote a family [in accordance with Rule 9 of the ICNP]; N.L. fem. pl. n. [Neo-Latin feminine plural noun] [in accordance with Rule 7] *Calderihabitantaceae* [grammatically, *Calderihabitans* is a present participle that can be used as an adjective and is here used as a noun; note that the family name was formed from the stem of the word: …*habitant–* as the genitive of *habitans* is *habitantis*], the *Calderihabitans* family)

2.7 Syllabification, Accentuation, and Pronunciation

The examples of etymology lines presented above include syllabification (showing how long words can be broken into syllables) and accentuation (the syllable to be stressed when the word is pronounced). For example, the specific epithet *kandeliicorticis* can be broken into six syllables, and the antepenultimate syllable (the third syllable from the end) is stressed: kan-de-li-i-**cor**-ti-cis. Although it is customary to add recommended syllabification and accentuation in the etymology lines, it is not a requirement given in Rule 27 of the ICNP. The code and its appendices do not provide guidelines for syllabification, accentuation, and pronunciation of names.

Classical Latin had no explicit rules on how to break up words into syllables. The nomenclature reviewers of the IJSEM routinely follow the recommendations given by Trüper (1999, 2001), which are not always identical to those given in authoritative Latin grammar textbooks such as Greenough et al. (1903). The way words are broken into syllables in English is quite different from most languages of mainland Europe, and this often causes confusion. The following guidelines are derived from Trüper's papers and from marello.org (2022):

- Every Latin word has as many syllables as it has vowels or diphthongs. In the division of words into syllables, a single consonant between two vowels is written and pronounced with the following vowel. Doubled consonants are separated. Examples: Sal.mo.nel.la.; Sa.li.sae.ta.
- Single consonants including x and z as well as ch, ph, and th belong to the subsequent syllable; qu, ch, ph, and th should never be separated in syllabification. Examples: Chlo.ro.fle.xus.; Rhi.zo.bi.um.; Me.thy.lo.pha.ga.
- If two or more consonants follow each other, the last one belongs to the following syllable. However, when b, p, d, t, g, and c are followed by l or r (sometimes by m or n), these sequences remain unseparated.
  Examples: Bac.te.ro.i.des.; Sim.pli.ci.spi.ra.; Mi.cro.bac.ter.; Ma.gne.to.coc.cus.
- Compound words are always broken between their composing elements.
  Examples: Spi.ril.lo.spo.ra. and not Spi.ril.los.po.ra.; Sul.fur.o.vum. and not Su.fu.ro.vum. Note also the example of *Acidianus*: A.cid.ia.nus. and not A.ci.

di.a.nus. *Acidianus* combines the Latin adjective *acidus*, "acidic," and *Ianus*, a mythical Roman figure with two faces looking in opposite directions. *Acidianus* means an acidic bifaced bacterium, reflecting the growth conditions and the metabolism of the organism.

The etymology line also gives the recommended accentuation: the accent-bearing syllable is followed by an accent sign instead of a period. For each word only one syllable is accentuated: the penultimate syllable (penult, the second syllable from the end) or the antepenultimate syllable (antepenult, the third syllable from the end). Examples: Mi.cro.lu.na'tus. (penult); Pa.ti.ri.i.so'ci.us. (antepenult).

The rules for accentuation are quite complex; only basic guidelines are given here. If the penult is long, the accent falls on it; otherwise, it falls on the antepenult. A syllable is long if it ends in two or more consonants or contains a long vowel. In Latin dictionaries, long vowels are often indicated by a "macron" (a diacritical mark given as a straight line: ā, ō) and short vowels by a "breve" (a curved symbol: ă, ŏ). In Greek, η (*eta*) and ω (*omega*) are always long, and ε (*epsilon*) and ο (*omicron*) are always short. If a syllable ends in at least one consonant, it counts as closed, otherwise as open. A closed syllable or one containing a long vowel or a diphthong counts as heavy, otherwise as light. Word stress falls on the penultimate syllable if it is heavy, else on the antepenultimate if available. Examples

heavy penultimate – long vowel or diphthong: Dei.no.coc'cus.; fer.men'tans.
heavy penultimate – short vowel with following consonant: Sta.nier.el'la.; haf.ni.en'se.
light penultimate – word stress on the antepenultimate syllable: Al.ka.li'phi.lus.; sub'ti.lis.

In view of the complexity of these guidelines, I happily agree with the statement by Trüper (2001) that "the practical sense of natural scientists should prevail and the present common usage of accentuation in bacterial names should be the guideline."

There are no "official" rules how to pronounce names of prokaryotic taxa. Pronunciation of Latin is strongly influenced by the speaker's mother tongue. Use of phonetic symbols in etymology lines is therefore not recommended. Native speakers of English and French often pronounce Latin according to the pronunciation rules of their languages, which are very different from, e.g., Italian and Spanish. The way most classical scholars pronounce Latin words also differs from church Latin. Trüper recommended to pronounce at least the vowels as they are pronounced in Spanish and Italian, languages whose pronunciations stayed close to their Latin origin (Trüper 1999; Trüper 2001).

Most microbiologists pronounce the "c" in *Acetobacter* and *Acinetobacter* as "s," although the Romans pronounced the letter c as the sound k (e.g., *Caesar* pronounced as "Kaesar" and *Cicero* as "Kikero"). *Acetobacter* is derived from that Latin *acetum*, "vinegar," and one could argue that for *Acinetobacter*, the "k" pronunciation is the correct one as the name is derived from the Greek verb κινεω (*kineo*, "to set into motion," "to move").

2.8 Prokaryotes Named After Persons

The names of many prokaryote genera and species honor famous microbiologists from the past and the present. Well-known examples are *Escherichia* (Theodor Escherich), *Salmonella* (Daniel Elmer Salmon), *Shigella* (Kiyoshi Shiga), *Methanococcus voltae* (Alessandro Volta), *Clostridium pasteurianum* (Louis Pasteur), and *Pyrodicium brockii* (Thomas Brock). Recommendation 10a (1) states "Refrain from naming genera and subgenera after persons unconnected with microbiology or at least with natural science," and Recommendation 12c (3) contains a similar statement for specific epithets. Genera and species should not be named to honor heads of state, politicians, etc., even if there is a precedent: *Streptomyces cavourensis*, honoring the Italian statesman Camillo Benso, Count of Cavour (1810–1861). It is good practice to ask the person to be honored by a scientific name for permission as long as she/he is alive.

The ways to include personal names in generic names and specific epithets of prokaryotes are standardized as presented in the tables below, derived from Appendix 9, Section D, of the ICNP. Generally the last name (family name) is used, but sometimes the first name (forename, given name) was used instead. A well-known example is the plant pathogenic genus *Erwinia*, named to honor Erwin Frink Smith. *Elioraea* honors Eliora Ron. *Staphylococcus arlettae*, named after Arlette van de Kerckhove, sounds better and is easier to pronounce than, e.g., *kerkhoveae* or *vandekerkhoveae*.

The latter example shows that prefixes common in many languages may be united to the name or may be omitted. Prefixes were united to the name in the genera *Deleya* (Josef De Ley), *Devosia* (Paul De Vos), *Leminorella* (Leon Le Minor), and *Vannielia* (Cornelis van Niel), who was also honored with the epithet *vannielii* in the genera *Methanococcus* and *Rhodomicrobium*. Prefixes were omitted in *Rochalimaea* (Henrique da Rocha-Lima), *Rickettsia prowazekii* (Stanislaus von Prowazek), *Leeuwenhoekiella* (Antonie van Leeuwenhoek), and *Spirillum itersonii* (Gerrit van Iterson). The Scottish "Mac" or "Mc" should be written "mac" and be united with the rest of the name: *Corynebacterium macginleyi* (Kenneth John McGinley) and *Legionella maceachernii* (Harold V. McEachern). The adjective Saint, San, Sankt, Santo, Santa, etc. as part of a family name may be omitted or united as in *Leptospira santarosai* (Carlos Santa Rosa) and *Leptospira saintgironsiae* (Isabelle Saint-Girons).

Generic names can be formed from personal names in three ways:

- By adding the ending *–a*, *–ea*, *–nia*, or *–ia*
- By adding a diminutive ending *–ella*, *–iella*, or *–nella*
- As the first word element of a compound word

The table below, modified from Table 2 of Appendix 9 of the 2022 revision of the ICNP, shows how generic names are formed from personal names using the endings *–a*, *–ea*, *–nia*, *–ia*, *–ella*, *–iella*, or *–nella*.

Personal name ending in	Person	Direct formation		Person	Diminutive formation	
		Add ending	Example		Diminutive ending	Example
–a	da Rocha Lima	*–ea*	*Rochalimaea*	Shiga	drop a, add *–ella*	*Shigella*
–e	Benecke	*–a*	*Beneckea*	Bruce	*–lla*	*Brucella*
	Hoppe	*–ia*	*Hoppeia*			
–i	Umezaki	*–a*	*Umezakia*	Terasaki	*–ella*	*Terazakiella*
–o	Beggiato	*–a*	*Beggiatoa*	Seino	*–nella*	*Seinonella*
	Kozako	*–nia*	*Kozakonia*			
–u	Simidu	*–ia*	*Simiduia*	Shimazu	*–ella*	*Shimazuella*
–y	Euzéby	*–a*	*Euzebya*	Bergey	*–ella*	*Bergeyella*
–er	Buchner	*–a*	*Buchnera*	Stanier	*–ella*	*Stanierella*
	Lister	*–ia*	*Listeria*	Turner	*–iella*	*Turneriella*
Any consonant	Nocard	*–ia*	*Nocardia*	Klebs	*–(i)ella*	*Klebsiella*
	De Vos	*–ia*	*Devosia*	Salmon	*–(i)ella*	*Salmonella*
	Escherich	*–ia*	*Escherichia*	Sneath	*–(i)ella*	*Sneathiella*

When naming new genera to honor persons, it may happen that names formed by adding *–ia* or *–(i)ella* to the person's name already exist in the botanical or zoological nomenclature, making the names illegitimate because of Principle 2 of the ICNP: a newly proposed generic name may not be a homonym of a genus that exists in botany or zoology. Combining the person's given name and surname generally solves the problem. Examples are *Andreprevotia* (André Prévot, earlier honored with the generic name *Prevotella*), *Elizabethkingia* (Elizabeth King), *Georgfuchsia* (Georg Fuchs, earlier honored with the name *Fuchsiella*), *Guyparkeria* (Cecil David "Guy" Parker), *Hansschlegelia* (Hans Schlegel, earlier honored with the name *Schlegelella*), *Hanstruepera* (Hans Trüper, earlier honored with the names *Truepera* and *Trueperella*), *Harryflintia* (Harry Flint), *Loktanella* (Tjhing-Lok Tan), *Marvinbryantia* (Marvin Bryant, earlier honored with the name *Bryantella*), *Owenweeksia* (Owen Weeks, earlier honored with the name *Weeksella*), *Williamwhitmania* (William Whitman), *Yinghuangia* (Ying Huang), *Yonghaparkia* (Yong-Ha Park), *Zhihengliuella* (Zhi-Heng Liu), *Zhizhongheella* (Zhi-Zhong He), and *Zunongwangia* (Zu-Nong Wang). A disadvantage is that such names tend to be long and often difficult to pronounce, contravening Recommendation 6(1) of the ICNP.

Examples how personal names can be included in compound generic names are shown in the table below, which is a simplified version of Table 2 of Appendix 9 of the 2022 revision of the ICNP. (m) and (f) refer to names of male and female persons, respectively.

Ending of name	Personal name	Compound generic name
–a	Ōhara (m)	*Oharaeibacter*

Ending of name	Personal name	Compound generic name
–e, –é	Pace (m)	*Candidatus* Paceibacter
–i	Terasaki	*Terasakiispira*
–o	Franco-Mora (m)	*Franconibacter*
–u	Le Testu (m)	*Letestuinema* (*Letestuiinema* would be preferred based on the current guidelines)
–y	Ráthay (m)	*Rathayibacter*
–er	Rubner (m)	*Rubneribacter*
Any other letter	Rummel (m) Young (f)	*Rummeliibacillus* *Youngiibacter*

Specific epithets can be formed from personal names in two ways: as adjectives that agree in gender with the generic name (Rule 12c(1)) or as a nouns in the genitive case (Rule 12c(3)). Adjectival specific epithets derived from personal names have the meaning of "pertaining to … (the person)." The gender of the adjective is determined by the gender of the generic name in accordance with Rule 12c(1) of the ICNP (so *–us*, *–a*, and *–um* for masculine, feminine, and neuter, respectively), not by the gender of the person in whose honor the organism was named, as shown in the table below.

Ending of name	Personal name (m = male, f = female)	Add the endings for gender			Examples
		Masculine	Feminine	Neuter	
Any consonant	Brock (m) Colwell (f) Pasteur (m)	*–ianus*	*–iana*	*–ianum*	*Thermus brockianus* *Alteromonas colwelliana* *Clostridium pasteurianum*
–a	Loya (m) Migula (m)	*–nus*	*–na*	*–num*	*Thalassomonas loyana* *Aneurinibacillus migulanus*
–e, –i, –o, –u, –y	Li (m) Love (f) Olley (f) Palleroni (m) Reverdatto (m)	*–anus*	*–ana*	*–anum*	*Cyclobacterium lianum* *Porphyromonas loveana* *Shewanella olleyana* *Pseudomonas palleroniana* *Trichormus reverdattoanus*

Below are examples of specific epithets formed as nouns in the genitive case:

Ending of name	Add for female person	Examples	Add for male person	Examples
–a	–e	Victoria, *victoriae*	–e	Volta, *voltae*
	–eae	Pamela Lee Oxley, *pamelaeae*	–i	Oshima, *oshimai*
	–iae	Zhihua Wu, *zhihuaiae*	–ei	Mukohata, *mukohataei*
			–ii	Vora, *voraii*

Ending of name	Add for female person	Examples	Add for male person	Examples
–e, –é	–ae	Curie, *curieae*	–i	Beveridge, *beveridgei*
–i	–ae	Yabuuchi, *yabuuchiae*	–i	Giovannoni, *giovannonii*
–o	–niae	Cato, *catoniae*	–nis	Hirano, *hiranonis*
–u	–iae	Plateau-Quénu, *quenuiae*	–ii	Brisou, *brisouii*
–y	–ae	Olley, *olleyae*	–i	De Ley, *deleyi*
–er	–ae	Miller, *millerae*	–i	Stutzer, *stutzeri*
any other letter	–iae	Gordon, *gordoniae*	–ii	Pfennig, *pfennigii*

The given name and the surname of the person can also be combined in the epithet: *Actinomyces tangfeifanii* (Feifan Tang), *Agrobacterium larrymoorei* (Larry Moore), *Agromyces badenianii* (Denian Ba), and *Salinibacterium hongtaonis* (Tao Hong).

It is not recommended to honor more than one person in a generic name or epithet (Appendix 9, 2022 revision of the ICNP), but there are a few precedents: *Azospirillum baldaniorum* ("of the Baldanis"), named in honor of José Ivo Baldani and Vera Divan Baldani, for their contributions to *Azospirillum* research; *Corynebacterium jeikeium*, based on the initial letters of the surnames of W.D. Johnson and D. Kaye, who described infections attributed to the taxon; and *Herbinix luporum* ("of wolves"), referring to Wolfgang Ludwig, Wolfgang Liebl, and Wolfgang Schwarz for their contributions to the taxonomy of Gram-positive bacteria.

Authors should refrain from naming bacteria after themselves or coauthors after each other in the same publication, as this is considered immodest by the majority of the scientific community, and it contravenes Recommendations 6 (7) and 12c (3) of the 2022 revision of the ICNP. There are a few cases of validly published names in which authors named organisms after themselves: *Eubacterium yurii* ("of George 'Yuri' Krywolap") and *Eubacterium yurii* subsp. *margaretiae* ("of Barbara S. Margaret"), *Legionella hackeliae* (honoring Meredith Hackel, who isolated the organism and coauthored the description paper), *Mycoplasma cottewii* ("of Gillian S. Cottew, one of the authors of the article describing the taxon"), and *Mycoplasma*/*Mesomycoplasma moatsii* (Kenneth E. Moats being a coauthor). For *Mycoplasma collis* ("of a hill"), the connection of the name with the author who described the taxon may not be immediately obvious. It was not isolated from a hill but from rats and mice, by Aureol C. Hill. The epithet of *Clostridium*/*Ruminiclostridium josui* ("of josu") is based on the first letters of the names of the authors who described the taxon: Jiraporn Sukhumavasi, Kunio Ohmiya, Shoichi Shimizu, and Kazue Ueno.

2.9 Animals and Plants in the Prokaryotic Nomenclature

Many prokaryotes are associated with plants and animals, often as symbionts or as pathogens, and this is reflected in many genus names and specific epithets. Here are examples of generic names in which the Latin or latinized name of the plant or animal is the first component of a compound name. As usual, connecting vowels are used: *-i-* after Latin and *-o-* after Greek:

Plants: *Arachidicoccus* (*Arachis*, genitive *arachidis*, "the peanut plant"), *Brassicibacter* (*brassica*, "cabbage"; genus name *Brassica*), *Camelliibacillus* (*Camellia*, genus name of the tea plant), *Oryzisolibacter* (*oryza*, "rice"; genus name *Oryza*), *Oryzomicrobium* (Greek *oryza*, "rice"), *Panacibacter* (*Panax*, genitive *panacis*, "the ginseng plant")

Animals: *Camelimonas* (*camelus*, "camel"; genus name *Camelus*), *Chelonobacter* (Greek *chelone*, "tortoise"), *Echinicola* (*echinus*, "sea urchin"; genus name *Echinus*), *Gryllotalpicola* (*Gryllotalpa*, a mole cricket genus), *Muribaculum* (*mus*, genitive *muris*, "mouse," genus name *Mus*), *Suicoccus* (*sus*, genitive *suis*, "hog"; genus name *Sus*)

The most common way to form specific epithets from plant or animal names is by using the genitive case of the Latin word and/or the generic name of the organism:

Plants

First declension: *Mesorhizobium acaciae*, "of the genus *Acacia*"; *Corynebacterium betae*, "of the sugar beet *Beta vulgaris*"; *Paenibacillus camelliae*, "of the tea plant *Camellia sinensis*"

Second declension, masculine and feminine: *Frankia alni*, "of the alder genus *Alnus*" (fem.); *Chryseobacterium lathyri*, "of the genus *Lathyrus*" (masc.); *Kurthia populi*, "of the poplar genus *Populus*" (fem.)

Second declension, neuter: *Agromyces allii*, "of the onion genus *Allium*"; *Variovorax gossypii*, "of the cotton genus *Gossypium*"; *Clavibacter tritici*, "of wheat, of the genus *Triticum*"

Third declension: *Bradyrhizobium arachidis*, "of the groundnut genus *Arachis*"; *Methylobacillus caricis*, "of the sedge genus *Carex*"; *Paenibacillus medicaginis*, "of the alfalfa genus *Medicago*"; *Salinicola tamaricis*, "of the genus *Tamarix*"

Fourth declension: *Deinococcus ficus*, "of the fig tree genus *Ficus*"

Animals

First declension: *Escherichia blattae*, "of the cockroach genus *Blatta*"; *Mycoplasma caviae*, "of the guinea pig genus *Cavia*"; *Limosilactobacillus gorillae*, "of the ape genus *Gorilla*"

Second declension, masculine: *Lactobacillus bombi*, "of the bumblebee genus *Bombus*"; *Streptococcus cameli*, "of a camel," "of the genus *Camelus*"; *Halomonas litopenaei*, "of the shrimp genus *Litopenaeus*"

Second declension, neuter: *Gordonia didemni*, "of the tunicate genus *Didemnum*"; *Undibacterium macrobrachii*, "of the shrimp genus *Macrobrachium*"; *Enterobacillus tribolii*, "of the flour beetle genus *Tribolium*"

Third declension: *Pasteurella anatis*, "of a duck," "of the genus *Anas*"; *Bombella apis*, "of a honeybee," "of the genus *Apis*"; *Bartonella bovis*, "of a cow," "of the genus *Bos*"; *Mycolicibacterium elephantis*, "of an elephant," "of the genus *Elephas*"; *Mycoplasma testudinis*, "of a tortoise," "of the genus *Testudo*"

Alternatively, the genitive of the plural number can be used:

First declension: *Acetobacter fabarum*, "of beans," from L. *faba*; *Flavobacterium phocarum*, "of seals," from L. *phoca*; *Mycoplasma gallinarum*, "of hens," from L. *gallina*; *Tissierella simiarum*, "of monkeys," from L. *simia*

Second declension: *Campylobacter cuniculorum*, "of rabbits," from L. *cuniculus*; *Helicobacter equorum*, "of horses," from L. *equus*; *Steptococcus porcorum*, "of pigs," from L. *porcus*

Third declension: *Bifidobacterium boum*, "of cattle," from L. *bos*; *Chlamydia avium*, "of birds," from L. *avis*; *Paracoccus suum*, "of pigs," from L. *sus*

Some epithets of prokaryotic genera were derived from the specific epithet rather than from the generic name of a plant or animal: *Methylophila carotae*, "of the carrot" (*Daucus carota*), and *Streptococcus caballi*, "of the horse" (*Equus caballus*). The nouns *carota*, "carrot," and *caballus*, "horse," are found in the Latin lexicon. However, the name *Pseudomonas avellanae* (of *Corylus avellana*, the hazel shrub) is problematic. The adjective *avellana* means "from Avella, a town in southern Italy." This refers to the tree, not to the type strain of the bacterium, which was isolated in Greece. In addition, there is a grammatical problem: *avellanae* is an adjective in the genitive case, and adjectives used as specific epithets must be in the nominative case (Rule 12c(1) of the ICNP). Better would therefore be *coryli* ("of the hazel," "of *Corylus*"), as used in *Streptomyces coryli*.

The genus name and the epithet of the plant or animal species can be combined in names of prokaryotes: *Nocardioides ginkgobilobae* ("of the ginkgo, *Ginkgo biloba*"), *Pseudomonas caricapapayae* ("of the papaya, *Carica papaya*"), and *Sphingomonas bovisgrunnientis* ("of the yak, *Bos grunniens*"). In the first two examples, the plant names were treated as single words, without a connecting vowel. For *Sphingomonas bovisgrunnientis*, both parts of the name have the genitive case (*bos*, genitive *bovis*; *grunniens*, genitive *grunnientis*). The latter form, concatenating the genitives of the components of the name, is recommended (Oren and Schink 2020b; Appendix 9 of the 2022 revision of the ICNP).

Names of higher taxa (family and higher) of plants and animals, which are always in the plural number, can be used in genitive case:

Plants

First declension: *Pectobacterium aroidearum* (of the plant subfamily Aroideae), *Pseudomonas brassicacearum* (of the plant family Brassicaceae)

Second declension, neuter: *Streptomyces bryophytorum* (of the phylum *Bryophyta*), *Paenibacillus marchantiophytorum* (of the liverwort division Marchantiophyta)

Animals

First declension: *Renibacterium salmoninarum* (of the fish subfamily Salmoninae), *Chlamydia muridarum* (of the mouse/hamster family Muridae)

Second declension, neuter: *Arcobacter bivalviorum* (of bivalves of the class Bivalvia), *Marinobacter bryozoorum* (of the moss animal phylum Bryozoa), *Winogradskyella poriferorum* (of the sponge phylum Porifera)

Third declension: *Actinobacillus anseriformium* (of the bird order Anseriformes)

Names of plants and animals can also be used in compound epithets in combination with other word elements:

Plants

Adjectives – Rule 12c(1): *Lactiplantibacillus fabifermentans*, "fermenting beans" (*Faba*); *Lysobacter erysipheiresistens*, "resisting *Erysiphe*" (powdery mildew)

Nouns in the nominative case in apposition – Rule 12c(2): *Pectobacterium cacticida*, "killer of cactus" (L. *cactus* from Gr. *kaktos*, "prickly plant"); *Rariglobus hedericola*, "inhabitant of *Hedera*" (ivy)

Nouns in the genitive case – Rule 12c(3): *Chryseobacterium arachidiradicis*, "of the root of *Arachis*, the peanut plant"; *Sporolactobacillus shoreicorticis*, "of the bark of the *Shorea* tree"

Animals

Adjectives – Rule 12c(1): *Streptococcus hyovaginalis*, "associated with pig vaginas," from Gr. *hus*, genitive *huos*, pig; *Streptococcus loxodontisalivarius*, "of elephant saliva" (genus *Loxodonta*, the African elephant)

Nouns in the nominative case in apposition – Rule 12c(2): *Blastomonas ursincola*, "inhabitant of bears" (genus *Ursus*); *Volucribacter psittacicida*, "killer of parrots" (genus *Psittacus*)

Nouns in the genitive case – Rule 12c(3): *Clostridium bovifaecis*, "of cow manure" (genus *Bos*, genitive *bovis*); *Streptococcus dentiloxodontae*, "of an elephant's tooth" (genus *Loxodonta*); *Nocardia otitidiscaviarum*, "of ear disease (otitis) of guinea pigs" (genus *Cavia*)

2.10 Geographical Names

Many genera and species were named after the geographical location where the organisms were isolated. The most often used geographical epithet is *massiliensis/massiliense* (pertaining to Marseille, France; Lagier et al. 2018), found in 239 species names (including effectively published names and synonyms, not including *Candidatus* taxa; LPSN, accessed 29 Nov 2022), and *koreensis/koreense* (pertaining to Korea), found 78 times. Trüper (2005) called this trend "localimania."

One reason why geographical epithets ending in *–ensis* or *–ense* became so popular is because no knowledge of Latin is necessary to correctly form such names. Authors only must ensure that the epithet agrees in gender with the generic name to comply with Rule 12c(1) of the ICNP. National and local pride are often an incentive to form such names: a scientist may gain esteem, and possibly even increased research funding, from the national and local government and national science agencies because he/she has named an organism after the country, city, or location. However, such naming of microorganisms can be misused for geopolitical aims, e.g., by naming organisms after areas of territorial dispute between neighboring countries. It must be stressed that prokaryote names formed from geographical locations have no geopolitical meaning (General Consideration 8 of the ICNP).

Here are a number of examples of genus names derived directly from names of geographical locations: *Basilea* (Basel, Switzerland), *Budvicia* (from *Budvicium*, the Latin name of the Cěské Buděјovice, Czechia), *Delftia* (Delft, the Netherlands, also in recognition of the Delft School and its contributions to microbiology), *Dokdonia* (Dokdo, an island in the East Sea near Korea), *Hafnia* (M.L. *Hafnia*, "Copenhagen"), *Massilia* (L. *Massilia*, "Marseille"), *Pragia* (Prague, Czechia), and *Scandinavium* (Scandinavia; the only case in which the neuter gender was chosen; all other names above are feminine).

The diminutive form in the feminine gender can also be used (ending *–ella*): *Dokdonella* (Dokdo, an island in the East Sea near Korea) and *Lusitaniella* (*Lusitania*, the Latin name of Portugal).

The following examples show the use of geographical locations in compound genus names: *Frisingicoccus* (*Frisinga*, the Medieval Latin name of Freising, Germany); *Turicibacter*, *Turicimonas* (*Turicum*, the Latin name of Zürich, Switzerland; and *Bavariicoccus* (*Bavaria* (Bayern), Germany).

Most geographical specific and subspecific epithets end in *–ensis* (masculine, feminine) or *–ense* (neuter). When the name of the locality ends in –a or –e or –en, these letters are dropped as shown for *jenensis* ("from Jena"), *aotearoense* ("from Aotearoa, the Maori name of New Zealand"), *californiensis* ("from California"), *drentensis* ("from Drente, the Netherlands"), and *bremensis* ("from Bremen, Germany"). When the locality's name ends in –o, the ending becomes *–nensis*/*–nense: Streptomyces sapporonensis* and *Streptoverticillum sapporonense* (Sapporo, Japan), *Aggregicoccus edonensis* (Edo, the old name of Tokyo), *Arcobacter ebronensis* (Ebro Delta, Spain), and *Marinomonas dokdonensis* (Dokdo, an island in the East Sea near Korea).

The *–ensis*/*–ense* ending can be used for epithets based on localities that had a classical or Medieval Latin name. Here are a few examples: *Vibrio albensis* (L. *Albis*, "the river Elbe, Labe"), *Corynebacterium argentoratense* (L. *Argentoratum*, "Strasbourg," France), *Paenibacillus barcinonensis* (L. *Barcino*, "Barcelona," Spain), *Mycobacterium basiliense* (L. *Basilea*, "Basel," Switzerland), *Neobacillus bataviensis* (L. *Batavia*, "the Netherlands"), *Arthrobacter gandavensis* (M.L. *Gandavum*, "Ghent," Belgium), *Desulfitobacterium hafniense* (M.L. *Hafnia*, "Copenhagen," Denmark), *Pseudomonas helmanticensis* (L. *Helmantica*, "Salamanca," Spain), *Helicobacter labacensis* (L. *Labacum*, "Ljubljana," Slovenia), *Legionella londiniensis* (L. *Londinium*, "London," England), *Haloferax lucentense* (L. *Lucentum*, "Alicante," Spain), *Staphylococcus lugdunensis* (L. *Lugdunum*, "Lyon," France), *Streptococcus lutetiensis* (L. *Lutetia*, "Paris," France), *Fancicolia* *mancuniensis* (L. *Mancunium*, "Manchester," England), *Brucella melitensis* (L. *Melita*, "Malta"), *Tetrathiobacter mimigardefordensis* (M.L. *Mimigardefordum*, "Münster," Germany), *Mycobacterium noviomagense* (L. *Noviomagus*, "Nijmegen," the Netherlands), *Megasphaera sueciensis* (N.L. *Suecia*, "Sweden"), *Chryseobacterium treverense* (L. *Augusta Treverorum*, "Trier," Germany), *Promicromonospora vindobonensis* (L. *Vindobona*, "Vienna," Austria), and *Mycobacterium ratisbonense* (L. *Ratisbona*, "Regensburg," Germany). Lists of Latin place names can be found at https://en.wikipedia.org/wiki/Category:Lists_of_Latin_place_names (List of Latin Place Names 2022)

When the location already had a name in classical Latin (L.) or Latin equivalents were in use in the Middle Ages ("Medieval Latin," M.L.) or later ("Neo-Latin," N.L.), those names are preferred, but the present-day name of the place can also be used. Thus, for Vienna, Austria, we have *Promicromonospora vindobonensis* (from L. *Vindobona*) and *Nitrososphaera viennensis* (N.L., "pertaining to Vienna"), and for Kiel, Germany, we have *Pseudomonas kilonensis* (M.L. *kilonensis* or *chilonensis*, "pertaining to Kiel") and *Pseudomonas kielensis* (N.L., "pertaining to Kiel"). When *Enterobacter turicensis* was transferred to the genus *Cronobacter*, *Cronobacter zurichensis* was proposed as nomen novum as the name *Cronobacter turicensis* already existed.

In classical Latin, many geographical epithets were based on adjectives of the first/second declension with –(*i*)*cus*, –(*i*)*ca*, and –(*i*)*cum* endings for masculine, feminine, and neuter, respectively. Examples are *Rhizobium aegyptiacus* and *Kocuria aegyptia* ("Egyptian"), *Mycobacterium africanum* ("African"), *Nocardioides antarcticus* ("Antarctic," "southern" in classical Latin), *Microvirga arabica* ("Arabian"), *Cryobacterium arcticum* ("Arctic," "northern" in classical Latin), *Campylobacter armoricus* ("pertaining to Brittany," from L. *Armoricae*), *Natrialba asiatica* ("from Asia"), *Marinisporobacter balticus* (M.L., "pertaining to the Baltic Sea"), *Pseudomonas bohemica* (M.L., pertaining to *Bohemia*), *Lonsdalea britannica* ("British"), *Amycolatopsis cappadoca* ("Cappadocian"), *Aneurinibacillus danicus* (M.L., "Danish"), *Bifidobacterium gallicum* ("of the Gauls," "French"), *Lactobacillus helveticus* ("Swiss"), *Yersinia hibernica* ("Irish"), *Nocardioides hungaricus* (M.L., "Hungarian"), *Lonsdalea iberica* ("Iberian," "Spanish"), *Flavobacterium indicum* ("Indian"), *Enterococcus italicus* ("Italian"), *Rhizobium lusitanum* ("Portuguese"), *Halothiobacillus neapolitanus* ("of Naples" (*Neapolis*), Italy), *Mesorhizobium norvegicum* (M.L., "Norwegian"), *Lentibacillus persicus* ("Persian"), *Pectobacterium polonicum* (M.L., "Polish"), and *Pseudomonas saxonica* ("from Saxony," Germany). Analogous to such examples from classical and medieval Latin, Neo-Latin adjectives were proposed to describe other locations, for example, *Ensifer americanus* ("American"), *Alteromonas australica* ("Australian"), *Bradyrhizobium australafricanum* ("South African," using the Latin adjective *australis*, "southern"; less correct is *Mycobacterium austroafricanum*), *Rickettsia japonica* ("Japanese"), *Desulfovibrio mexicanus* ("Mexican"), *Methylibium subsaxonicum* ("pertaining to Lower Saxony," Niedersachsen, Germany), *Thauera sinica* ("Chinese," from *Sinae*, "China"), and *Flavobacterium turcicum* ("Turkish").

The names of many localities (lakes, rivers, seas, islands, mountains, etc.) consist of two words, usually an adjective and a noun (e.g., Deep Lake, Black Sea, Rio Grande, Long Island, Blue Mountain) or two nouns (e.g., Death Valley, Loch Ness, Monte Carlo). Formation of specific epithets from such names by addition of the adjectival suffix *–ensis* or *–ense* could lead to formally correct but rather awkward names such as *deeplakensis* or *bluemountainense*. If the name of a locality lends itself to translation into Latin, specific epithets may be formed as genitive nouns, concatenating the translations of the two words without hyphenation. Examples are *lacusprofundi* ("of Deep Lake"), *marisnigri* ("of the Black Sea") (adjective + noun), and *vallismortis* ("of Death Valley") (two nouns).

The way the two word elements were concatenated was not always consistent. Therefore, the following proposal (Oren and Schink 2020b) was adopted in Appendix 9 of the 2022 revision of the ICNP: "Compound specific or subspecific epithets of prokaryotes based on localities can be formed by concatenating the genitives of the components if the name of the locality lends itself to translation into Latin. In these names, the basic noun comes first, and is followed by the determining word, which can be an adjective or a noun. Examples for a noun followed by an adjective: *marisnigri*, *lacusekhoensis*; for two nouns: *vallismortis*, *lacuslunae*." Note that use of the genitive forms (*maris*, *lacus*, *vallis*) is recommended rather than the stems followed by connecting vowel *-i-* as found, e.g., in *lacisaponensis*. Here is a non-exhaustive list of such geographical epithets in which two components of the original names were latinized and concatenated: *Aquirufa ecclesiirivi* (from L. *ecclesia*, "church," and L. *rivus*, "creek"; N.L. *ecclesiirivi*, "from Churchcreek, Kirchstaettbach, Obertrum near Salzburg, Austria"), *Polynucleobacter hirudinlacicola* (from L. *hirudo*, "leech," L. *lacus*, "lake" and L. suff. *–cola*, "inhabitant"; N.L. *hirudinilacicola*, "inhabitant of leech lake," Egelsee, Austria), *Simplicispira lacusdiani* ("of Dian Lake," China), *Halorubrum lacusprofundi* ("of Deep Lake," Antarctica), *Halanaerobium lacusrosei* ("of Rose Colored Lake," Retba Lake, Senegal), *Vibrio marisflavi* ("of the Yellow Sea"), *Haloarcula marismortui* ("of the Dead Sea"), *Methanoculleus marisnigri* ("of the Black Sea"), *Ruegeria marisrubri* ("of the Red Sea"), *Aquirufa nivalisilvae* (from L. *nivalis*, snow-covered, and L. *silva*, forest; N.L. *nivalisilvae*, "of the forest of Schneegattern," Lengau, Austria), *Nocardiopsis sinuspersici* ("of the Persian Gulf"), *Amycolatopsis speibonae* (from L. *spes*, hope, and L. *bonus*, good; N.L. *speibonae*, "of good hope," to indicate Cape Town, the Cape of Good Hope, South Africa), *Venenivibrio stagnispumantis* (from L. *stagnum*, pool, and L. *spumans*, foaming, frothing; N.L. *stagnispumantis*, "from a frothing pool," referring to Champagne Pool, Waiotapu, New Zealand), *Bacillus urbisdiaboli* (from L. *urbs*, city, and L. *diabolus*, devil; N.L. *urbisdiaboli*, "of Devil City," Mogui Cheng, Xinjiang, China), *Amantichitinum ursilacus* (from L. *ursus*, bear, and L. *lacus*, lake; N.L. *ursilacus*, "of Bear Lake," referring to Bärensee near Stuttgart, Germany), and *Haloarcula vallismortis* and *Actinopolyspora mortivallis* ("of Death Valley," California).

In a few cases, such compound epithets were formed as adjectives, using the geographical ending *–ensis* (masculine, feminine) or *–ense* (neuter): *Clostridium lacusfryxellense* ("pertaining to Lake Fryxell," Antarctica), *Nitrincola* *lacisaponensis* ("pertaining to Soap Lake," Washington), *Ruegeria lacuscaerulensis* ("pertaining to the Blue Lake," the Blue Lagoon geothermal lake, Iceland), and *Nesterenkonia lacusekhoensis* ("pertaining to Ekho Lake," Antarctica).

2.11 Chemical Elements and Compounds in the Prokaryotic Nomenclature

Section H of Appendix 9 of the ICNP, based in part on the recommendations by Buchanan (1960), presents guidelines for the use of names of chemical elements and compounds in names of prokaryotic taxa.

For some chemical elements, words existed in Latin, e.g., *calx* (genitive *calcis*), *carbo* (genitive *carbonis*), *cuprum*, *ferrum*, *phosphorus* (derived from Gr. *phosphoros*, the light bringer), *plumbum*, *silex* (genitive *silicis*), and *sulfur* (genitive *sulfuris*) for, respectively, calcium, carbon, copper, iron, phosphorus, lead, silica, and sulfur. Greek equivalents can also be used, e.g., *sideros* (iron) and *theion* (transliterated as *thion*) for iron and sulfur. Neo-Latin words ending in *–um* or *–ium* are used for the elements not found in the lexicon of classical Latin. Here are examples showing how names of the elements were used in prokaryotic genus names and epithets: *Deinococcus aluminii* ("of aluminum"), *Desulfuribacillus stibiiarsenatis* ("of antimony and arsenate"), *Arsenicibacter* ("a rod metabolizing arsenic"), *Pseudomonas arsenicoxydans* ("arsenic oxidizing"), *Lysinibacillus boronitolerans* ("boron-tolerating"), *Streptomyces cadmiisoli* ("of cadmium-rich soil"), *Mesorhizobium calcicola* ("inhabiting limestone"), *Desulfuromonas carbonis* ("of coal"), *Anaerosolibacter carboniphilus* ("coal-loving"), *Leucobacter chromiireducens* ("chromium-reducing"), *Maribacter cobaltidurans* ("cobalt-tolerating"), *Acidiplasma cupricumulans* ("copper accumulating"), *Hydrogeniiclostidium* (a hydrogen producing *Clostridium*), *Sulfuritalea hydrogenivorans* ("hydrogen devouring"), *Ferroplasma* (a ferrous ion oxidizing form), *Deferribacter* (a rod that reduces iron), *Sideroxyarcus* ("a curved iron-oxidizing organism"), *Sulfuriferula plumbiphila* ("lead loving"), *Halobaculum magnesiiphilum* ("magnesium-loving"), *Cumulibacter manganitolerans* ("manganese tolerating"), *Pseudodesufovibrio hydrargyri* ("of mercury"; L. *hydrargyrus*, from Gr. *hydrargyros*, referring to the capacity to methylate mercury), *Micromonospora nickelidurans* ("enduring nickel"), *Nitrogeniibacter* ("nitrogen rod"), *Geomonas nitrogeniifigens* ("nitrogen fixing"), *Azotobacter* (from Gr. prep. *a*-, "not," and Gr. n. *zoe*, "life") ("nitrogen rod"), *Thalassiella azotivora* ("nitrogen-eating"), *Candidatus* Methylomirabilis oxygeniifera (proposed correction of oxyfera) ("carrying oxygen"), *Gordonia phosphorivorans* ("phosphorus-devouring"), *Seleniivibrio* ("a vibrio that reduces selenate"), *Pseudomonas seleniipraecipitans* ("selenium precipitating"), *Silicibacter* ("a rod isolated from a silica-rich geothermal lake"), *Natrialba* ("sodium white"), *Streptomyces sodiiphilus* ("sodium ion-loving"), *Sulfurivirga* ("sulfur-oxidizing rod"), *Thiobacillus* ("sulfur rod"), *Halanaeroarchaeum sulfurireducens* ("sulfur-reducing"), *Desulfonatronum thiodismutans* ("sulfur dismutating"), *Geobacter uraniireducens* ("uranium-reducing"), and *Streptomyces zinciresistens* ("zinc-resisting").

In inorganic chemistry, the endings –i and –o are used to indicate different oxidation states of cations: –o for the reduced form such as ferro and cupro, –i for the oxidized forms such as ferri and cupri. The same convention is used in compound names where *ferro–* and *ferri–* indicate Fe(II) and Fe(III), respectively. Thus, *Ferroglobus* and *Mariprofundus ferrooxydans* are Fe(II) oxidizers, while *Ferrimonas* and *Pyrobaculum ferrireducens* are Fe(III) reducers. However, this was not always implemented consistently: *Ferribacterium*, *Ferrithrix*, and *Ferrimicrobium* are iron oxidizers. Here the authors used *-i-* as the connecting vowel as recommended following a Latin word element.

For "oxidizing," we both find the endings *–oxidans* and *–oxydans* in compound epithets. We have *Paracoccus sulfuroxidans* and *Thiovirga sulfuroxydans*, *Acidithiobacillus thiooxidans* and *Acidiferrobacter thiooxydans*, *Leptospirillum ferrooxidans* and *Mariprofundus ferrooxydans*, and *Pacificitalea manganoxidans* and *Aurantimonas manganoxydans*. The "i" spelling was used slightly more often than the "y" spelling. The "y" spelling reflects the origin from the Greek adjective *oxys*, "sharp," "pungent," and "acid."

The rules for the Latinization of names of oxyanions (nitrate, nitrite, phosphate, sulfite, etc.) are as follows. Anions ending in –ate and –ite are treated as masculine nouns of the third declension. The English ending –ate is latinized to *–as* (genitive *–atis*). Thus, nitrate becomes *nitras*, *nitratis*, and sulfate becomes *sulfas*, *sulfatis*. A genus name indicating the ability to reduce sulfate to sulfide may therefore start with *Desulfati–*: *Desulfatibacillum*, *Desulfatiferula*, *Desulfatiglans*, *Desulfatirhabdium*, and *Desulfatitalea*. However, the prefix *Desulfo–* is much widely used. The English ending –ite is latinized to *–is*, genitive *–itis*, so nitrite becomes *nitris*, *nitritis*. Names of reduced anions that in the chemical nomenclature end in –ide are latinized with the neuter ending *–idum*, genitive *–idi*. Thus, from sulfide we have *sulfidum* in names such as *Thermosulfidibacter*. Here are further examples showing how names of inorganic anions were latinized and used in the prokaryotic nomenclature: *Kocuria arsenatis* ("reducing arsenate"), *Intrasporangium chromatireducens* ("reducing chromate"), *Iodidimonas* ("iodide-oxidizing monad"), *Nitratifractor* ("nitrate breaker"), *Thioalkalivibrio nitratis* ("of nitrate"), *Hyphomicrobium nitrativorans* ("nitrate-consuming"), *Stenotrophomonas nitritireducens* ("nitrite-reducing"), *Candidatus* Halomonas phosphatis ("phosphate accumulating"), *Chryseobacterium phosphatilyticum* (dissolving phosphate), *Desulfotignum phosphitoxidans* ("phosphite-oxidizing"), *Thauera selenatis* ("of selenate"), *Sedimenticola selenatireducens* ("reducing selenate"), *Bacillus selenitireducens* ("reducing selenite"), *Desulfatiferula* ("a rod-shaped sulfate reducer"), *Desulfovibrio* ("a sulfate-reducing vibrio"), *Desulfonatronospira sulfatiphila* ("loving sulfate"), *Desulfitispora* ("sulfite reducing spore-forming bacterium"), *Acidianus sulfidivorans* ("sulfide-eating"), and *Desulfonispora thiosulfatigenes* ("thiosulfate-producing").

Latinized names of acids can be formed using the neuter word *acidum* followed by a neuter adjective: *acidum aceticum* ("acetic acid") and *acidum formicum* ("formic acid"). The latinized names of the anions end in *–as*, genitive *–atis*, e.g., *acetas*, genitive *acetatis*, "acetate." Here are examples to show how names of organic acids and their anions were used in the prokaryotic nomenclature.

- Acetic acid (*acidum aceticum*; *acetas*, genitive *acetatis*): *Acetatifactor* ("acetate maker"), *Proteiniphilum acetatigenes* ("acetate producing")
- Butyric acid (*acidum butyricum*; *butyras*, genitive *butyratis*): *Butyricicoccus* ("a coccus-producing butyrate"), *Desulfovibrio butyratiphilus* ("butyrate loving")
- Formic acid (*acidum formicum*; *formas*, genitive *formatis*): *Eubacterium formicigenerans* (formic acid-producing), *Bryantella formatexigens* (formate demanding)
- Lactic acid (*acidum lacticum*; *lactas*, genitive *lactatis*): *Lacticigenium* ("a bacterium that produces lactic acid"), *Streptomyces lacticiproducens* ("lactate producing")
- Oxalic acid (*acidum oxalicum*; *oxalas*, genitive *oxalatis*): *Oxalicibacterium* ("oxalic acid-utilizing rod"), *Azorhizobium oxalatiphilum* ("oxalate-loving")
- Propionic acid (*acidum propionicum*; *propionas*, genitive *propionatis*): *Propionicimonas* ("a propionic acid-producing monad"), *Paludibacter propionicigenes* ("propionic acid-producing")
- Pyruvic acid (*acidum pyruvicum*; *pyruvas*, genitive *pyruvatis*): *Pyruvatibacter* ("pyruvate-using rod"), *Corynebacterium pyruviciproducens* ("pyruvic acid-producing"), *Aminivibrio pyruvatiphilus* ("pyruvate-loving")
- Succinic acid (*acidum succinicum*; *succinas*, genitive *succinatis*): *Succinatimonas* ("succinate-producing monad"), *Basfia succiniciproducens* ("succinic acid-producing"), *Tepidiphilus succinatimandens* ("succinate-eating")

Latinized names of alcohols have the neuter gender and end in *–ol*, with the genitive ending *–is*, e.g., *ethanol*, genitive *ethanolis*:

- Alcohol: *Desulfotomaculum alcoholivorax* ("alcohol eating")
- Butanol: *Peribacillus butanolivorans* ("butanol-eating")
- Ethanol: *Ethanoligenens* ("ethanol-producing bacterium"), *Pediococcus ethanolidurans* ("ethanol resistant")
- Methanol: *Methylobacillus methanolivorans* ("methanol-eating")
- Phenol: *Paraburkholderia phenoliruptrix* ("breaker of phenol")

Latinized names of aldehydes, ethers, and esters have the neuter gender and end in *–al*, *–er*, and *–er*, respectively, with the genitive ending *–is*:

- Furfural: *Desulfovibrio furfuralis* ("of furfural")
- Ether: *Rhodococcus aetherivorans* ("devouring ether")
- Ester: *Paenibacillus esterisolvens* ("ester dissolving")

The authors who named *Paenibacillus etheri* ("of an ether") did not follow these guidelines; *etheris* is the preferred form.

The –yl ending for organic radicals is used in *methylo–* in numerous generic names and specific epithets belonging to organisms metabolizing methyl groups. Occasionally, methyl was latinized to *methylum*, resulting in the genus names *Methylibium* ("methyl life") and *Methylicorpusculum* ("methyl-using small body").

For a few chemicals, the names are latinized as feminine nouns of the first declension (ending *–a*, genitive *–ae*). An example is urea, e.g., in *Sporosarcina ureae*, of urea. Names of alkaloids, nucleic acid bases (adenine, thymine, etc.), and amino acids with English names ending in –ine (alanine, leucine, etc.) are latinized with the ending *–ina*, genitive *–inae*.

Most other names of chemicals that end in –ide (anions), –in, –ane, –ene, –one, –ol (for nonalcoholic compounds), –ose (sugars), –an (polysaccharides), and –ase (enzymes) are latinized as neuter nouns of the second declension ending in *–um*, genitive *–i*. The final –e is deleted and replaced by *–um*: *Pseudomonas benzenivorans* ("benzene-devouring"), *Thauera butanivorans* ("butane-devouring"), *Cellulosimicrobium* ("cellulose microbe") (note that *Cellulosi–* is preferred over *Cell–* as in *Cellvibrio* and *Cellulo–* as in *Cellulomonas*), *Pelobacter glucosidilyticus* ("glucoside-dissolving"), *Proteiniborus indolifex* ("indole-making") (note that in *Clostridium indolis* and *Lacrimispora indolis*, the word was treated as belonging to the third declension, as used for alcohols, although indole is not an alcohol), *Lactonifactor* ("lactone-maker"), *Bacillus mannanilytiucs* ("mannan-dissolving"), and *Methanimicrococcus* ("small methane coccus"); note that the prefix *Methano–* is nearly always used instead of *Methani–*: *Methanobacterium* ("methane rod"), *Methanothermococcus* ("methane producing thermophilic coccus"), *Cohnella laeviribosi* ("of L-ribose"), and *Actinomadura rifamycini* ("of rifamycin").

2.12 Names Originating from Languages Other than Latin and Classical Greek

According to Recommendation 6(3) of the ICNP, words from languages other than Latin or classical Greek should be avoided if equivalents exist in Latin or classical Greek or can be constructed by combining word elements from these two languages. Exceptions: names derived from typical local items, such as food, drink, or geographical localities for which no Latin or Greek names exist. Appendix 9 Section G(4) gives further explanations. National foods or fermentation products such as *sake*, *tofu*, and *kefir* do not have equivalent Latin names. They are latinized by forming a neuter noun by adding *–um* (*sakeum*, *tofuum*, *kefirum*). The genitive (*sakei*, *tofui*, *kefiri*) is then used as specific epithet, e.g., in *Lactobacillus odoratitofui* ("of stinky tofu").

Many studies were devoted to the fermentation of vegetables for the production of *kimchi* in Korea. We find *kimchii* as the epithet or word element in compound epithets in *Leuconostoc kimchii* ("of kimchi"), *Leuconostoc miyukkimchii* ("of *miyukkimchi*, a regional *kimchi* made of brown algae"), and *Leuconostoc mesenteroides* subsp. *jonggajibkimchii* ("of *jonggajibkimchi*, a commercial kind of kimchi"). An adjective was also used: *Lactobacillus kimchicus* ("pertaining to *kimchi*"). Use of the *–ensis* ending as in *Lactobacillus kimchiensis* is not recommended as the *–ensis/–ense* ending is generally reserved for "geographical" epithets.

We find the word *jeotgal*, a kind of Korean fermented seafood, in the generic names *Jeotgalibaca*, *Jeotgalibacillus*, and *Jeotgalicoccus* and as epithet, e.g., in *Bacillus jeotgali* ("of *jeotgal*") and in *Weissella jogaejeotgali* ("of *jogae jeotgal*"). Here are additional examples of the use of national foods in names of prokaryotic taxa: *Alkalibacterium kapii* and *Lentibacillus kapialis* ("of *ka-pi*, a Thai shrimp paste"), *Komagataeibacter nataicola* ("inhabitant of *nata de coco*, coconut gel produced by acetic acid bacteria in the Philippines"), *Lactobacillus futsaii* ("of *fu-tsai*, a traditional fermented mustard product in Taiwan"), *Lactobacillus suantsaii*, *Lactobacillus suantsaicola*, *Lactobacillus suantsaihabitans* ("of *suan-tsai* or inhabiting *suan-tsai*, a traditional fermented mustard green product in Taiwan"), *Lactobacillus plajomi* ("of *pla-jom*, a traditional fermented fish product in Thailand"), *Lactobacillus pobuzihii* ("of *pobuzihi*, fermented cummingcordia fruit in Japan"), *Lactobacillus sunkii* ("of *sunki*, a non-salted Japanese traditional pickle"), *Lysinibacter chungkukjangi* ("of *chungkukjang*, a traditional Korean fermented food"). and *Pectinatus sottacetonis* ("of *sottaceto* [Italian: "under vinegar"], Italian pickled cucumber"). Note that the last epithet was treated as a Latin noun of the third declension to give the *–onis* ending.

Here are a number of cases where the names of local drinks, nonalcoholic as well as alcoholic, were used to form generic names or specific epithets: *Lactobacillus kefiri* ("of kefir," Caucasian sour milk, from Turkish, *kefir*), *Bifidobacterium aquikefiri* ("of water kefir"), *Pueribacillus* (a bacillus from Chinese *pu'er* tea), *Paenibacillus pueri* ("of *pu'er* tea"), *Komagataeibacter kombuchae* ("of *kombucha*, a kind of fermented tea"), *Franconibacter daqui* ("of *daqu*, a starter culture used in the production of Chinese liquors"; an incorrectly formed variant is *Thermoactinomyces daqus*), *Lactobacillus sakei* ("of sake"), and *Corynebacterium nuruki* ("of *nuruk*, a fermentation starter used for Korean traditional rice wine").

When equivalents can be found in the classical languages, it is recommended to use those. For example, the Romans had a product named *garum* that resembled fermented fish sauces such as *nam-pla* known from the Thai cuisine. Accordingly, we have *Natrinema gari* ("of fermented fish sauce") and *Halomonas garicola* ("a dweller of fish sauce"). A nice translation of *makgeolli*, Korean rice wine vinegar, is found in *Acetobacter oryzoeni* (L. or Gr. *oryza*, "rice"; Gr. *oinos* (transliterated *oenos*), "wine"; N.L. gen. n. *oryzoeni*, "of rice wine").

The recommendation to use Latin and Greek equivalents when available was not always followed, as shown in the following examples: *Streptomyces nogalater* (from Spanish *nogal*, "walnut," and Latin *ater*, "black," referring to the production of an odor like that of black walnuts); *Ramlibacter* (from Arabic *raml*, "sand"); *Paenibacillus jamilae* ("of *jamila*," a word of Arabic origin used in Spain for residual water of olive oil production; *olivaquae* ("of olive water") would have been possible); *Halobacillus yeomjeoni* (from Korean *yeomjeon*, "a marine saltern"); *Gaetbulibacter*, *Gaetbulimicrobium*, and *Gramella gaetbulicola* (from Korean *gaetbul*, "a tidal flat"); *Mucilaginibacter gotjawali* and *Sphingomonas gotjawalisoli* (from Korean *gotjawal*, a volcanic forest); *Geobacillus toebii* (from Korean *toebi*, "farmland compost"); *Ehrlichia sennetsu* (from Japanese *sennetsu*, "glandular fever"; note that the epithet is not treated as Latin, contravening Rule 6 of the ICNP; an alternative would be *febriglandularum*, "of fever of the glands"); *Rhizobium etli* (from Nahuatl *etl*, "bean"); *Bradyrhizobium jicamae* (from the native Mexican *jicama* for the legume *Pachyrhizus*; *pachyrhizi* would be a simple alternative); *Streptomyces iakryus*, *Streptomyces tuirus*, and *Streptomyces capoamus* (from, respectively, *iakry*, *tuira*, and *capoama*, "green," "violet-blue," and "island" in the Amazonian Nheengatu language); *Bradyrhizobium paxllaeri* (from *paxllaer*, in the Mochica language referring to *Phaseolus lunatus* and from which the word *pallar* used in Peru for this crop is derived); *Bdellovibrio qatius* (from *q[q']a:yt* ("kill it") in the language of the Musqueam tribe of Coast Salish; a Latin alternative is *necator*, "the killer"); *Mycoplasma simbae* ("of a lion," from Swahili *simba*, "lion"); *Actinomadura kijaniata* (from Swahili *kijani*, "green"); *Ensifer terangae* (from *terenga*, "hospitality" in the West-African Wolof language, intended to mean that the species is found in different host plants); *Streptococcus waius* (from Maori *waiu*, "milk"); and *Flavobacterium akiainvivens* (from 'akia, the Hawaiian name of the shrub *Wikstroemia*; *wikstroemiae* would be a simple alternative).

2.13 Arbitrary Names and Acronyms

Rules 10a and 12c of the ICNP enable formation of "arbitrary" genus names and specific epithets. Section I of Appendix 9—Orthography provides further explanations. Rule 10a states that genus names "may be taken from any source and may even be composed in an arbitrary manner." They are treated as Latin nouns. Rule 10a gives as examples *Afipia*, *Desemzia*, *Waddlia*, or *Cedecea*, derived from the acronyms AFIP (Armed Forces Institute of Pathology), DSMZ (Deutsche Sammlung von Mikroorganismen und Zellkulturen), WADDL (Washington Animal Disease Diagnostic Laboratory), and CDC (Centers for Disease Control), respectively.

Here are a few additional examples of generic names formed from acronyms of the institutions where the studies were performed or of the funding agencies. When the acronyms consisted of consonants, vowels were added in an arbitrary way to make the word pronounceable: *Basfia* (BASF SE, Ludwigshafen, Germany); *Cnuella*, *Cnuibacter* (CNU, Capital Normal University, Beijing, PR China); *Deefgea* (DFG, Deutsche Forschungsgemeinschaft, the German Science Foundation); *Emticicia* (MTCC, Microbial Type Culture Collection, Chandigarh, India); *Ercella* (ERC, European Research Council); *Mameliella* (MMEL, Marine Microbial Ecology Laboratory, Xiamen, PR China); *Nubsella* (NUBS, Nihon University College of Bioresource Sciences, Fujisawa, Japan); *Stakelama* (State Key Laboratory of Marine Environment Science, Xiamen, PR China); and *Tistrella* (TISTR, Thailand Institute of Scientific and Technological Research, Bangkok, Thailand).

According to Rule 12c, "a specific epithet may be taken from any source and may even be composed arbitrarily. Example: *thetaiotaomicron* in *Bacteroides thetaiotaomicron* derived from a combination of the Greek letters *theta*, *iota* and *omicron*." This example is a rare case where the epithet was arbitrarily formed as a noun in the nominative case in apposition. Examples of genitive nouns ending in *–ae* (first declension, feminine), *–i* (second declension, masculine or neuter), or *–is* (third declension) are *Burkholderia unamae* (UNAM, Universidad Nacional Autónoma de México), *Brevundimonas nasdae* (NASDA, National Space Development Agency of Japan), *Citrobacter cronae* (CRONA, Chirurgie, Radiologie, Orthopädie, Neurologie, Anaesthesiologie clinics, Tübingen, Germany), *Muricauda iocasae* (IOCAS, Institute of Oceanology, Chinese Academy of Sciences, Qingdao, PR China), *Rhodoferax bucti* (BUCT, Beijing University of Chemical Technology, Beijing, PR China), *Vibrio cidicii* (CDC, Centers for Disease Control and Prevention, Atlanta, Georgia, USA), *Adhaeribacter swui* (SWU, Seoul Women's University, Seoul, Republic of Korea), *Bacteroides kribbi* (Korean Research Institute of Biosciences and Biotechnology, Daejeon, Republic of Korea), *Profundibacter amoris* (Arctic Mid-Ocean Ridge System, where the isolate was found; the epithet is homonymous with L. gen. n. *amoris*, "of love"), and *Dysosmobacter welbionis* (Walloon Excellence in Life Sciences and Biotechnology, Louvain, Belgium).

Examples of arbitrarily formed adjectives based on acronyms are *Sphingomonas buctiana* (BUCT, Beijing University of Chemical Technology, Beijing, PR China) and *Geobacter icigianus* (ICIG, Institute for Cytology and Genetics, Novosibirsk, Russia). In the past, such adjectives were often formed with the geographical ending *–ensis* (masculine, feminine) or *–ense* (neuter). Rule 12c gives as example *Bacillus safensis*, arbitrarily derived from SAF (the Spacecraft-Assembly Facility at the Jet Propulsion Laboratory, Pasadena, CA, USA). Additional examples are *Pseudomonas asuensis* (ASU, Arizona State University), *Bacillus cecembensis* (CCMB, Centre for Cellular and Molecular Biology, Hyderabad, India), *Bacillus encimensis* (NCIM, National Collection of Industrial Microorganisms, Pune, India), and *Lysobacter niastensis* (NIAST, National Institute of Agricultural Science and Technology, Suwon, Republic of Korea). Use of the *–ensis/–ense* endings is no longer recommended for newly formed names of this type (Oren et al. 2017).

2.14 Final Comments

In his 1999 review *How to Name a Prokaryote*, Trüper wrote: "Latin and Greek offer a wealth of word elements and ways to form words that are by far not exhausted yet and will do so for a long time in the future, although scientists have not always been careful or reasonable in their 'creations.' By mixing Greek and Latin elements, by dropping syllables, repositioning letters, contractions and arbitrary formations the antique wealth of words has been partially widely changed." At the time those words were written, the numbers of new validly published names of genera and species added annually were around 50 and 225, respectively (1995–1999). Since then, the rate at which new taxa of prokaryotes are described has increased greatly: for the period 2017–2021, these numbers were around 175 and 960, respectively (data from LPSN).

I concur with Trüper's statement that the options provided by the dictionaries of Latin and classical Greek, together with the ways Latin and Greek word elements can be combined to form names that meet the demands of the ICNP, are far from being exhausted. By combining a small number of word elements following the guidelines given in Appendix 9 of the ICNP, large numbers of correctly formed names can be created that are not overly long and awkward and relate to the properties of the organisms and/or the environments from which they were recovered. Computer programs that generate names from lists of Latin and Greek word elements can be helpful (Pallen et al. 2021; Telatin 2022), but the machine-generated names are dull and uninteresting. With a little effort, more attractive names that may boost interest in nomenclature can be formed by creative use of the Latin and classical Greek lexicons (Oren 2020), so that there will be a meaning and a story behind the scientific names of prokaryotic genera and species.

References

Buchanan RE (1960) Chemical terminology and microbiological nomenclature. Int Bull Bacteriol Nomencl Taxon 10:16–22. Reprinted (1994) Int J Syst Bacteriol 44: 588–590

Buchanan RE, Holt JG, Lessel EF (1966) Index Bergeyana. An annotated alphabetic listing of names of the taxa of the bacteria. Williams & Wilkins, Baltimore

Catalogue of Life (2022) Annual release 2022. https://www.catalogueoflife.org/2022/08/30/release. Accessed 20 Nov 2022

Deshmukh UB, Oren A (2022a) Proposal of *Membranihabitans* gen. nov. as a replacement name for the illegitimate prokaryotic generic name *Membranicola* Li *et al.* 2016. Int J Syst Evol Microbiol 72:005576

Deshmukh UB, Oren A (2022b) Proposal of *Holzapfeliella* gen. nov. and *Litorivicinus* gen. nov. as a replacements names for the illegitimate prokaryotic generic names *Holzapfelia* Zheng *et al.* 2020 and *Litoricola* Kim *et al.* 2007, respectively. Int J Syst Evol Microbiol 72:005688

Deshmukh UB, Oren A (2023) Proposal of *Christiangramia* gen. nov., *Neomelitea* gen. nov., and *Nicoliella* gen. nov. as replacement names for the illegitimate prokaryotic generic names *Gramella* Nedashkovskaya *et al.* 2005, *Melitea* Urios *et al.* 2008, and *Nicolia* Oliphant *et al.* 2022, respectively. Int J Syst Evol Microbiol 73:005806

Diggle J (editor-in-chief), Fraser BL, James P et al (eds) (2021) The Cambridge Greek Lexicon. Cambridge University Press, Cambridge

Euzéby JP, Boemare NE (2000) The modern Latin word *rhabdus* belongs to the feminine gender, inducing necessary corrections according to Rules 65(2), 12c(1) and 13b of the Bacteriological Code (1990 Revision). Int J Syst Evol Microbiol 50:1691–1692

Google Translate (2022). https://translate.google.com. Accessed 26 Nov 2022

Greenough JB, Kittredge GL, Howard AA et al (eds) (1903) Allen and Greenough's new Latin grammar for scholars and colleges. Ginn & Co., The Athenæum Press, Boston and London. https://www.mtnbrook.k12.al.us/cms/lib/AL01901445/Centricity/Domain/1258/NEW%20LATIN%20GRAMMAR%20OCR.pdf. Accessed 20 Nov 2022

Hobern D, Barik SK, Christidis L et al (2021) Towards a global list of accepted species VI: The Catalogue of Life checklist. Org Divers Evol 21:677–690

Lagier J-C, Bilen M, Cadoret F et al (2018) Naming microorganisms: the contribution of the IHU Méditerranée Infection, Marseille, France. New Microbes New Infect 26:S89–S95

Lewis CT, Short C (1891) A new Latin dictionary (Harper's Latin Dictionary). Clarendon Press, New York, Harper & Brothers/Oxford. https://ia801904.us.archive.org/35/items/LewisAndShortANewLatinDictionary/lewisandshort.pdf. Accessed 20 Nov 2022

Liddell HG, Scott R (1996) A Greek-English lexicon. Ninth edition with revised supplement. Clarendon Press, Oxford. Eighth edition (1882): American Book Company, New York. https://ia802608.us.archive.org/19/items/greekenglishlex00liddrich/greekenglishlex00liddrich.pdf. Accessed 20 Nov 2022

Linnaeus C (1751) *Philosophia Botanica*. G. Kiesewetter, Stockholm

List of Latin Place Names (2022). https://en.wikipedia.org/wiki/Category:Lists_of_Latin_place_names. Accessed 30 Nov 2022

LPSN - List of Prokaryotic names with Standing in Nomenclature (2022) lpsn.dsmz.de; bacterio.net. Accessed 20 Nov 2022

MacAdoo TO (1993) Nomenclatural literacy. In: Goodfellow M, O'Donnell AG (eds) Handbook of new bacterial systematics. Academic Press, London, pp 339–360

Marello.org (2022). https://marello.org/tools/syllabifier. Accessed 20 Nov 2022

Morwood J (ed) (2005) The pocket Oxford classical greek dictionary. Oxford University Press, Oxford

Morwood J, Taylor J (2002) The pocket Oxford classical greek dictionary. Oxford University Press, Oxford

Numen (2022) The Latin Lexicon. https://latinlexicon.org/index.php. Accessed 20 Nov 2022

Oren A (2011) How to name new taxa of prokaryotes? In: Rainey FA, Oren A (eds) Taxonomy of prokaryotes. Methods in microbiology, vol 38. Elsevier/Academic Press, Amsterdam, pp 438–463

Oren A (2019) How to name new taxa of archaea and bacteria. In: Bergey's manual of systematics of archaea and bacteria. John Wiley & Sons, in association with Bergey's Manual Trust

Oren A (2020) Prokaryotic names: the bold and the beautiful. FEMS Microbiol Lett 367:fnaa096

Oren A, Schink B (2016) Formation of names of genera of prokaryotes that end on *–oides* or *–opsis*: a proposal for addenda to Rule 65(2) and to Appendix 9 of the International Code of Nomenclature of Prokaryotes. Int J Syst Evol Microbiol 66:2452–2453

Oren A, Schink B (2020a) Use of Greek in prokaryotic nomenclature: proposal to change Principle 3, Recommendation 6, Rule 7, Rule 65 and Appendix 9 of the International Code of Nomenclature of Prokaryotes. Int J Syst Evol Microbiol 70:3559–3560

Oren A, Schink B (2020b) Further guidelines for the formation of compound specific and subspecific epithets. A proposal to emend Appendix 9 of the International Code of Nomenclature of Prokaryotes. Int J Syst Evol Microbiol 70:3561–3562

Oren A, Vandamme P, Schink B (2016) Notes on the use of Greek roots in the genus and species names of prokaryotes. Int J Syst Evol Microbiol 66:2129–2140

Oren A, Garrity GM, Schink B et al (2017) 'Localimania' revisited: guidelines for the formation of specific epithets for names of prokaryotes based on names of institutions or their acronyms. A proposal for emendation of Appendix 9 to the International Code of Nomenclature of Prokaryotes. Int J Syst Evol Microbiol 67:1618–1619

Oren A, Chuvochina M, Schink B (2019) The use of Greek and Latin prepositions and prefixes in compound names: proposed emendation of Appendix 9 of the International Code of Nomenclature of Prokaryotes. Int J Syst Evol Microbiol 69:1831–1832

Oren A, Arahal DR, Göker M et al (eds) (2023) International code of nomenclature of prokaryotes. Prokaryotic Code (2022 Revision). Int J Syst Evol Microbiol 73:005585

Pallen MJ, Telatin A, Oren A (2021) The next million bacterial names. Trends Microbiol 29:289–298

Parker CT, Tindall BJ, Garrity GM (eds) (2019) International code of nomenclature of prokaryotes. Prokaryotic code (2008 revision). Int J Syst Evol Microbiol 69:S1–S111

SeqCode Publications (2022). https://disc-genomics.uibk.ac.at/seqcode/page/publications. Accessed 30 Oct 2022

Simpson DP (1968) Cassell's Latin dictionary. Latin English, English-Latin, 5th edn. Wiley Publishing, Inc., Hoboken NJ

Skerman VBD, Sneath PHA, McGowan V (1980) Approved lists of bacterial names. Int J Syst Bacteriol 30:225–420

Telatin A (2022) Gan: the Great Automatic Nomenclator. https://github.com/telatin/gan. Accessed 20 Nov 2022

Tindall BJ, Euzéby JP (2006) Proposal of *Parvimonas* gen. nov. and *Quatrionicoccus* gen. nov. as replacements for the illegitimate, prokaryotic, generic names *Micromonas* Murdoch and Shah 2000 and *Quadricoccus* Maszenan et al. 2002, respectively. Int J Syst Evol Microbiol 56:2711–2713

Trüper HG (1996) Help! Latin! How to avoid the most common mistakes while giving Latin names to newly discovered prokaryotes. Microbiología 12:473–475

Trüper HG (1999) How to name a prokaryote? Etymological considerations, proposals and practical advice in prokaryote nomenclature. FEMS Microbiol Rev 23:231–249

Trüper HG (2001) Etymology in nomenclature of prokaryotes. In: Boone DR, Castenholz RW (eds) Bergey's manual of systematic bacteriology, 2nd edn, vol 1. The Archaea and the deeply branching and phototrophic Bacteria. Springer-Verlag, New York, pp 89–99

Trüper HG (2005) Is 'localimania' becoming a fashion for prokaryote taxonomists? Int J Syst Evol Microbiol 55:1753

Trüper HG, Euzéby JP (2009) International code of nomenclature of prokaryotes. Appendix 9: Orthography. Int J Syst Evol Microbiol 59:2107–2113

Woodhouse SC (1910) English-Greek dictionary. Routledge & Kegan Paul, London. https://dcthree.github.io/woodhouse/# and https://perseids-project.github.io/woodhouse-js. Accessed 20 Nov 2022

[illegible]

Chapter 3
Technologies Promoting Genome-Based Taxonomy

Bhagwan Narayan Rekadwad, Zheng-Han Lian, Jian-Yu Jiao, and Wen-Jun Li

3.1 Introduction

Prokaryotic genomes are essential for thorough studies on genetic and functional similarity among microorganisms and give novel insights about new characters in a species (Konstantinidis and Tiedje 2005). Getting the whole genome of type strains becomes a critical requirement to sustain and/or for progressive evaluation of microbial species, but obtaining genome sequences of all type strains is still a challenging task (Collins and Fink 1995; Nurk et al. 2021; Reardon 2021). On the other hand, one can sequence the whole genome in a few hours (Gautam et al. 2019). The speed and quality of whole-genome sequencing have significantly improved, reducing sequencing time (Saunders et al. 2012; Tyler et al. 2018; Sereika et al. 2022). Ultimately, there is a question mark over the quality of the obtained genome and its error-free completeness. Hence, the incorporation of well-curated genomes and high-end computational tools has boosted the credibility of today's prokaryotic taxonomy in this modern genomic era (Chun and Rainey 2014). Culturing methods have avoided or failed to achieve 100% isolation of microbial taxa in a pure-culture form (Connon and Giovannoni 2002; Martiny 2019; Youseif et al. 2021). In contrast, bulk DNA extracted from samples proved that sequencing and computer analysis tools have considerably improved and are able to recover most of the sequence

B. N. Rekadwad
Department of Microbiology and Biotechnology, Yenepoya Research Centre, Yenepoya (Deemed to be University), Mangalore, Karnataka, India

Z.-H. Lian · J.-Y. Jiao · W.-J. Li (✉)
State Key Laboratory of Biocontrol, Guangdong Provincial Key Laboratory of Plant Resources and Southern Marine Science and Engineering Guangdong Laboratory (Zhuhai), School of Life Sciences, Sun Yat-Sen University, Guangzhou, China
e-mail: liwenjun3@mail.sysu.edu.cn

W.-J. Li et al. (eds.), *Modern Taxonomy of Bacteria and Archaea*,
https://doi.org/10.1007/978-981-99-5720-0_3

data in an uncultured way. More than >80% of microbial taxa are represented by uncultured species (Hugenholtz et al. 2016).

A shotgun technique is commonly used to identify genome sequences, in which sequence reads are produced from random locations in the genome using one of a limited number of sequencing technologies (Heather and Chain 2015). Second-generation sequencing generates a huge number of short sequence reads, whereas third-generation sequencing generates fewer but significantly longer sequence reads (Slatko et al. 2018; Adewale 2020). Additionally, recognizing specific bases is more inaccurate when reading long sequences. Today's next-generation technologies are now overcoming this issue (Amarasinghe et al. 2020). These issues are often solved using a hybrid strategy that combines long read data with short read data or a self-correction that uses a consensus approach (Morisse et al. 2021). Although certain WGS investigations are done directly on the sequence read data (e.g., detecting SNPs by mapping the reads to a reference sequence), assembly software is often employed to build a genome assembly that will be utilized in downstream research such as cgMLST, AMR gene identification, and SNP detection (Kwong et al. 2015; Fu et al. 2019; Mahmoud et al. 2019; Valiente-Mullor et al. 2021). There are significantly more assembly software options available than sequencing technologies, making it impossible to examine objectively (Pereira et al. 2020).

The way that microbial systematics describe novel species using genetic and environmental information is undergoing a revolutionary transformation. For instance, a solid framework for the evolutionary and taxonomy relationships of bacteria in the *Idiomarinaceae* family has been developed using comparisons based on single copies of orthologous proteins from core genomes, digital DNA-DNA association (dDDH), and average nucleotide identity (AAI) (Liu et al. 2018). The great majority of the unseen microbial world on Earth needs a close tap to reveal their genetic similarity or diversity, metabolomics profiles, and potential industrious use (Rosselló-Móra and Whitman 2019; Romalde et al. 2019). Various algorithms, software solutions, and service websites offer free or paid analyses and comparative information to a microbial taxonomist. Therefore, sequencing technology, algorithms, software solutions, and services become invaluable for "microbial systematics" and "microbiome analyses" research. This manuscript aims to offer a wide overview of genome-sequencing technologies, algorithms, software solutions, and services for prokaryotic genomes based on their documented use in genome databases.

3.2 DNA Discovery to Genomic Era

The discovery of the DNA double helix (Watson and Crick 1953) has pulled the world into the genomic era (Fig. 3.1). Watson and Crick (1953) showed the world that nucleotide bases could be put in a certain order and that two strands of DNA could be wound around each other to form a nucleotide double helix. Nucleotide base pairs have been arranged, and their sequence arrangement may be read or

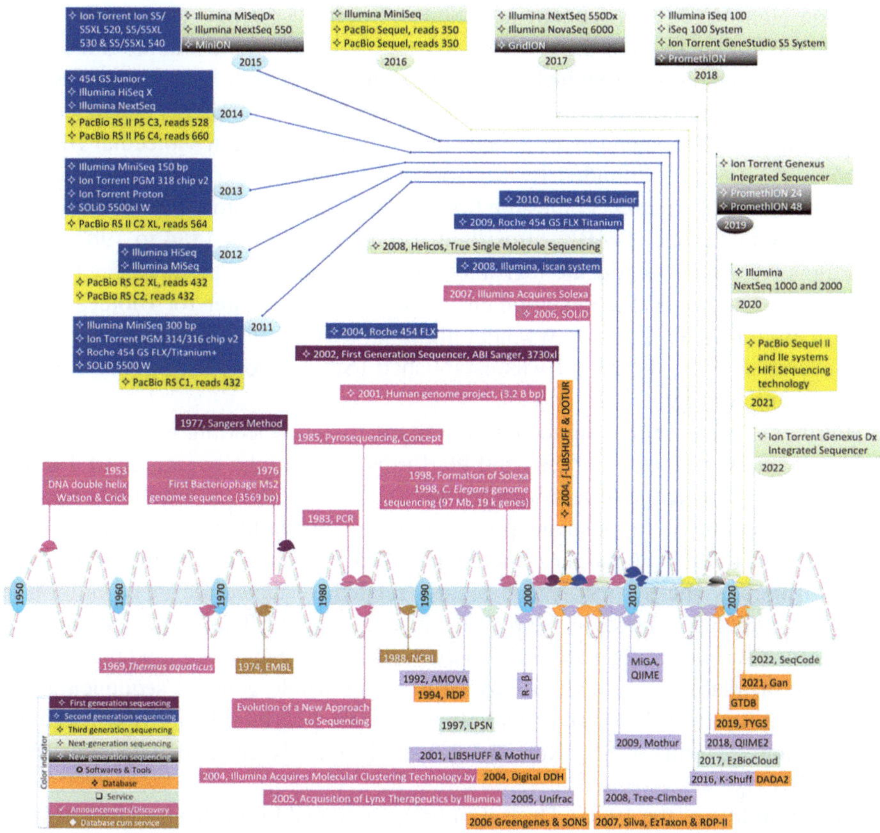

Fig. 3.1 Overview of sequencing technologies, algorithms, software solutions, and services

identified. This concept led to the development of a core DNA reading technology called "sequencing," which permits the reading and counting of base pairs (Koonin and Galperin 2003).

Since about the early 1970s, microbial taxonomists and ecologists have employed most culture-dependent approaches to describe bacteria and have effectively created several keys to identify and classify a strain. Given the time and effort required, genome sequencing was not widely used for strain verification and authentication until the publication of the 3569-nucleotide-long RNA sequence of Bacteriophage MS2 (Fiers et al. 1976). This entire MS2 RNA sequence was obtained through two-dimensional fractionation of radiolabeled phosphate (^{32}P) in nucleotides (Sanger et al. 1965). *T. aquaticus* YT-1 (NCBI Accession No. X58343) was discovered and is a thermophile capable of growing at 70–75 °C. Thomas D. Brock discovered it in Yellowstone National Park in a hot spring (Brock and Freeze 1969; Bateson et al. 1990). The discovery of YT-l paved the way for the discovery of polymerase chain reaction, which was invented by Kary B. Mullis, who was awarded the Nobel Prize in Chemistry in 1993 (Saiki et al. 1988; Gelfand 1989; Logan 2021).

Soon after MS2 RNA sequencing, "Sanger" described a new method for determining nucleotide sequence that employs the 2′, 3′-dideoxy and arabino-nucleoside moieties of standard deoxynucleoside triphosphates, which function as specific chain-terminating inhibitors of DNA polymerase (Sanger et al. 1977). Concurrent with the above phenomena is the establishment of the European Molecular Biology Laboratory (EMBL), founded by Sir John C. Kendrew, who served as the first Director General, which becomes the first and premier laboratory for basic molecular biology in Europe (EMBL, https://www.embl.org/). From 1884 to 1987, many meetings were held to educate legislators and their staffs on the need for funding genomic research and the development of the National Center for Biotechnology Information (NCBI, https://www.ncbi.nlm.nih.gov/books/NBK148949/). Finally, in November 1988, the NCBI was established (Smith 2008; 2013), which offers various services (Table 3.1).

Further to that, sequencing technology began to zoom, leading to the creation of new or alternative methodologies and platforms. Around the beginning of 1986, a fundamental knowledge of DNA polymerase's function was gained. It involves the examination of the pyrophosphate produced after a nucleotide is incorporated into a DNA strand. In this method, enzyme-coupled processes and bioluminescence were used to monitor the pyrophosphate release related to nucleotide incorporation in real time, popularly known as the pyrosequencing method. This is the first and only de novo DNA sequencing alternative to the well-known Sanger method that is commercially available. The method may be used for whole-genome sequencing, tag sequencing (up to 100 bases), and single-nucleotide polymorphism (SNP) analysis (Nyrén and Lundin 1985; Ronaghi et al. 1998; Nyrén 2001; Nyrén 2007; Nyrén 2015).

In the middle of the 1990s, Dr. Balasubramanian and his colleagues investigated the synthesis of DNA by a DNA polymerase enzyme. DNA polymerases are well-known molecular catalysts for precisely transferring nucleotide monophosphates from a developing strand of DNA to an RNA or DNA template. In collaboration with his colleague David Klenerman, a laser spectroscopy expert, Dr. Balasubramanian advanced this investigation. Their collaboration led them down the path of using a single-molecule fluorescence detection technique to observe a DNA polymerase interact with a DNA substrate. To examine the integration of fluorescently tagged deoxynucleotide monophosphates (dNTPs) by a DNA polymerase employing radiolabeled dNTPs as a co-substrate, tests were conducted in which discrete single molecules of DNA were immobilized on a surface (Balasubramanian 2011). Concurrently with the evolution of sequencing technologies, a 97 Mb genome of *Caenorhabditis elegans* has been sequenced that contains 19,000 different genes (The C. elegans Sequencing Consortium 1998).

Table 3.1 Data analysis and retrieval tools are offered by the National Center for Biotechnology Information (NCBI, http://www.ncbi.nlm.nih.gov)

Year	Service	Description
1990	BLAST	A sequence comparison algorithm for quick sequence similarity search
1991	Entrez	Search and retrieval system to find information
1992	GenBank	A nucleotide sequence databases
1993	Network	Network Entrez bought Entrez to the Internet
1994	NCBI website	dbEST, dbSTS, and BLAST service mounted
1995	Genomes	Annotations, maps, sequences, and other genomic data are all part of this new database
	Blankit	Facilitate sequence submissions to GenBank
1996	OMIM	Directory of human genes and genetic disorders
1997	PubMed	Free access to bibliographic retrieval system
1998	NIH disease-based service	Cancer Genome Anatomy Project (CGAP) identifies genes expressed in malignant phases
1999	Human genome	Complete sequence of first human chromosome and working draft of the entire human genome
	Suite of genomic resources	LocusLink (genetic loci descriptors) RefSeq (human reference sequences – nonredundant set) dbSNP (human genetic variation)
2000	PubMed Central	Digital archives of biomedical and life sciences full-text
	GEO	Gene Expression Omnibus (GEO) – public microarray data repository
2001	Bookshelf	Books and publications in the domains of medicine and healthcare are available for free
2002	WGS	Whole-genome shotgun sequence available
2003	DTDs	Document type definitions (DTDs) used to archive and exchange journal content
	Gene	Gene database
2004	PubChem	Chemical structure of macromolecules
2005	NIH Public Access	Manuscript submission system for submission of published papers to PubMed Central
	My NCBI	User can customize service of NCBI databases
2006	dbGAP	Genotypes and Phenotypes (dbGaP) database, which includes genome-wide association studies (GWAS)
2007	Genome Reference Consortium	Provide reference genome of model organisms/type strains
2008	Discovery Initiative	A service – automated methods – for data mining through query
	Public access becomes mandatory	Publication that receives NIH funding made available in PubMed Central after embargo period
	1000 genome project	Mapping of human genetic variations

(continued)

Table 3.1 (continued)

Year	Service	Description
2010	dbVar	Massive genomic variation data and the phenotypic information that goes along with it
	My Bibliography	Referencing tool
2011	PubMed Health	The service provides information on sickness prevention and treatment, focusing on clinical efficacy studies. NLM shuttered it on October 31, 2018, to consolidate consumer health information in PubMed, MedlinePlus, and Bookshelf
2012	GTR	Genetic Testing Registry (GTR) explains genetic tests to doctors, scientists, patients, and others. The database lists valid, beneficial genetic tests
2013	ClinVar	ClinVar gathers sequencing and health data. The database comprises external and internal sources such as dbSNP, dbVar, dbGap, and Gene Reviews
	PubReader	NCBI makes a new way to show PMC articles on a desktop computer, laptop, or tablet

3.2.1 Sequencing Technologies and Platforms for DNA and RNA Based Sequencing

Since the 1990s through 2022, sequencing technologies can be divided into five key categories: first generation, second-generation, third-generation, and next-generation and new-generation technologies (see Fig. 3.1). Based on Sanger's method, the first generation sequencing (ABI Sanger, 3730XL, 2002) is the best sequencing technology currently available (1977). Numerous researchers, including myself, believe that this is the best sequencing method, especially for 16S rRNA gene sequencing and multi-locus sequence typing (MLST), i.e., multi-locus sequence analysis (MLSA) of housekeeping genes and other genes, due to its low cost, ease of operation, speed, and 99.999% accuracy.

About 24 years have passed since David Klenerman and Shankar Balasubramanian had the idea for research that would eventually allow for quick and complete human genome sequencing. Discoveries made during fundamental exploratory research conducted in our facilities in the mid- to late-1990s served as the main inspiration for Solexa sequencing, and the unintentional direction of these observations toward DNA sequencing was a by-product of these observations. The University of Cambridge was the site of the first concepts and proof-of-concept tests. We later created the first commercial sequencing system at Solexa Limited, a business we started in 1998. Early in 2007, Illumina, Inc. purchased Solexa, and after making more upgrades, numerous new sequencing systems emerged (Balasubramanian 2015), including Roche 454 FLX in 2004 (Mardis 2008; Balzer et al. 2010) and SOLiD in 2006 (Valouev et al. 2008; Cloonan et al. 2008; McKernan et al. 2009; Pettersson et al. 2009). Moreover, Illumina acquired Molecular Clustering Technology in 2004 and Lynx Therapeutics in 2005 (Illumina Inc. 2022).

As the years passed, more high-throughput technologies were discovered and termed "second-" and "third"-generation sequencing, more commonly known as "next-generation" or "new-generation" sequencing technologies. Second-generation technologies have launched competition among sequencing platforms and have a greater number of sequencing platforms, namely, Roche 454 FLX (2004), Illumina (2008), and the iscan system (2009), followed by Roche 454 GS Junior in 2010, Illumina MiniSeq 300 bp, Ion Torrent PGM 314/316 chip v2, Roche 454 GS FLX/Titanium+, SOLiD 5500 W in 2011, Illumina HiSeq and MiSeq in 2012, Illumina MiniSeq 150 bp, Ion Torrent PGM 318 chip v2, Ion Torrent Proton, SOLiD 5500xl W in 2013, 454 GS Junior+, Illumina HiSeq X, Illumina NextSeq in 2014, and Ion Torrent Ion S5/S5XL 520, S5/S5XL530, and S5/S5XL 540 in 2015. Similarly, third-generation technologies are also in use, and PacBio dominates all of them including PacBio RS C1, reads 432 in 2011, PacBio RS C2 XL, reads 432 and PacBio RS C2, reads 432 in 2012, PacBio RS II C2 XL, reads 564 in 2013, PacBio RS II P5 C3, reads 528 and PacBio RS II P6 C4, reads 660 in 2014, PacBio Sequel, reads 350 and PacBio Sequel, reads 350 in 2016, and PacBio Sequel II and IIe systems and HiFi sequencing technology in 2021.

In the middle of the twenty-first century, next-generation sequencing became more popular than the second and third generations. These high-throughput sequencing technologies started to obtain the exact order of nucleotides, from a few base pairs to a complete genome, epigenome, and transcriptome. Platforms offering more accurate and quality of sequence are generated by next-generation sequencing platforms including Illumina MiSeqDx and Illumina NextSeq 550 in 2015, Illumina MiniSeq in 2016, Illumina NextSeq 550Dx and Illumina NovaSeq 6000 on 2017, Illumina iSeq 100, iSeq 100 System and Ion Torrent GeneStudio S5 System in 2018, Ion Torrent Genexus Integrated Sequencer in 2019, Illumina NextSeq 1000 and 2000 in 2020, and Ion Torrent Genexus Dx Integrated Sequencer in 2022 in an almost half-decade. Precisely, the preparation of DNA or RNA samples, read length, and the data analysis choices employed are the primary differences between next-generation sequencing (NGS) technologies. MinION in 2015, GridION in 2017, PromethION 24 in 2018, and PromethION 48 in 2017 all emerged as new-generation sequencing platforms that competed with one another for competitive analyses and services.

3.3 Algorithms, Software Solutions, Webservers, and Services for Genome-Based Taxonomy and Sequence Analysis

In the quest for cutting-edge technologies to analyze genome-based variation and function and the evolution of sequencing platforms, another development was happening in the shape of diverse software solutions, algorithms, databases, etc. Technology advancements and the accessibility of a wide range of datasets have fueled researchers' interest, which has driven them to investigate the origins, functions, and interactions of the microbial communities present in ecosystems. Initially, the majority of investigations were restricted to estimating diversity levels. These are now utilized to compare the beta-diversity within and between groups in order to gauge their relative diversity. The Simpson index, Shannon-Weaver index, and richness index are the three most commonly used diversity indices. These indices enable statistical comparisons between various communities, but not among them (Bent and Forney 2008). The range of information and the foundation of the relationship were therefore constrained. Operational taxonomic units (OTUs) were therefore found to be necessary for the assessment of the range of species dominance, richness, and evenness indices and to understand how ecosystems work (Saitou and Nei 1987; Hill et al. 2003).

In meantime, algorithms for comparing gene sequence libraries and an explanation for structural and compositional variations began to be included in the analyses. LIBSHUFF was one of the first algorithms for this type of analysis (Singleton et al. 2001). Several analysis tools were developed, including AMOVA (Excoffier et al. 1992), ∫-LIBSHUFF, DOTUR, (Schloss et al. 2004; Schloss 2008), Mothur (Schloss et al. 2009), UniFrac (Lozupone and Knight 2005; Lozupone et al. 2006), simplex SONS (Glover and Klingman 1981), PSON (Xiao et al. 2011), TreeClimber, ∫-LIBSHUFF, Library Compare, Metastats (Schloss 2008), and many others. Recently, the structure and makeup of microbial communities were compared simultaneously using a sophisticated algorithm called K-shuff. The Intra-k function and the Cross-k function, which increase our power to statistically comprehend the observed differences between the contrasted microbial communities, were also presented by K-shuff as two new diversity indices. While some of them make use of the whole set of sequencing data, others rely on clustering to assess the degree of diversity as well as the phylogenetic composition of the variety in a particular environment (Jangid et al. 2016). As a result, users rely on various algorithms to interpret the data and provide relevant results (Rifaie et al. 2022). In addition, more sophisticated analysis platforms, software, web servers, and databases are in place to fulfill the intended goal.

3.3.1 The Divisive Amplicon Denoising Algorithm (DADA)

The Divisive Amplicon Denoising Algorithm (DADA) infers the sample genotypes and error factors that formed a metagenome dataset in the absence of training data (Rosen et al. 2012). It is a method for fixing amplicon errors that is based on a model. DADA finds small differences in 454-sequenced amplicon data while giving out a few false positives.

There are several advantages of using DADA:

- DADA is more accurate than other techniques, especially for detecting near, actual diversity. DADA uses both read abundance and sequence distance.

- It is based on a straightforward probabilistic model; therefore, its parameters are simple to interpret. Using DADA, you may determine which sequences DADA believes to be true.
- DADA will infer your dataset's error parameters.
- DADA is designed to be fast and minimize sequence alignments.

3.3.2 DADA2

In DADA2, putting similar sequences together makes it less likely that mistakes will be mistaken for biological variation. DADA2 uses an updated model of Illumina amplicon errors that takes quality into account. The error model is used to partition the amplicon reads into groups that reflect the sample's composition (Methods). DADA2 does not rely on external references and can be used on any genetic locus. The DADA2 R package implements the full amplicon workflow, including filtering, dereplication, sample composition, and merged paired-end reads output (Callahan et al. 2016). However, OTUs don't make the most of the quality of modern sequencing because they don't allow for fine-scale variation to be resolved. In the sequenced region, the strains in the balanced, HMP, and extreme communities were different by as little as one nucleotide (nt). Sequence quality was different, with Balanced having the best quality (mean $Q = 35.9$ forward and 33.5 reverse) and Extreme, Moderate, and HMP having the worst quality (32.3/28.7) (Kondratenko et al. 2020). The sequences that had an exact BLAST hit (100% identity and 100% coverage) were called "Exact," while the ones that had the best hit but had one mismatch or gap were called "One Off." DADA2 did a good job of separating sequence variants that were only different by one nucleotide and were found in as few as two reads. By every measure except the number of reference strains found in the merged Extreme dataset, UPARSE was the most accurate OTU method that was tested. Mothur and QIIME made errors that were both chimeric and not chimeric. DADA2's main denoising algorithm was slower than UPARSE but roughly the same speed, and it could easily handle Illumina samples. For the filtered balanced forward reads, UPARSE took 9 s, DADA2 21 s, QIIME (uclust) 17 s, Mothur 2 min and 26 s, and MED 2 min and 34 s. DADA found small differences in 454-sequenced amplicon data with low false-positive rates. Rather than building OTUs from scratch, DADA uses a model-based approach to fix amplicon errors (Callahan et al. 2016). To fix this issue, the DADA2 variant was developed. DADA2, an open-source R package available at https://github.com/benjjneb/dada2, builds upon and refines the DADA algorithm.

The DADA2 software can be installed using the following code through Bioconductor:

```
if (!requireNamespace("BiocManager", quietly = TRUE))
 install.packages("BiocManager")
BiocManager::install("dada2", version = "3.11")
```

Refer to https://benjjneb.github.io/dada2/dada-installation.html for more details.

3.3.3 *Quantitative Insights into Microbial Ecology (QIIME)*

QIIME is an open-source bioinformatics pipeline for analyzing microbiome sequencing data. QIIME is intended to guide users from raw sequencing data obtained on Illumina or other platforms to publication-quality visuals and statistics. This comprises demultiplexing and quality filtering, OTU selection, taxonomic assignment, and phylogenetic reconstruction, as well as studies and visualizations of diversity. Studies involving billions of sequence reads from tens of thousands of samples have utilized QIIME (Caporaso et al. 2010).

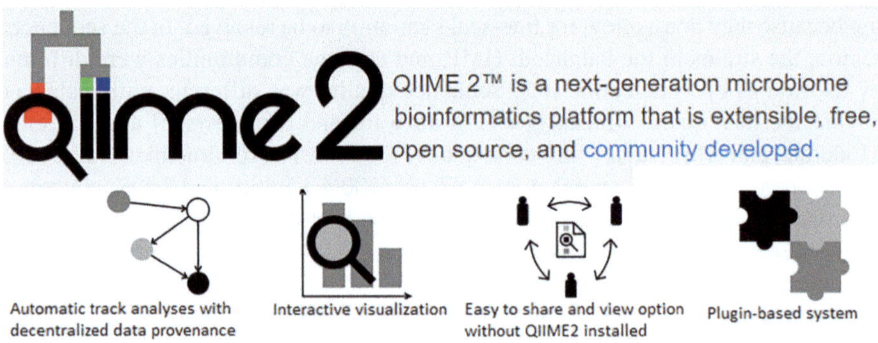

QIIME 2 supplanted QIIME 1 on January 1, 2018. QIIME 1 is currently no longer supported, since all development and support efforts for QIIME are now concentrated on QIIME 2. QIIME 1 users should make the switch to QIIME 2. If you are unfamiliar with QIIME, you should begin by studying QIIME 2, not QIIME 1 (http://qiime.org/). QIIME 2™ is an extensible, free, open-source, and community-developed microbiome bioinformatics platform of the next generation (https://qiime2.org/). QIIME 2 automatically tracks your analyses with decentralized data provenance – no more guesswork on what commands were executed – it enables you to explore your data with beautiful visualizations that provide new perspectives in an interactive manner; it makes it simple to share results with your team, even those members who do not have QIIME 2 installed; and it has a plugin-based system that houses all of your favorite microbiome methods (Bolyen et al. 2019).

Natively, qiime2 can be installed in a `conda` environment and `miniconda` package manager, which is a best way to install qiime2. While installation, if conda environment is already available, simply update it using command `conda install wget` or use command `conda install wget` to install a fresh conda. Once conda environment is created, install qiime 2 Core 2022.8, which is a recent version as of today.

```
conda update condawget https://data.qiime2.org/distro/core/
qiime2-2022.8-py38-linux-conda.yml
conda env create -n qiime2-2022.8 --file qiime2-2022.8-py38-linux-
conda.yml
```

For any update if announced in the future, one can simply replace `qiime2-2022.8` in above two commands with latest version similar format.

Once qiime2 is installed, activate qiime2 using the following command:

```
conda activate qiime2-2022.8
```

A successful installation can be checked by running the following command:

```
qiime -help
```

The qiime2 can be reactivated, version overview, and deactivated.

Activate qiime2: `conda activate qiime2-2022.8.`
Version check: `conda info.`
Deactivate qiime2: `conda activate qiime2-2022.8.`

3.3.4 The Ribosomal Database Project (RDP)

Concurrently with the development of algorithms, several software solutions, web servers, and services have been developed from 1994 to 2022. After the discovery of the 16S rRNA gene (Sogin et al. 1971), followed by its use as a marker gene for species delineation (Woese et al. 1990; Patel 2001), the Ribosomal Database Project (RDP) has launched its services and software through an online repository of ribosome-related information. The ribosomal RNA gene sequences used in the RDP alignments come from GenBank, EMBL, and individuals who sent them directly to the RDP. Phylogenetic trees, rRNA secondary structure diagrams, and other software for processing, evaluating, and displaying alignments and trees are all included in the service. FTP has provided access to the ribosomal sequence data. RDP servers have provided access to programs such as Convert Aln, DNArates, Editor_AE2, Editor GDE, EPSFilter, fastDNAml, Graphic Converter, Readseq, SeqEdit, Subalign, and TreeTool (Maidak et al. 1994). In 2004, RDP was modified in response to the significant growth in 16S rRNA gene sequences and the necessity for high-volume rRNA analysis tools as the Ribosomal Database Project, i.e.,

RDP-II (Cole et al. 2005). RDP-II databases (http://rdp.cme.msu.edu/) are continuously updated on a regular basis to provide services for analyses (Cole et al. 2011), such as Classifier (Wang et al. 2007), FunGene (Fish et al. 2013), FrameBot (Wang et al. 2013), the latest data and high-throughput software tools (Cole et al. 2014), and Xander (Wang et al. 2015). Currently, RDP contains programmers such as the Hierarchy Browser, Classifier, Probe Match, FunGene, MIxS GoogleSheets, Library Compare, Sequence Match, RDPipeline, Aligner, and Tree Builder (http://rdp.cme.msu.edu/).

3.3.5 Greengenes

A 16S rRNA gene database (http://greengenes.lbl.gov) addresses the shortcomings of public repositories by offering chimaera screening, standard alignment, and taxonomic classification based on multiple published taxonomies. It was discovered that curators' taxonomic nomenclature differs even at the phylum level. Putative chimaeras were found in 3% of environmental sequences and 0.2% of isolate-derived records. Environmental sequences from archaea and bacteria were classified into 100 phylum-level lineages (DeSantis et al. 2006). With the use of an enhanced "taxonomy to tree" approach, we applied the taxonomies of Greengenes, the NCBI, and CyanoDB to a tree topology. In addition, the NCBI taxonomy provides clear rank information to group names for greater user orientation and consistency in classification. The revised Greengenes taxonomy provides essential infrastructure for a wide range of megasequencing initiatives ranging from single organisms to the entire planet's genomics (McDonald et al. 2012).

3.3.6 SILVA

A multidisciplinary Microbial Genomics and Bioinformatics Research Group (Bremen, Germany) developed SILVA, which provides updated 16S/18S small subunit and 23/28S large subunit RNA "Parc" databases on the website as well as ARB reference files. These datasets can be used for a variety of analyses and comparisons of sequences for phylogenetic reconstructions, detection and quantification of microbial diversity, and so on (Pruesse et al. 2007; Quast et al. 2013; Yilmaz et al. 2014). The SILVA SSU 138.1 update release (https://www.arb-silva.de/) currently contains 9, 469, 124 (SSU Parc), 510, 508 (SSU Ref NR 99), 1, 312, 534 (LSU Parc), 95, 286 (SSU Ref NR 99), and NR 99 (LSU Ref). SINA is the new SILVA (web) aligner. SINA (SILVA INcremental Aligner) is a novel aligner that can steadily align hundreds of thousands of sequences using a chosen SEED alignment (Pruesse et al. 2012).

3.3.7 *EzBioCloud*

EzBioCloud (https://www.ezbiocloud.net/) provides Microbial Taxonomy, Systematics, and Microbiome Analyses (former name: EzTaxon) to researchers worldwide in the field of microbiology for bioinformatics. These include identification of prokaryotes based on curated 16S rRNA from type strains (16S-based ID), genomics (whole-genome analysis), identifying bacterial isolates using genomes (genome-based ID/TruBac ID), microbiome taxonomic profiling (16S-based MTP), microbiome taxonomic profiling-beta (shotgun-based MTP), and comparative genomics along with the EzCOVID19 database for detection and characterization of SARS-CoV-2 (Yoon et al. 2017).

3.4 Digital DDH

The bacteria and archaea pragmatic species concept was previously based on DNA-DNA hybridization (DDH). This technique allows the taxonomist to obtain an estimate of the overall similarity between the genomes of two strains. It is time-consuming and error-prone, and it cannot be used to incrementally build up a comparative database. Recent advances in genome-sequencing technology necessitate the use of bioinformatics methods to replace wet-lab DDH with in silico genome-to-genome comparison (Auch et al. 2010; Rekadwad et al. 2016). On a digital DDH basis, inferring intergenomic distances efficiently and determining high-scoring segment pairs or maximally unique matches perform well (Auch et al. 2010; Meier-Kolthoff et al. 2013). The investigated distance functions, which can cope with heavily reduced genomes and repetitive sequence regions, outperform previously described ones in terms of correlation with and error ratios when emulating DDH. Some distance formulas are very robust against missing fractions of genomic information, according to simulations of incompletely sequenced genomes. DDH values have a better correlation with 16S rRNA gene sequence distances than digitally derived genome-to-genome distances (Auch et al. 2010).

3.4.1 *MiGA*

A system for managing and processing data for microbial genomes and metagenomes is called MiGA. Its primary objective is to offer a consistent framework for genome-based taxonomy classification and diversity investigations, though its foundation can be used to other fields as well. MiGA provides following services which are exclusively available for all users.

TypeMat

Search the type material genomes for the nearest genomes from formally designated species to determine taxonomy and uniqueness.

NCBI Prok

Use the entire- and chromosome-level NCBI Genome database (prokaryotes) to find closely similar complete genomes.

Preproc

Submit your genomic or metagenomic data to this project to preprocess it without further taxonomic analysis.

Clades

Using average nucleotide identity, query clade-specific databases for high resolution within well-characterized species.

Environments

The MiGA database feature genome collections for use in environmental genomics

MAGs

Metagenome-assembled genomes (MAGs) recovered from different binning projects

WB Bins HQ

Population genomes (MAGs) collected from the Chattahoochee River, other lakes, and estuaries (Southeast USA)

GROS

Genome repository of oil systems provides a curated collection of MAGs, SAGs, and isolates from habitats affected by crude oil.

TARA MAGs (Delmont) and TARA MAGs (Tully)

MAGs retrieved from the TARA Oceans metagenome

RefSoil genomes

Manually curated list of complete genomes from NCBI RefSeq associated with soil environments

Bio-Gas Microbiome

MAGs from 134 samples collected from biogas plants and lab-scale reactors in seven countries

More information about getting started with MiGA, interfaces, deployment examples, and workflows can be found at http://microbial-genomes.org/ (Rodriguez-R et al. 2018).

3.4.2 Type Strain Genome Server (TYGS)

TYGS offers completely automated analyses developed by the Leibniz Institute DSMZ (German Collection of Microorganisms and Cell Cultures GmbH, Braunschweig, Germany). The TYGS pipeline includes extraction of the 16S rDNA gene sequence, determination of the nearest type strains, and cutting-edge phylogenomics analysis, all of which are relevant to modern microbial taxonomy (https://tygs.dsmz.de/). TYGS provides categorization and identification of prokaryotic strains based on the genome without overestimating phylogenetic certainty. In addition to the phylogenomic technique, TYGS provides digital DNA-DNA hybridization values, changes in G + C content, and other information. A type strain is the nomenclatural type of a species in microbiological taxonomy. Comparing novel strains to type strains determines whether a strain belongs to a certain species. There are presently 22,000 microbial type strain genomes cataloged (recent update) in the extensive TYGS database. The TYGS web service with the List of Prokaryotic Names with Standing in Nomenclature (LPSN) uses a number of well-known methods (GGDC, GBDP) from the past few years and adds a whole new level of fast data analysis, including taxonomic nomenclature, which makes the analysis fast, accurate, and easy to understand (Meier-Kolthoff and Göker 2019; Meier-Kolthoff et al. 2022).

3.4.3 The Genome Taxonomy Database (GTDB)

Axenic bacteria and archaea genomes (Mukherjee et al. 2017), single-amplified genomes (SAGs), and metagenome-assembled genomes (MAGs) (Parks et al. 2017; Alneberg et al. 2018) improved the genomic pool. The GTDB is a genome-based taxonomy of 317,542 bacterial and archaeal genomes. Unlike NCBI-classified representatives, over a third (3620; 35.1%) have a different species assignment in the GTDB. The majority of these changes (2473; 68.3%) were due to modifying the generic name of the species as a result of resolving polyphyletic genera or normalizing genera by relative evolutionary divergence (RED). GTDB established species borders and proposed species groups using average nucleotide identity (ANI) criteria and proposed 148 phyla according to relative evolutionary divergence. The GTDB website provides a taxonomic framework for bacterial and archaeal genomes, encouraging uncultivated species study and scientific communication (Parks et al. 2020), and the GTDB-Tk is a software toolkit (https://github.com/Ecogenomics/GTDBTk) for assigning objective taxonomic classifications to bacterial and archaeal genomes based on the Genome Database Taxonomy (GTDB).

3.4.4 The Great Automatic Nomenclator (GAN)

In an era of discovery through culturomics, genomics, and metagenomics, the bacteria and archaea require new names. Taxonomic naming is a time-consuming process that requires Latin, Greek, and ICNP criteria. Millions of multi-omics names can be easily generated. Innovative approach, GAN, has three features: mass generation of names before they are connected to taxa, combinatorial concatenation of Latin and Greek roots, and computerized name formation (Pallen et al. 2021).

3.4.5 The SeqCode

In genome taxonomy databases, around 50% of microbial phyla lack a cultivated representative. Although these genome sequences indisputably distinguish microbial species, they cannot be formally designated under the International Code of Nomenclature of Prokaryotes since a pure-culture strain is required. The SeqCode program is geared toward developing a system for naming these uncultivated archaea and bacteria. The SeqCode is a new nomenclature code based on the use of DNA sequence type information as the basis for proof (https://www.isme-microbes.org/). The SeqCode implementation date would be January 1, 2022. The SeqCode Registry (https://disc-genomics.uibk.ac.at/seqcode/) has additional information regarding the SeqCode.

3.5 Conclusion

With the improvement of technical methods, microbial classification has shifted from morphological observation to chemical classification and then to molecular classification. With the reduction of the cost of sequencing technology and the improvement of genomic and metagenomic analyses, microbial classification at the genome level will become an important and efficient tool for cultivated and uncultivated microbial taxonomy. Given recent changes and the different services offered by different sequencing platforms, databases, software, algorithms, etc., it is necessary to use hybrids, which involve combining technologies and picking the right criteria for customized research in this growing genomic era.

References

Adewale BA (2020) Will long-read sequencing technologies replace short-read sequencing technologies in the next 10 years? Afr J Lab Med 9:1340

Alneberg J, Karlsson CMG, Divne AM, Bergin C, Homa F, Lindh MV, Hugerth LW, Ettema TJG, Bertilsson S, Andersson AF, Pinhassi J (2018) Genomes from uncultivated prokaryotes: a comparison of metagenome-assembled and single-amplified genomes. Microbiome 6:173

Amarasinghe SL, Su S, Dong X, Zappia L, Ritchie ME, Gouil Q (2020) Opportunities and challenges in long-read sequencing data analysis. Genome Biol 21:30

Auch AF, von Jan M, Klenk HP, Göker M (2010) Digital DNA-DNA hybridization for microbial species delineation by means of genome-to-genome sequence comparison. Stand Genomic Sci 2:117–134

Balasubramanian S (2011) Sequencing nucleic acids: from chemistry to medicine. Chem Commun 47:7281–7286

Balasubramanian S (2015) Solexa sequencing: decoding genomes on a population scale. Clin Chem 61:21–24

Balzer S, Malde K, Lanzén A, Sharma A, Jonassen I (2010) Characteristics of 454 pyrosequencing data—enabling realistic simulation with flowsim. Bioinformatics 26:i420–425

Bateson MM, Thibault KJ, Ward DM (1990) Comparative analysis of 16S ribosomal RNA sequences of Thermus species. Syst Appl Microbiol 13:8–13

Bent SJ, Forney LJ (2008) The tragedy of the uncommon: understanding limitations in the analysis of microbial diversity. ISME J 2:689–695

Bolyen E, Rideout JR, Dillon MR, Bokulich NA, Abnet CC, Al-Ghalith GA, Alexander H, Alm EJ, Arumugam M, Asnicar F, Bai Y, Bisanz JE, Bittinger K, Brejnrod A, Brislawn CJ, Brown CT, Callahan BJ, Caraballo-Rodríguez AM, Chase J, Cope EK, Da Silva R, Diener C, Dorrestein PC, Douglas GM, Durall DM, Duvallet C, Edwardson CF, Ernst M, Estaki M, Fouquier J, Gauglitz JM, Gibbons SM, Gibson DL, Gonzalez A, Gorlick K, Guo J, Hillmann B, Holmes S, Holste H, Huttenhower C, Huttley GA, Janssen S, Jarmusch AK, Jiang L, Kaehler BD, Kang KB, Keefe CR, Keim P, Kelley ST, Knights D, Koester I, Kosciolek T, Kreps J, Langille MGI, Lee J, Ley R, Liu YX, Loftfield E, Lozupone C, Maher M, Marotz C, Martin BD, McDonald D, McIver LJ, Melnik AV, Metcalf JL, Morgan SC, Morton JT, Naimey AT, Navas-Molina JA, Nothias LF, Orchanian SB, Pearson T, Peoples SL, Petras D, Preuss ML, Pruesse E, Rasmussen LB, Rivers A, Robeson MS, Rosenthal P, Segata N, Shaffer M, Shiffer A, Sinha R, Song SJ, Spear JR, Swafford AD, Thompson LR, Torres PJ, Trinh P, Tripathi A, Turnbaugh PJ, Ul-Hasan S, van der Hooft JJJ, Vargas F, Vázquez-Baeza Y, Vogtmann E, von Hippel M, Walters W, Wan Y, Wang M, Warren J, Weber KC, Williamson CHD, Willis AD, Xu ZZ, Zaneveld JR, Zhang Y, Zhu Q, Knight R, Caporaso JG (2019) Reproducible, interactive, scalable and extensible microbiome data science using QIIME 2. Nat Biotechnol 37:852–857

Brock TD, Freeze H (1969) *Thermus aquaticus* gen. n. and sp. n., a nonsporulating extreme thermophile. J Bacteriol 98:289–297

Callahan B, McMurdie PJ, Rosen MJ, Han AW, Johnson AJ, Holmes SP (2016) DADA2: high-resolution sample inference from Illumina amplicon data. Nat Methods 13:581–583

Caporaso JG, Kuczynski J, Stombaugh J, Bittinger K, Bushman FD, Costello EK, Fierer N, Peña AG, Goodrich JK, Gordon JI, Huttley GA, Kelley ST, Knights D, Koenig JE, Ley RE, Lozupone CA, McDonald D, Muegge BD, Pirrung M, Reeder J, Sevinsky JR, Turnbaugh PJ, Walters WA, Widmann J, Yatsunenko T, Zaneveld J, Knight R (2010) QIIME allows analysis of high-throughput community sequencing data. Nat Methods 7:335–336

Chun J, Rainey FA (2014) Integrating genomics into the taxonomy and systematics of the bacteria and archaea. Int J Syst Evol Microbiol 64:316–324

Cloonan N, Forrest AR, Kolle G, Gardiner BB, Faulkner GJ, Brown MK, Taylor DF, Steptoe AL, Wani S, Bethel G, Robertson AJ, Perkins AC, Bruce SJ, Lee CC, Ranade SS, Peckham HE, Manning JM, McKernan KJ, Grimmond SM (2008) Stem cell transcriptome profiling via massive-scale mRNA sequencing. Nat Methods 5:613–619

Cole JR, Chai B, Farris RJ, Wang Q, Kulam SA, McGarrell DM, Garrity GM, Tiedje JM (2005) The Ribosomal Database Project (RDP-II): sequences and tools for high-throughput rRNA analysis. Nucleic Acids Res 33:D294–296

Cole JR, Wang Q, Chai B, Tiedje JM (2011) The Ribosomal Database Project: sequences and software for high-throughput rRNA analysis. In: de Bruijn FJ (ed) Handbook of molecular microbial ecology I: metagenomics and complementary approaches, J. Wiley & Sons, Inc, Hoboken, NJ

Cole JR, Wang Q, Fish JA, Chai B, McGarrell DM, Sun Y, Brown CT, Porras-Alfaro A, Kuske CR, Tiedje JM (2014) Ribosomal Database Project: data and tools for high throughput rRNA analysis. Nucleic Acids Res 42:D633–642

Collins FS, Fink L (1995) The human genome project. Alcohol Health Res World 19:190–195

Connon SA, Giovannoni SJ (2002) High-throughput methods for culturing microorganisms in very-low-nutrient media yield diverse new marine isolates. Appl Environ Microbiol 68:3878–3885

DeSantis TZ, Hugenholtz P, Larsen N, Rojas M, Brodie EL, Keller K, Huber T, Dalevi D, Hu P, Andersen GL (2006) Greengenes, a chimera-checked 16S rRNA gene database and workbench compatible with ARB. Appl Environ Microbiol 72:5069–5072

Excoffier L, Smouse PE, Quattro JM (1992) Analysis of molecular variance inferred from metric distances among DNA haplotypes: application to human mitochondrial DNA restriction data. Genetics 131:479–491

Fiers W, Contreras R, Duerinck F, Haegeman G, Iserentant D, Merregaert J, Min Jou W, Molemans F, Raeymaekers A, Van den Berghe A, Volckaert G, Ysebaert M (1976) Complete nucleotide sequence of bacteriophage MS2 RNA: primary and secondary structure of the replicase gene. Nature 260:500–507

Fish JA, Chai B, Wang Q, Sun Y, Brown CT, Tiedje JM, Cole JR (2013) FunGene: the functional gene pipeline and repository. Front Microbiol 4:291

Fu S, Wang A, Au KF (2019) A comparative evaluation of hybrid error correction methods for error-prone long reads. Genome Biol 20:26

Gautam SS, Kc R, Leong KW, Mac Aogáin M, O'Toole RF (2019) A step-by-step beginner's protocol for whole genome sequencing of human bacterial pathogens. J Biol Methods 6:e110

Gelfand DH (1989) Taq DNA polymerase. In: Erlich HA (ed) PCR technology. Palgrave Macmillan, London. https://doi.org/10.1007/978-1-349-20235-5_2

Glover F, Klingman D (1981) The simplex SON algorithm for LP/embedded network problems. In: Klingman D, Mulvey JM (eds) Network models and associated applications. Mathematical programming studies, vol 15. Springer, Berlin, Heidelberg

Heather JM, Chain B (2015) The sequence of sequencers: the history of sequencing DNA. Genomics 107:1–8

Hill TC, Walsh KA, Harris JA, Moffett BF (2003) Using ecological diversity measures with bacterial communities. FEMS Microbiol Ecol 43:1–11

Hugenholtz P, Skarshewski A, Parks DH (2016) Genome-based microbial taxonomy coming of age. Cold Spring Harb Perspect Biol 8:a018085

Illumina Inc. (2022) History of sequencing by synthesis—The evolution of the next-generation sequencing technology powering Illumina instruments. https://sapac.illumina.com/science/technology/next-generation-sequencing/illumina-sequencing-history.html (Assessed on 2022-07-17 time 12:51:51)

Jangid K, Kao MH, Lahamge A, Rathbun SL, Williams MA, Whitman WB (2016) K-shuff: a novel algorithm for characterizing structural and compositional diversity in gene libraries. PLoS One 11:e0167634

Kondratenko Y, Korobeynikov A, Lapidus A (2020) CDSnake: Snakemake pipeline for retrieval of annotated OTUs from paired-end reads using CD-HIT utilities. BMC Bioinformatics 21(Suppl 12):303

Konstantinidis KT, Tiedje JM (2005) Towards a genome-based taxonomy for prokaryotes. J Bacteriol 187:6258–6264

Koonin EV, Galperin MY (2003) Sequence—evolution—function: computational approaches in comparative genomics. Kluwer Academic, Boston

Kwong JC, McCallum N, Sintchenko V, Howden BP (2015) Whole genome sequencing in clinical and public health microbiology. Pathology 47:199–210

Liu Y, Lai Q, Shao Z (2018) Genome-based analysis reveals the taxonomy and diversity of the family Idiomarinaceae. Front Microbiol 9:2453

Logan DW (2021) Hot to touch: the story of the 2021 Nobel prize in physiology or medicine. Dis Model Mech 14:dmm049352

Lozupone C, Hamady M, Knight R (2006) UniFrac—an online tool for comparing microbial community diversity in a phylogenetic context. BMC Bioinformatics 7:371

Lozupone C, Knight R (2005) UniFrac: a new phylogenetic method for comparing microbial communities. Appl Environ Microbiol 71:8228–8235

Mahmoud M, Gobet N, Cruz-Dávalos DI, Mounier N, Dessimoz C, Sedlazeck FJ (2019) Structural variant calling: the long and the short of it. Genome Biol 20:246

Maidak BL, Larsen N, McCaughey MJ, Overbeek R, Olsen GJ, Fogel K, Blandy J, Woese CR (1994) The ribosomal database project. Nucleic Acids Res 22:3485–3487

Mardis ER (2008) The impact of next-generation sequencing technology on genetics. Trends Genet 24:133–141

Martiny AC (2019) High proportions of bacteria are culturable across major biomes. ISME J 13:2125–2128

McDonald D, Price M, Goodrich J, Nawrocki EP, DeSantis TZ, Probst A, Andersen AL, Knight R, Hugenholtz P (2012) An improved Greengenes taxonomy with explicit ranks for ecological and evolutionary analyses of bacteria and archaea. ISME J 6:610–618

McKernan KJ, Peckham HE, Costa GL, McLaughlin SF, Fu Y, Tsung EF, Clouser CR, Duncan C, Ichikawa JK, Lee CC, Zhang Z, Ranade SS, Dimalanta ET, Hyland FC, Sokolsky TD, Zhang L, Sheridan A, Fu H, Hendrickson CL, Li B, Kotler L, Stuart JR, Malek JA, Manning JM, Antipova AA, Perez DS, Moore MP, Hayashibara KC, Lyons MR, Beaudoin RE, Coleman BE, Laptewicz MW, Sannicandro AE, Rhodes MD, Gottimukkala RK, Yang S, Bafna V, Bashir A, MacBride A, Alkan C, Kidd JM, Eichler EE, Reese MG, De La Vega FM, Blanchard AP (2009) Sequence and structural variation in a human genome uncovered by short-read, massively parallel ligation sequencing using two-base encoding. Genome Res 19:1527–1441

Meier-Kolthoff JP, Auch AF, Klenk HP, Klenk HP, Markus Göker M (2013) Genome sequence-based species delimitation with confidence intervals and improved distance functions. BMC Bioinformatics 14:60

Meier-Kolthoff JP, Carbasse JS, Peinado-Olarte RL, Göker M (2022) TYGS and LPSN: a database tandem for fast and reliable genome-based classification and nomenclature of prokaryotes. Nucleic Acids Res 50:D801–D807

Meier-Kolthoff JP, Göker M (2019) TYGS is an automated high-throughput platform for state-of-the-art genome-based taxonomy. Nat Commun 10:2182

Morisse P, Marchet C, Limasset A, Lecroq T, Lefebvre A (2021) Scalable long read self-correction and assembly polishing with multiple sequence alignment. Sci Rep 11:761

Mukherjee S, Seshadri R, Varghese NJ, Eloe-Fadrosh EA, Meier-Kolthoff JP, Göker M, Coates RC, Hadjithomas M, Pavlopoulos GA, Paez-Espino D, Yoshikuni Y, Visel A, Whitman WB, Garrity GM, Eisen JA, Hugenholtz P, Pati A, Ivanova NN, Woyke T, Klenk HP, Kyrpides NC (2017) 1,003 reference genomes of bacterial and archaeal isolates expand coverage of the tree of life. Nat Biotechnol 35:676–683

Nurk S, Koren S, Rhie A, Rautiainen M, Bzikadze AV, Mikheenko A, Vollger MR, Altemose N, Uralsky L, Gershman A, Aganezov S, Hoyt SJ, Diekhans M, Logsdon GA, Alonge M, Antonarakis SE, Borchers M, Bouffard GG, Brooks SY, Caldas GV, Chen NC, Cheng H, Chin CH, Chow W, de Lima LG, Dishuck PC, Durbin R, Dvorkina T, Fiddes IT, Formenti G, Fulton RS, Fungtammasan A, Garrison E, Grady PGS, Graves-Lindsay TA, Hall IM, Hansen NF, Hartley GA, Haukness M, Howe K, Hunkapiller MW, Jain C, Jain M, Jarvis ED, Kerpedjiev P, Kirsche M, Kolmogorov M, Korlach J, Kremitzki M, Li H, Maduro VV, Marschall T, McCartney AM, McDaniel J, Miller DE, Mullikin JC, Myers EW, Olson ND, Paten B, Peluso P, Pevzner PA, Porubsky D, Potapova T, Rogaev EI, Rosenfeld JA, Salzberg SL, Schneider VA, Sedlazeck FJ, Shafin K, Shew CJ, Shumate A, Sims Y, Smit AFA, Soto DC, Sovic I, Storer JM, Streets A, Sullivan BA, Thibaud-Nissen F, Torrance J, Wagner J, Walenz BP, Wenger A, Wood

JMD, Xiao C, Yan SM, Young AC, Zarate S, Surti U, McCoy RC, Dennis MY, Alexandrov IA, Gerton JL, O'Neill RJ, Timp W, Zook JM, Schatz MC, Eichler JE, Miga KH, Phillippy AM (2021) The complete sequence of a human genome. Science 376:44–53

Nyrén P (2001) Method for sequencing DNA based on the detection of the release of pyrophosphate and enzymatic nucleotide degradation. Patents: US 6 258 568BI and WO98/28440

Nyrén P (2007) The history of pyrosequencing®. In: Walker JM, Marsh W (eds) Pyrosequencing protocols, methods in molecular biology. Humana Press, pp 1–13

Nyrén P (2015) The history of pyrosequencing(®). Methods Mol Biol 1315:3–15

Nyrén P, Lundin A (1985) Enzymatic method for continuous monitoring of inorganic pyrophosphate synthesis. Anal Biochem 151:504–509

Pallen MJ, Telatin A, Oren A (2021) The next million names for archaea and bacteria. Trends Microbiol 29:289–298

Parks DH, Chuvochina M, Chaumeil PA, Rinke C, Mussig AJ, Hugenholtz P (2020) A complete domain-to-species taxonomy for bacteria and archaea. Nat Biotechnol 38:1079–1086

Parks DH, Rinke C, Chivochina M, Chaumeil PA, Woodcroft BJ, Evans PN, Hugenholtz P, Tyson GW (2017) Recovery of nearly 8,000 metagenome-assembled genomes substantially expands the tree of life. Nat Microbiol 2:1533–1542

Patel JB (2001) 16S rRNA gene sequencing for bacterial pathogen identification in the clinical laboratory. Mol Diagn 6:313–321

Pereira R, Oliveira J, Sousa M (2020) Bioinformatics and computational tools for next-generation sequencing analysis in clinical genetics. J Clin Med 9:132

Pettersson E, Lundeberg J, Ahmadian A (2009) Generations of sequencing technologies. Genomics 93:105–111

Pruesse E, Peplies J, Glöckner FO (2012) SINA: accurate high-throughput multiple sequence alignment of ribosomal RNA genes. Opens external link in new window. Bioinformatics 28:1823–1829

Pruesse E, Quast C, Knittel K, Fuchs BM, Ludwig W, Peplies J, Glockner FO (2007) SILVA: a comprehensive online resource for quality checked and aligned ribosomal RNA sequence data compatible with ARB. Nucleic Acids Res 35:7188–7196

Quast C, Pruesse E, Yilmaz P, Gerken J, Schweer T, Yarza P, Peplies J, Glöckner FO (2013) The SILVA ribosomal RNA gene database project: improved data processing and web-based tools. Opens external link in new window. Nucleic Acids Res 41(D1):D590–D596

Reardon S (2021) A complete human genome sequence is close: how scientists filled in the gaps. Nature 594:158–159

Rekadwad BN, Gonzalez JM, Khobragade CN (2016) Genomic analysis of a marine bacterium: bioinformatics for comparison, evaluation, and interpretation of DNA sequences. Biomed Res Int 7215379:7. https://doi.org/10.1155/2016/7215379

Rifaie S, Patil V, Jangid K (2022) Advances in sequencing technology, databases, and analyses tools for the assessment of microbial diversity. In: Gunjal A, Shinde S (eds) Microbial diversity and ecology in hotspots, pp 317–347

Rodriguez-R LM, Gunturu S, Harvey WT, Rosselló-Mora R, Tiedje JM, Cole JR, Konstantinidis KT (2018) The microbial genomes atlas (MiGA) webserver: taxonomic and gene diversity analysis of archaea and bacteria at the whole genome level. Nucleic Acids Res 46(W1):W282–W288

Romalde JL, Balboa S, Ventosa A (2019) Editorial: microbial taxonomy, phylogeny and biodiversity. Front Microbiol 10:1324

Ronaghi M, Uhlén M, Nyrén P (1998) A sequencing method based on real-time pyrophosphate detection. Science 281:363–365

Rosen MJ, Callahan BJ, Fisher DS, Homes SP (2012) Denoising PCR-amplified metagenome data. BMC Bioinformatics 13:283

Rosselló-Móra R, Whitman WB (2019) Dialogue on the nomenclature and classification of prokaryotes. Syst Appl Microbiol 42:5–14

Saiki RK, Gelfand DH, Stoffel S, Scharf SJ, Higuchi R, Horn GT, Mullis KB, Erlich HA (1988) Primer-directed enzymatic amplification of DNA with a thermostable DNA polymerase. Science 239:487–491

Saitou N, Nei M (1987) The neighbor-joining method: a new method for reconstructing phylogenetic trees. Mol Biol Evol 4:406–425

Sanger F, Brownlee GG, Barrell BG (1965) A two-dimensional fractionation procedure for radioactive nucleotides. J Mol Biol 13:373–398

Sanger F, Nicklen S, Coulson AR (1977) DNA sequencing with chain-terminating inhibitors. Proc Natl Acad Sci U S A 74:5463–5467

Saunders CJ, Miller NA, Soden SE, Dinwiddie DL, Noll A, Alnadi NA, Andraws N, Patterson ML, Krivohlavek LA, Fellis J, Humphray S, Saffrey P, Kingsbury Z, Weir JC, Betley J, Grocock RJ, Margulies EH, Farrow EG, Artman M, Safina NP, Petrikin JE, Hall KP, Kingsmore SF (2012) Rapid whole-genome sequencing for genetic disease diagnosis in neonatal intensive care units. Sci Transl Med 4:154ra135

Schloss P (2008) Evaluating different approaches that test whether microbial communities have the same structure. ISME J 2:265–275

Schloss PD, Larget BR, Handelsman J (2004) Integration of microbial ecology and statistics: a test to compare gene libraries. Appl Environ Microbiol 70:5485–5492

Schloss PD, Westcott SL, Ryabin T, Hall JR, Hartmann M, Hollister EB, Lesniewski RA, Oakley BB, Parks DH, Robinson CJ, Sahl JW, Stres B, Thallinger GG, Van Horn DJ, Weber CF (2009) Introducing mothur: open-source, platform-independent, community-supported software for describing and comparing microbial communities. Appl Environ Microbiol 75:7537–7541

Sereika M, Kirkegaard RH, Karst SM, Michaelsen TY, Sørensen EA, Wollenberg RD, Albertsen M (2022) Oxford Nanopore R10.4 long-read sequencing enables the generation of near-finished bacterial genomes from pure cultures and metagenomes without short-read or reference polishing. Nat Methods 19:823–826

Singleton DR, Furlong MA, Rathbun SL, Whitman WB (2001) Quantitative comparisons of 16S rRNA gene sequence libraries from environmental samples. Appl Environ Microbiol 67:4374–4376

Slatko BE, Gardner AF, Ausubel FM (2018) Overview of next-generation sequencing technologies. Curr Protoc Mol Biol 122:e59

Smith K (2013) A brief history of NCBI's formation and growth, the NCBI handbook [internet], 2nd edn. National Center for Biotechnology Information, US, Bethesda, MD

Smith KA (2008) Laws, leaders, and legends of the modern National Library of Medicine. J Med Libr Assoc 96:121–133

Sogin M, Pace B, Pace NR, Woese CR (1971) Primary structural relationship of p16 to m16 ribosomal RNA. Nat New Biol 232:48–49

The C. elegans Sequencing Consortium (1998) Genome sequence of the nematode *C. elegans*: a platform for investigating biology. Science 282:2012–2018

Tyler AD, Mataseje L, Urfano CJ, Schmidt L, Antonation KS, Mulvey MR, Corbett CR (2018) Evaluation of Oxford Nanopore's MinION sequencing device for microbial whole genome sequencing applications. Sci Rep 8:10931

Valiente-Mullor C, Beamud B, Ansari I, Francés-Cuesta C, García-González N, Mejía L, Ruiz-Hues P, González-Candelas F (2021) One is not enough: on the effects of reference genome for the mapping and subsequent analyses of short-reads. PLoS Comput Biol 17:e1008678

Valouev A, Ichikawa J, Tonthat T, Stuart J, Ranade S, Peckham H, Zeng K, Malek JA, Costa G, McKernan K, Sidow A, Fire A, Johnson SM (2008) A high-resolution, nucleosome position map of C. elegans reveals a lack of universal sequence-dictated positioning. Genome Res 18:1051–1063

Wang Q, Fish JA, Gilman M, Sun Y, Brown CT, Tiedje JM, Cole JR (2015) Xander: employing a novel method for efficient gene-targeted metagenomic assembly. Microbiome 3:32

Wang Q, Garrity GM, Tiedje JM, Cole JR (2007) Naïve Bayesian classifier for rapid assignment of rRNA sequences into the new bacterial taxonomy. Appl Environ Microbiol 73:5261–5267

Wang Q, Quensen JF III, Fish JA, Lee TK, Sun Y, Tiedje JM, Cole JR (2013) Ecological patterns of nifH genes in four terrestrial climatic zones explored with targeted metagenomics using FrameBot, a new informatics tool. MBio 4:e00592–13

Watson JD, Crick FHC (1953) Molecular structure of nucleic acids. Nature 171:737–738

Woese CR, Kandler O, Wheelis ML (1990) Towards a natural system of organisms: proposal for the domains archaea, bacteria, and eucarya. Proc Natl Acad Sci U S A 87:4576–4579

Xiao T, Yuan C, Huang Y (2011) PSON: a parallelized SON algorithm with MapReduce for mining frequent sets. Fourth International Symposium on Parallel Architectures, Algorithms and Programming, pp. 252–257

Yilmaz P, Parfrey LW, Yarza P, Gerken J, Pruesse E, Quast C, Schweer T, Peplies J, Ludwig W, Glöckner FO (2014) The SILVA and "all-species living tree project (LTP)" taxonomic frameworks. Opens external link in new window. Nucleic Acids Res 42:D643–D648

Yoon SH, Ha SM, Kwon S, Lim J, Kim Y, Seo H, Chun J (2017) Introducing EzBioCloud: a taxonomically united database of 16S rRNA and whole genome assemblies. Int J Syst Evol Microbiol 67:1613–1617

Youseif SH, Abd El-Megeed FH, Humm EA, Maymon M, Mohamed AH, Saleh SA, Hirsch AM (2021) Comparative analysis of the cultured and total bacterial community in the wheat rhizosphere microbiome using culture-dependent and culture-independent approaches. Microbiol Spectr 9:e0067821

Chapter 4
Phylogenetics and Phylogenomics for Inferring Relationships of Bacteria and Archaea Phyla

Bhagwan Narayan Rekadwad, Zheng-Han Lian, Jian-Yu Jiao, and Wen-Jun Li

4.1 Introduction

The ocean and soil are home to a vast array of bacteria and archaea, the most ancient group of microorganisms involved in metabolic and biogeochemical activities (Louca et al. 2019). Evolutionary studies on microorganisms have shed light on the selective pressures that influence the diversity and function of microbial lineages of all origins and speciation (Ley et al. 2006; Braga et al. 2018) to define their taxonomic position and ecological roles. Continuously developing sequencing technology and sequence analysis tools has escalated the scope of evolutionary genomics dramatically in recent years. A couple of decades ago, as is often said historically, all gene banks and databases consisted of a small collection of manually read and corrected gene sequences derived from cultivable microorganisms. The genomic representation of lineages across the tree of life (Falkowski 1998; Kasting 2002; Mende et al. 2017; Martinez-Gutierrez and Aylward 2022) has grown as sequencing technologies and evidence-based tools and databases have improved (Falkowski 1998; Kasting 2002; Mende et al. 2017; Martinez-Gutierrez and Aylward 2022).

B. N. Rekadwad
Division of Microbiology and Biotechnology, Yenepoya Research Centre, Yenepoya (Deemed to be University), Mangalore, Karnataka, India

Z.-H. Lian · J.-Y. Jiao · W.-J. Li (✉)
State Key Laboratory of Biocontrol, Guangdong Provincial Key Laboratory of Plant Resources and Southern Marine Science and Engineering Guangdong Laboratory (Zhuhai), School of Life Sciences, Sun Yat-Sen University, Guangzhou, China
e-mail: liwenjun3@mail.sysu.edu.cn

W.-J. Li et al. (eds.), *Modern Taxonomy of Bacteria and Archaea*,
https://doi.org/10.1007/978-981-99-5720-0_4

4.2 Sequencing, Phylogenetics, and Phylogenomics

Sequencing technology and sequence analysis tools play a critical role in evolutionary microbiology by enabling researchers to study the genetic makeup of microorganisms and how it has changed over time. With this information, researchers can gain insights into the evolution and diversity of microbial life, as well as the mechanisms that drive evolutionary change (Bryant et al. 2012; Donkor 2013). One key application of sequencing technology in evolutionary microbiology is the reconstruction of phylogenetic trees (Munjal et al. 2018). These trees represent the evolutionary relationships among different microorganisms, based on their genetic similarity. By comparing the genetic sequences of different microorganisms, researchers can determine which ones are closely related and which ones are more distantly related. This information can be used to study the evolution of different groups of microorganisms, as well as the mechanisms that drive evolutionary change, such as natural selection and genetic drift. Sequencing technology and sequence analysis tools also play a crucial role in the study of horizontal gene transfer, which is the process by which microorganisms acquire new genetic material from other organisms. This is a common mechanism of evolution in microbial communities, and it can be studied by comparing the genetic sequences of different microorganisms to identify instances of horizontal gene transfer. This information can help researchers understand how microbial communities evolve and adapt to changing environments. Indeed, sequencing technology and sequence analysis tools are essential tools in evolutionary microbiology, providing researchers with the ability to study the genetic makeup of microorganisms and how it has changed over time. This information can be used to gain insights into the evolution and diversity of microbial life, as well as the mechanisms that drive evolutionary change.

The accuracy of Sanger sequencing and advances in sequencing technologies generate sequences with great ease and inexpensively, and the quality of the sequence is reliable (Ko et al. 2022). These highly specific sequences, which are rich in information from ancestral or inherited DNA, aid in the rapid inference of evolutionary relationships. In the case of difficult to cultivate, slow growing, rarely discovered, and uncultivated species, morphological features or known characteristics have traditionally been used to infer evolutionary relationships (Handelsman 2004; Robinson et al. 2021). However, the authenticity and reliability of morphology-based phylogenetic analysis and even mislabeled sequence data (Kozlov et al. 2016) are always questionable in the case of bacteria and archaea. Therefore, inferring phylogenetic trees from genuine molecular or sequence data has become a prerequisite to distinguishing multiple independent evolutionary lineages or genes from similar strains (Jukes and Cantor 1969; Bauldauf 2003; Baum et al. 2005; Shakya et al. 2020). This approach is usually termed "molecular phylogenetics" (Brown 2002; Yang and Rannala 2012; Tkach et al. 2021). Phylogenetics improves our understanding of how genes, genomes, species, and, more broadly, molecular sequences evolve. In addition to helping us discover how the sequences came to be as they are, phylogenetics teaches us overarching principles that let us foresee future changes in the sequences. In addition to being fundamentally important, this is also very practical for a wide range of uses.

Phylogenetics, also known as taxonomy, is the study of evolutionary relationships among biological entities, most commonly species, individuals, or genes (Pace et al. 2012; Brower and Schuh 2021). Observation of sequence, reconstruction of evolutionary history, in-depth learning of evolutionary processes, and development of a better evolutionary model are major components of phylogenetics. Phylogeneticists typically investigate the evolutionary relationship or history of species or genes of interest or the evolution of sequences or describe the evolution of sequences using a mathematical model (EMBL, https://www.ebi.ac.uk/). One can reconstruct a phylogenetic tree by analyzing the sequences of nucleotides or proteins and combining this information with our knowledge of how these sequences have evolved over time, which is described using an evolutionary model. This not only provides more information about the evolutionary processes that are at work on sequences but also enables us to infer evolutionary events that took place in the past. As a result, we might be able to improve our understanding of how the process of evolution works and construct more accurate mathematical models of evolution (Jukes and Cantor 1969; Bauldauf 2003; Sterpetti 2021; Styles et al. 2021).

Phylogenomics is a method for predicting uncharacterized genes through evolutionary analysis (Eisen 1998). Phylogenomics aims to reconstruct organisms' evolutionary histories by using whole genome sequences (WGS) or large scaffolds of genomes. The abundance of genomic data for a range of species has facilitated molecular phylogenetic inference for many groups, leading to the development of numerous computer algorithms and software that implement the associated methods (Patané et al. 2018). In this era of rapid growth in genomics and continuously updating tools for evolutionary analysis, more curated genomic data has provided opportunities for updating microbial phylogenies that are frequently challenged by disagreement due to topological differences and distances between individual genes.

Zhu et al. (2019) have built a reference phylogenomic tree from 10, 575 whole genomes of bacteria and archaea evenly sampled from the 86,200 nonredundant genomes that were available in NCBI GenBank and RefSeq24 as of March 7, 2017 (Haft et al. 2018; Zhu et al., 2019). These 10,575 were evenly sampled and reconstructed topology using ASTRAL based on a 381-marker gene set and multiple strategies (Fig. 4.1a). Several quantifiable factors, including taxon and site sampling, non-vertical evolution, amino acid substitution (heterogeneity and saturation), and the influence of excluding candidate phyla radiation taxa, were used to confirm the results obtained. Taxonomic units covering phyla, classes, orders, families, genera, and species as mentioned on the Y-axis (Fig. 4.1b). Almost 96.7% of genomes were medium to high quality comprising scaffolds (47%), contigs (31.6%), complete genomes (19.9%), and chromosomes (1.6%) (Fig. 4.1c, d). Most of extracted and selected genomes are sourced from mono-isolates (62.1%), genomes from environmental samples (27%), multispecies (5.7%), and multi-isolates (4.5%) followed by single cells and other sources (0.7%) (Fig. 4.1e) (Zhu et al., 2019).

Moreover, phylogenomics and comparative genomics can resolve inconsistency in the naming and classification of *Bacillus* polyphyly. The genus *Bacillus* is a

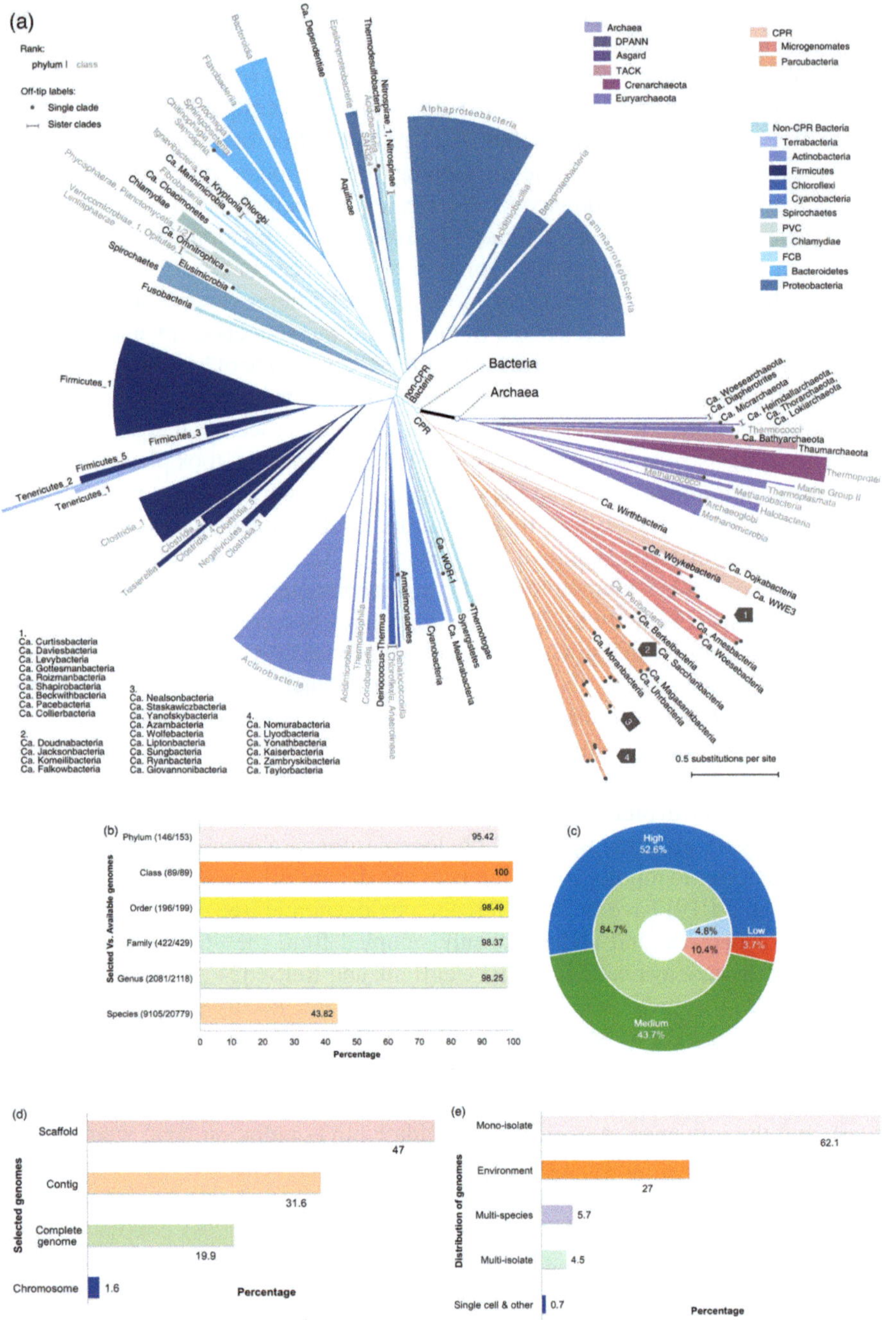

Fig. 4.1 (**a**) The construction of 10,575 genomes for archaea, candidate phyla radiation (CPR), and non-CPR bacteria has provided an entirely new perspective at the evolutionary relationships between bacteria and archaea. (**b**) Taxonomic units covering phyla, classes, orders, families, genera, and species as mentioned on Y-axis. (**c**) Quality of genomes. (**d**) Definition of genomes. (**e**) Genome source (Modified, rearranged indicators and represented by adopting graphics/contents from Zhu et al., 2019; Nature Communications; https://www.nature.com/articles/s41467-019-13443-4#rightslink)

group of Gram-positive, spore-forming bacteria that are commonly found in soil and aquatic environments. However, recent studies have revealed that the current taxonomic classification of *Bacillus* species is polyphyletic, meaning that it includes multiple distinct evolutionary lineages that are not closely related to each other. This situation can lead to confusion and inconsistency in the naming and classification of *Bacillus* species. To address this issue, a phylogenomic and comparative genomic framework has been proposed for resolving the polyphyly of the genus *Bacillus*. This framework involves the use of advanced sequencing technology and sequence analysis tools to study the genetic makeup of *Bacillus* species and their evolutionary relationships. Based on this analysis, it has been proposed to split the genus *Bacillus* into six new genera: *Peribacillus*, *Cytobacillus*, *Mesobacillus*, *Neobacillus*, *Metabacillus*, and *Alkalihalobacillus*. These new genera would be based on the evolutionary relationships among different *Bacillus* species, providing a more accurate and consistent taxonomic classification for these bacteria. Thus, the proposed phylogenomic and comparative genomic framework provides a new approach for resolving the polyphyly of the genus *Bacillus* and improving the taxonomic classification of *Bacillus* species. This approach has the potential to improve our understanding of the diversity and evolution of these bacteria, as well as their roles in different environments (Patel and Gupta 2020).

As of today, a total of 15 scientifically accepted microbial tree of life have been constructed in a span of more than 40 years, i.e., from 1977 to 2019 and so on (Table 4.1). Most of the microbial trees of life have used RAxML for reconstruction

Table 4.1 A summary on trees of microbial life (1977–2019) (Modified from Zhu et al., 2019)

Sr. no.	Date (dd/mm/yy)	Domain (s)	Total taxa	Markers	Selected taxa	Method
1	01/11/77	A, B, E	13	SSU	N/A	Comparative
2	20/08/96	A, B, E	64	SSU	64	fastDNAml
3	03/03/06	A, B, E	191	31	181	PhyML
4	09/08/08	A, B	6727	SSU	9975	RAxML
5	13/10/08	B	578	31	578	PhyML
6	23/12/08	A, B, E	40	45	40	P4
7	20/05/13	A, B	2,03,452	SSU	2,03,452	FastTree
8	25/04/13	A, B	841	24	840	BUCKy
9	14/07/13	B	2229	38	2228	RAxML
10	14/08/13	A, B	3737	400	3139	RAxML
11	11/04/16	A, B, E	3083	16	3080	RAxML
12	12/06/17	A, B	1003	56	1039	RAxML
13	17/10/17	B	12,400	SSU	926	RAxML
14	27/08/18	B	21,943	120	21,547	FastTree
15	02/12/19	A, B	10,575	381	10,575	RAxML, ASTRAL
16	29/03/15	A, B, E	>55, 000	10–20 genes	∞	ExaML v3

Note: A = archaea; B = bacteria; E = eukarya

inferred either from nucleotides or amino acids. Of these 15, the tree of life reconstructed by Carl Woese and Fox in 1977 and Zhu et al. follows a comparative and RAxML/ASTRAL approach (Zhu et al., 2019). All branches of the microbial tree of life have selected and used representative genomes based on core genome selection or inputs from statistical analysis tools. But there has been a lot of debate about the current microbial tree of life for the last few decades. In 2019, a new view of the microbial tree of life from 10,575 genomes will be shown. This is very different from Carl Woese and Fox's approach, which was to compare all genomes.

Moreover, a strategic selection method leads to preferential loss of some important genes through the deletional bias and during the processes of demuxing and preventive fixation (Martinez-Gutierrez and Aylward 2021). Information about the loss of genes remains unclear, and how many lineages have been lost is a million-dollar question, which is contradictory to the approach of Carl Woese.

4.3 Discoveries and Tools for Phylogenetics and Phylogenomics

4.3.1 The Primary Kingdoms: Ribosomal RNA Based Phylogenetic Analysis

Phylogenetic analysis based on existing ribosomal RNA sequence data revealed that life has descended from one of the following: (i) the eubacteria, (ii) the archaebacteria, and (iii) the eukaryotes. To this day, no one knows for sure if the ancestor of all living things was a prokaryote or not. If the former statement isn't true, then urkingdom definitely represents a separate evolution at the prokaryotic level (Woese and Fox 1977). After the 1990s, new algorithms and platforms for sequencing, phylogenetics, and phylogenomics started to develop. Different resources have been heavily used, and in microbial phylogenetics and phylogenomics, there has been a paradigm shift that has led to the creation of tools that can handle a lot of data.

4.3.2 fastDNAml

fastDNAml is a powerful tool for constructing phylogenetic trees of DNA sequences using the maximum-likelihood method. This method is widely used in evolutionary biology, as it allows researchers to infer the evolutionary relationships between different species based on the similarity of their DNA sequences. fastDNAml is particularly useful for large datasets, as it is able to quickly and accurately analyze large amounts of data and construct high-resolution phylogenetic trees. Additionally, the software is user-friendly and easy to use, making it accessible to researchers with a wide range of backgrounds and expertise. The program can be downloaded from the Ribosomal Database Project (RDP) using anonymous FTP. fastDNAml can be used on Unix systems and workstations based on Felsenstein's dnaml v3.3 and uses a method called maximum likelihood (ML). Several improvements, including changes to the algorithms, make the program run much faster and use less memory, making it possible to build big trees with massive genomics data at that time (Olsen et al. 1994). Hence, fastDNAml is an indispensable tool for anyone interested in studying the evolution of DNA sequences.

4.3.3 P4

P4 is a phylogenetic tool for maximum-likelihood and Bayesian analysis of sequences that was created using Python. It employs a heterogeneous model, meaning that the model's properties might vary depending on the data or the tree. Compositional heterogeneity refers to the variation in the relative proportions of different components within a system. For example, in the context of biological systems, compositional heterogeneity can refer to the variation in the relative abundances of different types of molecules, such as proteins or nucleic acids, within a cell or tissue. Modeling compositional heterogeneity is important for understanding the complex dynamics of biological systems and can provide insight into how these systems evolve and function. Foster 2004 describe a method for modeling compositional heterogeneity using a probabilistic model based on Dirichlet distributions. This method allows for the incorporation of various sources of information, such as measurement error and uncertainty, and can provide a more accurate and comprehensive view of the compositional heterogeneity within a system. P4 is able to read data in a few popular phylogenetic formats, such as FastA, ClustalW, Nexus, Phylip, and pir/nbrf, but it cannot read additional bioinformatics formats (e.g., EMBL, genbank). P4 can read trees in Phylip or Nexus format (Foster 2004).

4.3.4 BUCKy

BUCKy is a C++ program for reconciling gene trees and species trees using Bayesian concordance analysis. It uses Bayesian inference to accurately estimate the evolutionary history of a set of genes, taking into account the known phylogeny of the species in which the genes are found. By integrating information from both the gene trees and the species tree, BUCKy can provide a more accurate and comprehensive view of the evolutionary relationships between the genes. BUCKy is particularly useful in cases where the gene trees and species tree have conflicts or discrepancies, as it can help resolve these inconsistencies and provide a more accurate picture of the evolution of the genes. BUCKY, unlike other tools, does not recognize that all genes have a similar topology. Instead, it counts how many genes have the same tree and merges their outputs to make a single tree with a high level of taxonomic resolution (Ané et al. 2007; Larget et al. 2010). Thus, BUCKy is a valuable tool for researchers studying the evolution of genes and their relationships to the species in which they are found.

4.3.5 ASTRAL

ASTRAL-Pro is a tool for inferring species trees based on quartet-based phylogenetic inference. This means that it uses information from the relationships between pairs of species (i.e., quartets) to infer the overall species tree. Meaning, ASTRAL is one of the most commonly used techniques for reconstructing species trees from gene sequences. It is scalable to very big datasets and can handle any inputs commonly used in practice, including missing data, polytomies, multiple species, etc. (10 K leaves). To do this, ASTRAL-Pro first generates a set of quartets from the input data. This can be done in several different ways, depending on the specific assumptions and constraints of the analysis. For example, ASTRAL-Pro can use a maximum-likelihood or Bayesian approach to infer the most likely quartets based on the input data. Once the quartets have been inferred, ASTRAL-Pro uses a specialized algorithm to combine them into a single species tree. This algorithm considers the compatibility of the quartets, as well as any additional constraints or assumptions that have been specified, to generate the most parsimonious species tree. This tree is then output by ASTRAL-Pro, along with any relevant statistics or results from the analysis. With massive phylogenetic or phylogenomic trees, it has high accuracy and is a robust algorithm. As per the data represented by Zhu et al. in 2019, one of the most effective methods for reconstructing species trees and determining the evolutionary distance between the domains of bacteria and archaea is ASTRAL, which combines CS theory, statistical analysis methods, heuristic techniques, efficient implementation, parallelism, and extensive software support (Mirarab et al. 2014; Zhang et al. 2018, 2020). Overall, ASTRAL-Pro is a powerful tool for inferring species trees based on quartet-based phylogenetic inference and can be used to analyze large datasets and complex evolutionary scenarios.

4.3.6 PhyML

A phylogeny program based on the maximum-likelihood idea is called PhyML. Because of its ease of use and decent accuracy/speed trade-off, PhyML has seen extensive use since its initial release. PhyML can be used to search the tree (Guindon and Gascuel 2003), and enhanced PhyML version 3.0 achieves a user-defined intensity. New methods are implemented. With its user interface, PhyML 3.0 can perform a nonparametric branch test that is similar to the Shimodaira-Hasegawa test. It also has a few new evolutionary models that can be used to make reliable, usable, and still fairly fast and accurate datasets (Guindon et al. 2010).

4.3.7 FastTree and FastTree 2

The fastDNAml was created to be used as a tool for construction of phylogenetic trees of DNA sequences using maximum likelihood. It implements the Fitch algorithm for the maximum-likelihood (ML) estimation, and it is written in C++ programming language. The FastTree algorithm is a tool for phylogenetic tree construction. It is used to find the minimum evolution tree for a set of DNA/RNA or amino acid sequences. FastTree uses the profile sequence alignment method to compute the distance matrix and then uses a heuristic search algorithm to efficiently find the optimal tree. This tool is particularly useful when there are many sequences and many taxa, as it can quickly find an approximate answer without using too much memory or computer time. FastTree uses alignments of nucleotide (N) or amino acid sequences to generate phylogenetic trees with a maximum-likelihood approach. FastTree can align up to one million sequences in a reasonable amount of time and RAM. In 17 hours and 2.4 GB of memory, FastTree generated a tree with support values comprising 158,022 unique 16S rRNAs (Price et al. 2009). Similarly, the FastTree 2 also includes maximum-likelihood NNIs and subtree pruning and regrafting (SPR) with minimum evolution. FastTree 2 is a program that calculates approximately maximum-likelihood (ML) trees for large alignments. It can handle input files that are too large to be processed by traditional methods, making it ideal for use with large genome sequences. The program is designed for alignment lengths of about 10,000 nucleotides or less but can handle larger inputs if the fast mode is used. The algorithm has been designed to be scalable so that it can run on a cluster of machines in parallel (Price et al. 2010).

4.3.8 Randomized Axelerated Maximum Likelihood (RAxML)

RAxML is a popular program for phylogenetic data from large datasets using maximum likelihood that is known for its speed and accuracy. It can be used to infer phylogenetic trees from large datasets and can handle a wide range of evolutionary models. The version 8 of RAxML includes several new features and improvements that make it even more effective at analyzing large phylogenies. For example, it includes support for parallel computing, which allows it to take advantage of multiple processors to speed up the analysis. It also includes support for new evolutionary models, such as the GTRGAMMA model, which can provide more accurate estimates of phylogenetic relationships. Additionally, the version 8 of RAxML includes several new features for post-analysis of phylogenies, such as support for generating consensus trees and calculating bootstrap support values. RAxML now supports data on multistate morphological features, binary and secondary structures of RNA, as well as coarse-grain, fine-grain, and coarse-grain/fine-grain parallelization via MPI, PThreads, and MPI/PThreads, respectively. Moreover, RAxML offers a plethora of sets of trees in their post-analysis functions, two algorithms-preparation and analysis-for next-generation sequencing data along with optimized x86 vector intrinsics to accelerate analysis. Overall, RAxML version 8 is a powerful tool that is well-suited for phylogenetic analysis and post-analysis of large phylogenies (Stamatakis 2014).

4.3.9 ExaML v3.0

Next-generation sequencing generated a massive amount of data, which may not always be fully understood or analyzed properly using the aforementioned bioinformatics tools and algorithms. Hence, to resolve this issue, ExaMl version 3.0 (v3.0) offers phylogenetic analyses that can grow at an unprecedented pace. ExaML v3.0 possesses dedicated codes for inferring phylogenies with whole genome (WG) and whole transcriptome (WT) data using supercomputers. ExaML (v. 3) offers new models and data types as well as a production-ready implementation of the Intel MIC architecture, which required a lot of reengineering. We also show a new way to align I/O method in parallel.

4.3.9.1 Parallel I/O Optimization in Examl v3.0

To read in the input alignment, the parallel I/O optimization method (hereafter IO) can be optimized in two ways. A specialized parser component is responsible for performing the initial evaluation and conversion of a plain-text PHYLIP file to a binary file format. Because the parser may be run on a standard server, it does not require a substantial amount of time from a supercomputer to complete tasks such as verifying the format, searching for duplicate taxon names, compressing site patterns, and so on. The raw sequences are kept in partition order and are included in the binary alignment file. The file also contains global data information such as the alignment length, data kinds, model settings, and partition boundaries. In other words, the sequences for all taxa in partition 1 are saved first, followed by all taxa in partition 2, etc. Since it is an unpublished binary format, each ExaML process is able to simultaneously access only the alignment parts on which it will infer likelihood tree.

The codes are available via HTTS (https://github.com/stamatak/ExaML.git) and GitHub CLI (gh repo clone stamatak/ExaML) (Kozlov et al. 2015). ExaML also uses checkpointing, SSE3, AVX vectorization, and techniques that save memory, just like RAxML-Light. The RAxML-Light tool is four times slower than ExaML v3.0 (Stamatakis et al. 2012). This is the best advantage of using ExaML v3.0 for big data computing over other algorithms and tools.

References

Ané C, Larget B, Baum DA, Smith SD, Rokas A (2007) Bayesian estimation of concordance among gene trees. Mol Biol Evol 24(2):412–426

Bauldauf SL (2003) Phylogeny for the faint of heart: a tutorial. Trends Genet 19:346–351

Baum DA, Smith SD, Donovan SS (2005) Evolution: the tree-thinking challenge. Science 310:979–980

Braga LPP, Soucy SM, Amgarten DE, da Silva AM, Setubal JC (2018) Bacterial diversification in the light of the interactions with phages: the genetic symbionts and their role in ecological speciation. Front Ecol Evol 6:6

Brower AVZ, Schuh RT (2021) Biological systematics: principles and applications. Cornell University Press, Ithaca, NY. ISBN13: 9781501752780, ISBN10: 1501752782

Brown TA (2002) Molecular phylogenetics, In Genomes, 2nd Edition. Garland Science. Bookshelf ID: NBK21122

Bryant J, Chewapreecha C, Bentley SD (2012) Developing insights into the mechanisms of evolution of bacterial pathogens from whole-genome sequences. Future Microbiol 7(11):1283–1296

Donkor ES (2013) Sequencing of bacterial genomes: principles and insights into pathogenesis and development of antibiotics. Genes 4(4):556–572

Eisen JA (1998) Phylogenomics: improving functional predictions for uncharacterized genes by evolutionary analysis. Genome Res 8:163–167

Falkowski PG (1998) Biogeochemical controls and feedbacks on ocean primary production. Science 281:200–206

Foster PG (2004) Modeling compositional heterogeneity. Syst Biol 53:485–495

Guindon S, Dufayard JF, Lefort V, Anisimova M, Hordijk W, Gascuel O (2010) New algorithms and methods to estimate maximum-likelihood phylogenies: assessing the performance of PhyML 3.0. Syst Biol 59(3):307–321

Guindon S, Gascuel O (2003) A simple, fast, and accurate algorithm to estimate large phylogenies by maximum likelihood. Syst Biol 52(5):696–704

Haft DH, DiCuccio M, Badretdin A, Brover V, Chetvernin V, O'Neill K, Li W, Chitsaz F, Derbyshire MK, Gonzales NR, Gwadz M, Lu F, Marchler GH, Song JS, Thanki N, Yamashita RA, Zheng C, Thibaud-Nissen F, Geer LY, Marchler-Bauer A, Pruitt KD (2018) RefSeq: an update on prokaryotic genome annotation and curation. Nucleic Acids Res 46(D1):D851–D860

Handelsman J (2004) Metagenomics: application of genomics to uncultured microorganisms. Microbiol Mol Biol Rev 68:669–685

Jukes TH, Cantor CR (1969) Evolution of protein molecules. In: Munro HN (ed) Mammalian protein metabolism, 3rd edn. Academic Press, New York, pp 21–132

Kasting JF (2002) Life and the evolution of Earth's atmosphere. Science 296:1066–1068

Ko KKK, Chng KR, Nagarajan N (2022) Metagenomics-enabled microbial surveillance. Nat Microbiol 7:486–496

Kozlov AM, Aberer AJ, Stamatakis A (2015) ExaML version 3: a tool for phylogenomic analyses on supercomputers. Bioinformatics 31(15):2577–2579

Kozlov AM, Zhang J, Yilmaz P, Glöckner FO, Stamatakis A (2016) Phylogeny-aware identification and correction of taxonomically mislabeled sequences. Nucleic Acids Res 44:5022–5033

Larget B, Kotha SK, Dewey CN, Ané C (2010) BUCKy: gene tree/species tree reconciliation with the Bayesian concordance analysis. Bioinformatics 26(22):2910–2911

Ley RE, Peterson DA, Gordon JI (2006) Ecological and evolutionary forces shaping microbial diversity in the human intestine. Cell 124:837–848

Louca S, Mazel F, Doebeli M, Parfrey LW (2019) A census-based estimate of Earth's bacterial and archaeal diversity. PLoS Biol 17:e3000106

Martinez-Gutierrez CA, Aylward FO (2021) Genome size distributions in bacteria and archaea are strongly linked to phylogeny. bioRxiv. https://doi.org/10.1101/2021.12.15.472816

Martinez-Gutierrez CA, Aylward FO (2022) Evolutionary genomics of marine bacteria and archaea. In: Stal LJ, Cretoiu MS (eds) The marine microbiome. The microbiomes

of humans, animals, plants, and the environment, vol 3. Springer, Cham. https://doi.org/10.1007/978-3-030-90383-1_7
Mende DR, Bryant JA, Aylward FO, Eppley JM, Nielsen T, Karl DM, DeLong EF (2017) Environmental drivers of a microbial genomic transition zone in the ocean's interior. Nat Microbiol 2:1367–1373
Mirarab S, Reaz R, Bayzid MS, Zimmermann T, Swenson MS, Warnow T (2014) ASTRAL: genome-scale coalescent-based species tree estimation. Bioinformatics 30(17):i541–i548
Munjal G, Hanmandlu M, Srivastava S (2018) Phylogenetics algorithms and applications. Ambient Commun Comput Syst 904:187–194
Olsen GJ, Matsuda H, Hagstrom R, Overbeek R (1994) fastDNAmL: a tool for construction of phylogenetic trees of DNA sequences using maximum likelihood. Comput Appl Biosci 10(1):41–48
Pace NR, Sapp J, Goldenfeld N (2012) Phylogeny and beyond: scientific, historical, and conceptual significance of the first tree of life. PNAS 109:1011–1018
Patané JSL, Martins J, Setubal JC (2018) Phylogenomics. In: Setubal J, Stoye J, Stadler P (eds) Comparative genomics. Methods in molecular biology, vol 1704. Humana Press, New York, NY. https://doi.org/10.1007/978-1-4939-7463-4_5
Patel S, Gupta RS (2020) A phylogenomic and comparative genomic framework for resolving the polyphyly of the genus Bacillus: proposal for six new genera of *Bacillus* species, *Peribacillus* gen. nov., *Cytobacillus* gen. nov., *Mesobacillus* gen. nov., *Neobacillus* gen. nov., *Metabacillus* gen. nov. and *Alkalihalobacillus* gen. nov. Int J Syst Evol Microbiol 70(1):406–438
Price MN, Dehal PS, Arkin AP (2009) FastTree: computing large minimum evolution trees with profiles instead of a distance matrix. Mol Biol Evol 26(7):1641–1650
Price MN, Dehal PS, Arkin AP (2010) FastTree 2—approximately maximum-likelihood trees for large alignments. PLoS One 5(3):e9490
Robinson SL, Piel J, Sunagawa S (2021) A roadmap for metagenomic enzyme discovery. Nat Prod Rep 38:1994–2023
Shakya M, Ahmed SA, Davenport KW, Flynn MC, Lo CC, Chain PSG (2020) Standardized phylogenetic and molecular evolutionary analysis applied to species across the microbial tree of life. Sci Rep 10:1723
Stamatakis A (2014) RAxML version 8: a tool for phylogenetic analysis and post-analysis of large phylogenies. Bioinformatics 30(9):1312–1313
Stamatakis A, Aberer AJ, Goll C, Smith SA, Berger SA, Izquierdo-Carrasco F (2012) RAxML-light: a tool for computing TeraByte phylogenies. Bioinformatics 28(15):2064–2066
Sterpetti F (2021) Mathematical explanations in evolutionary biology or naturalism? A challenge for the statisticalist. Found Sci. https://doi.org/10.1007/s10699-021-09818-w
Styles KM, Brown AT, Sagona AP (2021) A review of using mathematical modeling to improve our understanding of bacteriophage, bacteria, and eukaryotic interactions. Front Microbiol 12:724767
Tkach N, Nobis M, Schneider J, Becher H, Winterfeld G, Jacobs SWL, Röser M (2021) Molecular phylogenetics and micromorphology of Australasian Stipeae (Poaceae, Subfamily Pooideae), and the interrelation of whole-genome duplication and evolutionary radiations in this Grass Tribe. Front Plant Sci 11:630788
Woese CR, Fox GE (1977) Phylogenetic structure of the prokaryotic domain: the primary kingdoms. PNAS 74(11):5088–5090
Yang Z, Rannala B (2012) Molecular phylogenetics: principles and practice. Nat Rev Genet 13:303–314
Zhang C, Rabiee M, Sayyari E, Mirarab S (2018) ASTRAL-III: polynomial time species tree reconstruction from partially resolved gene trees. BMC Bioinformatics 19:153

Zhang C, Scornavacca C, Molloy EK, Mirarab S (2020) ASTRAL-Pro: quartet-based species-tree inference despite paralogy. Mol Biol Evol 37(11):3292–3307

Zhu Q, Mai U, Pfeiffer W, Janssen S, Asnicar F, Sanders JG, Belda-Ferre P, Al-Ghalith GA, Kopylova E, McDonald D, Kosciolek T, Yin JB, Huang S, Salam N, Jiao JY, Wu Z, Xu ZZ, Cantrell K, Yang Y, Sayyari E et al (2019) Phylogenomics of 10,575 genomes reveals evolutionary proximity between domains Bacteria and Archaea. Nat Commun 10(1):5477. https://doi.org/10.1038/s41467-019-13443-4

Chapter 5
Phenotypes and Genotypes

Manik Prabhu Narsing Rao and Wen-Jun Li

5.1 Introduction

Prokaryotes have been known since Leeuwenhoek's observations (Leeuwenhoek 1677). The development of solid media for the separation of single cells by streaking bacteria aided scientists in carefully isolating specific microorganisms and studying them in their environment (Koch 1877). The retrieval of prokaryotes and studying their useful information helped to develop the species concept for prokaryotes (Rosselló-Mora and Amann 2001). In the 1870s, Ferdinand Cohn conducted the first studies on bacterial classification and classified bacteria (at that time as the member of domain plant) into six different genera based on morphologic traits (cell shape) (Cohn 1875; Trüper and Schleifer 2006). Based on phenotypic traits, different methods of classification have been published, making taxonomic studies difficult due to a lack of minimal standards and common agreement on classification (Woese 1992). This status was changed with the definition of a reference guideline for bacterial classification in the first edition of Bergey's *Manual of Determinative Bacteriology*, which allowed microbiologists to combine the criteria they used (Bergey et al. 1923). Soon, based on phenotypic traits, huge volumes of phenotypic data were analyzed on electronic platforms to derive meaningful relationships among a large number of microorganisms, and this system of analysis was called numerical taxonomy (Sokal 1966; Sneath and Sokal 1973). Although numerical

M. P. Narsing Rao
Instituto de Ciencias Aplicadas, Facultad de Ingeniería, Universidad Autónoma de Chile, Sede Talca, Talca, Chile

W.-J. Li (✉)
State Key Laboratory of Biocontrol, Guangdong Provincial Key Laboratory of Plant Resources and Southern Marine Science and Engineering Guangdong Laboratory (Zhuhai), School of Life Sciences, Sun Yat-Sen University, Guangzhou, People's Republic of China
e-mail: liwenjun3@mail.sysu.edu.cn

W.-J. Li et al. (eds.), *Modern Taxonomy of Bacteria and Archaea*,
https://doi.org/10.1007/978-981-99-5720-0_5

taxonomy improved bacteria identification, it did not take phylogenetic analyses into account (Schleifer 2009; Ferraz Helene et al. 2022). With the evolution of numerical taxonomy, new techniques were developed that took into account physiologic and biochemical traits, such as chromatographic and electrophoretic methods for defining chemotaxonomic markers (Rosselló-Mora and Amann 2001; Ferraz Helene et al. 2022). It was soon discovered that many standard phenotypic tests produce data with low reproducibility and unacceptable margins of error (Kim et al. 2015; Sutcliffe 2015).

The idea that bacteria might be best categorized by comparing their genomes was supported throughout the early 1960s (Rosselló-Mora and Amann 2001). Zuckerkandl and Pauling predicted that macromolecules could act as a molecular clock to infer evolutionary relationships (Zuckerkandl and Pauling 1965). This inspired Woese to begin his search for a molecular chronometer that could serve as the foundation of an evolutionary framework for all life, and he established ribosomal RNA genes as standard phylogenetic markers in taxonomic studies (Woese and Fox 1977). Over time, several other DNA-based classification methods have been developed (Lee et al. 1956; Brenner et al. 1969). The prokaryotic species delineation today is based on a polyphasic approach that includes phenotypic, chemotaxonomic, and genomic data analysis (Colwell 1970). The present chapter describes phenotype and genome-based characterization.

5.2 Phenotype

The phenotype is best defined as the observable characteristics that result from an organism's gene expression, which can be greatly influenced by environmental or other factors (Rosselló-Móra and Amann 2015). The observable characteristics include morphological, physiological, and biochemical features. Phenotypic-based characterization is the most time-consuming and difficult task in classification, and the techniques should be standardized to avoid subjective observations (Rosselló-Mora and Amann 2001). While analyzing the phenotypic characters of new strain, it is essential to use both new strain and the reference strains (Rosselló-Mora and Amann 2001).

5.2.1 *Morphological, Microscopic, and Biochemical Characters*

The morphological features of the bacteria include colony characters, Gram staining, cellular shape, endospores, flagella, etc. The colony characters include colony form (circular or irregular), colony margin (entire, irregular, undulate, lobate, curled, or filiform), colony texture (smooth, rough, or wrinkled), colony size (punctiform, small, or large), colony color (white, brown, yellow, etc.), colony elevation (flat, raised, umbonate, convex, or pulvinate), and colony opacity (transparent, opaque, translucent, or iridescent) (Aneja 2010; Sousa et al. 2013).

Gram staining is one of the oldest methods of distinguishing between cell-wall components (peptidoglycan type and teichoic acids) and allows bacteria to be classified into two basic types of cells: Gram-positive and Gram-negative (Gram 1884).

Bacteria and archaea cells vary greatly between species, so better methods for visualizing prokaryotic cellular organization and structure are being developed (Kämpfer and Glaeser 2013). Light and electron microscopy can be used to determine the shape of the cell. Some of the shapes are round or ellipsoidal, rods, long, filamentous branched, comma, and spiral. The cell arrangement is also an important factor. Some cells form chains, while others form grapelike clusters and cubic packets (Salton and Kim 1996). Some bacteria (such as *Bacillus*) were classified based on their capability to sporulate (Fritze 2004). The formation of spores inside the cell (endospores) or outside the cell (exospores) should be noted. In the case of endospore-forming organisms, the location of the spores, the formation of sporangia, and the shape of the spores should be noted (Tindall et al. 2010).

Some organisms may move quickly, while others move slowly, and thus motility must be accurately described. Flagella-mediated motility should be distinguished from gliding or other forms of motility (Tindall et al. 2010). Depending on the number and position of flagella, they were further characterized as monotrichous (having one flagellum), amphitrichous (single flagellum at both ends), lophotrichous (numerous flagella as a tuft), and peritrichous (flagella distributed all over the cell) flagella (Ruan et al. 2012).

The physiological and biochemical traits include data on growth at various temperatures, salt concentrations, pH values, oxygen requirement (aerobic/anaerobic), growth in the presence of various substances (like antimicrobial agents), and data on the presence or absence of various enzyme activities (oxidase, catalase, lipase, urease, gelatin liquefaction, starch hydrolysis, nitrate reduction, and others) and metabolization of compounds (carbon and nitrogen sources). The detailed physiological and biochemical trait evaluation methods were mentioned by Tindall et al. (2007) and Qinyuan et al. (2016). Further, commercial kits (analytical profile index; API) and easy-to-use systems (GENIII MicroPlate; Biolog) are also available for the rapid detection of biochemical characters (Washington 2nd et al. 1971; Holmes et al. 1994).

5.3 Genotype

Despite the widespread use of phenotypic-based classification, many tests produce results with low reproducibility and unacceptable error margins, which cannot be regarded as sound scientific practice (Sutcliffe 2015). For instance, it has recently been demonstrated that determining G + C content in wet lab experiments is unreliable when compared to determining this value using high-quality genome sequence data (Kim et al. 2015). In this way, genotypic-based classification was made, which involved comparing genetic material (Tang et al. 1998).

5.3.1 DNA GC Content and DNA-DNA Hybridization

In the early 1960s, the advancement of molecular techniques enabled the incorporation of methodologies such as DNA guanine and cytosine content (GC mol%) and DNA-DNA hybridization (DDH) for taxonomic studies (Brenner et al. 1969; Stackebrandt and Liesack 1993). The GC content of DNA is still an important character and is required for the description of the new taxon. Variation in percent GC content should not exceed 3% within a well-defined species and 10% within a well-defined genus (Stackebrandt and Liesack 1993; Prakash et al. 2007). Furthermore, greater similarity in DNA base composition does not necessarily imply that the two strains are closely related. It should be emphasized that the DNA base composition is a valuable attribute and should be used to distinguish between unrelated bacteria (Kämpfer and Glaeser 2013).

DDH is based on a comparison of the whole genomes of two bacterial species (Brenner et al. 1969). It is determined by heating the DNA of the two strains being compared. Heating converts double-stranded DNA to single-stranded DNA. The DNA preparations are then mixed in equal parts and slowly cooled. Thereafter, the DNA will begin to renature into double-stranded DNA. When DNA from two strains is mixed, it renatures less perfectly than DNA from the same strain. It was noticed that phenotypically coherent microorganisms could be considered a single species if they shared DDH values above 70%. An ad hoc committee later recommended this cutoff value as an approximate threshold for circumscribing species (Wayne et al. 1987). Although DDH has been used as the gold standard in prokaryote taxonomy, it is being criticized for being labor-intensive and prone to experimental errors (Konstantinidis and Tiedje 2005).

5.3.2 16S rRNA Gene Sequencing

The groundbreaking study of Woese and Fox (1977) established ribosomal RNA genes as standard phylogenetic markers in taxonomic studies. The 16S rRNA gene sequence has been considered the gold standard technique for prokaryotic classification, and currently, 98.7% 16S rRNA gene sequence similarity was recommended as the cutoff for delineating species (Chun et al. 2018). Furthermore, there is growing interest in the use of other genes in prokaryotic taxonomy, such as *rpoB*, *gyrA*, and others, as an alternative or supporting molecular marker (Sutcliffe 2015).

5.3.3 *DNA Fingerprinting and Multilocus Sequence Analysis*

DNA fingerprinting can be used to differentiate closely related taxa, particularly at the subspecies and/or strain level, using different methods such as amplified fragment-length polymorphism PCR (AFLP), random amplified polymorphic DNA (RAPD) analysis, repetitive element primed PCR (rep-PCR) using different primers that bind to highly conserved, repetitive DNA sequences distributed all over the genome, including repetitive extragenic palindromic (REP)-PCR, enterobacterial repetitive intergenic consensus sequences (ERIC)-PCR, BOX-PCR (derived from the boxA element), and (GTG)5-PCR, and ribotyping (Tindall et al. 2010; Kämpfer and Glaeser 2013).

Multilocus sequence analysis (MLSA) was introduced as an intermediate resolution between the 16S rRNA gene and genome-based approaches when complete genome sequences were not yet available for more in-depth genomic comparisons (Oren and Garrity 2014; Glaeser and Kämpfer 2015). It is a method for genotypic characterization of a specific group of prokaryotes, most commonly strains of a single species; species of a genus, as well as closely related genera, are investigated. It contains partial sequences of 5 to 12 housekeeping genes that were investigated and compared as individual data sets or combined to form concatenated sequences that were then used for phylogenetic analysis (Kämpfer and Glaeser 2013; Glaeser and Kämpfer 2015).

5.4 Genome-Based Analysis

The first prokaryote genome to be sequenced was *Haemophilus influenzae* in 1995 (Fleischmann et al. 1995), but the high cost and time-consuming process of genome sequencing prevented significant progress in this area. However, advancements in this field, as well as the introduction of high-throughput sequencing methods, have made it possible to obtain low-cost and high-quality bacterial genome sequences (Liu et al. 2019). Minimal standards for the use of genome data (like genome completeness, quantity, and others) in prokaryote taxonomy have also been proposed (Chun et al. 2018).

Overall genome relatedness indices (Chun and Rainey 2014) were described to estimate sequence similarity and to replace DDH. Several overall genome relatedness indices methods were developed, the most common of which are digital DNA-DNA hybridization (dDDH) (Auch et al. 2010) and average nucleotide identity (ANI) (Konstantinidis and Tiedje 2005). The accepted species boundaries for ANI and dDDH values were 95–96 (Goris et al. 2007; Richter and Rosselló-Móra 2009) and 70% (Meier-Kolthoff et al. 2013), respectively. A detailed section on this topic is discussed in Chap. 7 of this book.

References

Aneja KR (2010) Experiments in microbiology plant pathology and biotechnology, 4th edn, New Delhi. New age International (p) Ltd

Auch AF, von Jan M, Klenk HP, Göker M (2010) Digital DNA-DNA hybridization for microbial species delineation by means of genome-to-genome sequence comparison. Stand Genomic Sci 2:117–134. https://doi.org/10.4056/sigs.531120

Bergey DH, Harrison FC, Breed RS, Hammer BW, Hantoon FM (eds) (1923) Bergey's manual of determinative bacteriology, 1st edn. Williams and Wilkins, Baltimore

Brenner DJ, Fanning GR, Rake AV, Johnson KE (1969) Batch procedure for thermal elution of DNA from hydroxyapatite. Anal Biochem 28:447–459. https://doi.org/10.1016/0003-2697(69)90199-7

Chun J, Oren A, Ventosa A, Christensen H, Arahal DR, da Costa MS, Rooney AP, Yi H, Xu XW, De Meyer S, Trujillo ME (2018) Proposed minimal standards for the use of genome data for the taxonomy of prokaryotes. Int J Syst Evol Microbiol 68:461–466. https://doi.org/10.1099/ijsem.0.002516

Chun J, Rainey FA (2014) Integrating genomics into the taxonomy and systematics of the bacteria and archaea. Int J Syst Evol Microbiol 64:316–324. https://doi.org/10.1099/ijs.0.054171-0

Cohn F (1875) Untersuchungen Über Bacterien: I. JU Kern

Colwell R (1970) Polyphasic taxonomy of bacteria. In: Proceedings of the first international conference an culture collections, October 7–11, 1968. University of Tokyo Press, pp 421–436

Ferraz Helene LC, Klepa MS, Hungria M (2022) New insights into the taxonomy of bacteria in the genomic era and a case study with rhizobia. Int J Microbiol 2022:4623713. https://doi.org/10.1155/2022/4623713

Fleischmann RD et al (1995) Whole-genome random sequencing and assembly of *Haemophilus influenzae* Rd. Science 269:496–512

Fritze D (2004) Taxonomy of the genus bacillus and related genera: the aerobic endospore-forming bacteria. Phytopathology 94:1245–1248. https://doi.org/10.1094/phyto.2004.94.11.1245

Glaeser SP, Kämpfer P (2015) Multilocus sequence analysis (MLSA) in prokaryotic taxonomy. Syst Appl Microbiol 38:237–245. https://doi.org/10.1016/j.syapm.2015.03.007

Goris J, Konstantinidis KT, Klappenbach JA, Coenye T, Vandamme P, Tiedje JM (2007) DNA-DNA hybridization values and their relationship to whole-genome sequence similarities. Int J Syst Evol Microbiol 57:81–91. https://doi.org/10.1099/ijs.0.64483-0

Gram HC (1884) über die isolierte Farbung der Schizomyceten in Schnittund Trocken praparaten. Fortschr Med 2:185–189. (In German)

Holmes B, Costas M, Ganner M, On SL, Stevens M (1994) Evaluation of biolog system for identification of some gram-negative bacteria of clinical importance. J Clin Microbiol 32:1970–1975. https://doi.org/10.1128/jcm.32.8.1970-1975.1994

Kämpfer P, Glaeser SP (2013) Prokaryote characterization and identification. In: Rosenberg E, DeLong EF, Lory S, Stackebrandt E, Thompson F (eds) The prokaryotes: prokaryotic biology and symbiotic associations. Springer, Berlin Heidelberg, Berlin, Heidelberg, pp 123–147

Kim M, Park SC, Baek I, Chun J (2015) Large-scale evaluation of experimentally determined DNA G+C contents with whole genome sequences of prokaryotes. Syst Appl Microbiol 38:79–83. https://doi.org/10.1016/j.syapm.2014.11.008

Koch R (1877) Untersuchungen uber Bakterien: VI. Verfahren zur Conserviren und Photographiren der Bakterien. Beitrage zur Biologie der Pflanzen 2

Konstantinidis KT, Tiedje JM (2005) Genomic insights that advance the species definition for prokaryotes. Proc Natl Acad Sci U S A 102:2567–2572. https://doi.org/10.1073/pnas.0409727102

Lee KY, Wahl R, Barbu E (1956) Contenu en bases puriques et pyrimidiques des acides désoxyribonucléiques des bactéries. In: Annales de l institut pasteur. Masson editeur 120 blvd saint-germain, 75280 Paris 06, France, pp 212–224

Leeuwenhoek AV (1677) Observations, communicated to the publisher by Mr. Antony van Leewenhoeck, in a dutch letter of the 9th Octob. 1676. here English'd: concerning little animals by him observed in rain-well-sea-and snow water; as also in water wherein pepper had lain infused. Phil Trans 12:821–831. https://doi.org/10.1098/rstl.1677.0003

Liu GH, Narsing Rao MP, Dong ZY, Wang JP, Che JM, Chen QQ, Sengonca C, Liu B, Li WJ (2019) Genome-based reclassification of *bacillus okuhidensis* as a later heterotypic synonym of *bacillus halodurans*. Int J Syst Evol Microbiol 69:3599–3602. https://doi.org/10.1099/ijsem.0.003666

Meier-Kolthoff JP, Auch AF, Klenk HP, Göker M (2013) Genome sequence-based species delimitation with confidence intervals and improved distance functions. BMC Bioinformatics 14:60. https://doi.org/10.1186/1471-2105-14-60

Oren A, Garrity GM (2014) Then and now: a systematic review of the systematics of prokaryotes in the last 80 years. Antonie Van Leeuwenhoek 106:43–56. https://doi.org/10.1007/s10482-013-0084-1

Prakash O, Verma M, Sharma P, Kumar M, Kumari K, Singh A, Kumari H, Jit S, Gupta SK, Khanna M, Lal R (2007) Polyphasic approach of bacterial classification - an overview of recent advances. Indian J Microbiol 47:98–108. https://doi.org/10.1007/s12088-007-0022-x

Qinyuan L, Xiu C, Yi J, Chenglin J (2016) Cultural, physiological, and biochemical identification of actinobacteria. In: Dharumadurai D, Yi J (eds) Actinobacteria. IntechOpen, Rijeka. p Ch. 4

Richter M, Rosselló-Móra R (2009) Shifting the genomic gold standard for the prokaryotic species definition. Proc Natl Acad Sci U S A 106:19126–19131. https://doi.org/10.1073/pnas.0906412106

Rosselló-Mora R, Amann R (2001) The species concept for prokaryotes. FEMS Microbiol Rev 25:39–67. https://doi.org/10.1111/j.1574-6976.2001.tb00571.x

Rosselló-Móra R, Amann R (2015) Past and future species definitions for bacteria and archaea. Syst Appl Microbiol 38:209–216. https://doi.org/10.1016/j.syapm.2015.02.001

Ruan J, Kato T, Santini CL, Miyata T, Kawamoto A, Zhang WJ, Bernadac A, Wu LF, Namba K (2012) Architecture of a flagellar apparatus in the fast-swimming magnetotactic bacterium MO-1. Proc Natl Acad Sci U S A 109:20643–20648. https://doi.org/10.1073/pnas.1215274109

Salton MRJ, Kim KS (1996) Structure. In: Baron S (ed) Medical Microbiology, 4th edn. University of Texas Medical Branch at Galveston, Galveston (TX). Chapter 2

Schleifer KH (2009) Classification of bacteria and archaea: past, present and future. Syst Appl Microbiol 32(8):533–542. https://doi.org/10.1016/j.syapm.2009.09.002

Sneath PHA, Sokal RR (1973) Numerical Taxonomy: The Principles and Practice of Numerical Classification. WF Freeman & Co., San Francisco

Sokal RR (1966) Numerical taxonomy. Sci Am 215:106–117

Sousa AM, Machado I, Nicolau A, Pereira MO (2013) Improvements on colony morphology identification towards bacterial profiling. J Microbiol Methods 95:327–335. https://doi.org/10.1016/j.mimet.2013.09.020

Stackebrandt E, Liesack W (1993) Nucleic acid and classification. In: Goodfellow M, O'Donnell AG (eds) Handbook of new bacterial systematics. Academic Press Ltd, London, pp 151–194

Sutcliffe IC (2015) Challenging the anthropocentric emphasis on phenotypic testing in prokaryotic species descriptions: rip it up and start again. Front Genet 6:218. https://doi.org/10.3389/fgene.2015.00218

Tang YW, Ellis NM, Hopkins MK, Smith DH, Dodge DE, Persing DH (1998) Comparison of phenotypic and genotypic techniques for identification of unusual aerobic pathogenic gram-negative bacilli. J Clin Microbiol 36:3674–3679. https://doi.org/10.1128/jcm.36.12.3674-3679.1998

Tindall BJ, Rosselló-Móra R, Busse HJ, Ludwig W, Kämpfer P (2010) Notes on the characterization of prokaryote strains for taxonomic purposes. Int J Syst Evol Microbiol 60:249–266. https://doi.org/10.1099/ijs.0.016949-0

Tindall BJ, Sikorski J, Smibert RA, Krieg NR (2007) Phenotypic characterization and the principles of comparative systematics. In: Methods for general and molecular microbiology, pp 330–393

Trüper HG, Schleifer KH (2006) Prokaryote characterization and identification. In: Dworkin M, Falkow S, Rosenberg E, Schleifer K-H, Stackebrandt E (eds) The prokaryotes: volume 1: symbiotic associations, biotechnology, applied microbiology. Springer New York, New York, NY, pp 58–79

Washington JA 2nd, Yu PK, Martin WJ (1971) Evaluation of accuracy of multitest micromethod system for identification of Enterobacteriaceae. Appl Microbiol 22:267–269. https://doi.org/10.1128/am.22.3.267-269.1971

Wayne LG et al (1987) Report of the Ad Hoc Committee on Reconciliation of Approaches to Bacterial Systematics. Int J Syst Evol Microbiol 37:463–464. https://doi.org/10.1099/00207713-37-4-463

Woese CR (1992) Prokaryote systematics: the evolution of a science. The Prokayotes:3–18

Woese CR, Fox GE (1977) Phylogenetic structure of the prokaryotic domain: the primary kingdoms. Proc Natl Acad Sci U S A 74:5088–5090. https://doi.org/10.1073/pnas.74.11.5088

Zuckerkandl E, Pauling L (1965) Molecules as documents of evolutionary history. J Theor Biol 8:357–366. https://doi.org/10.1016/0022-5193(65)90083-4

Chapter 6
Chemotaxonomy and Genomes

Bao-Zhu Fang, Jian-Yu Jiao, Shuang Wang, and Wen-Jun Li

6.1 Introduction

Chemotaxonomy was first used by organic chemists to investigate naturally occurring substances in plants. With the advancement of natural product chemistry, many botanists and chemists realized they can identify and categorize plants based on their chemical contents. Likewise, botany and chemistry become increasingly separated due to the rise in the specialization of their respective subject experts. With the growing knowledge of the structure of natural products and their occurrence in plants, the potentialities of "chemotaxonomy" had become increasingly obvious. Since 1960, chemotaxonomy, which was often employed in defining novel species

B.-Z. Fang
State Key Laboratory of Desert and Oasis Ecology, Key Laboratory of Ecological Safety and Sustainable Development in Arid Lands, Xinjiang Institute of Ecology and Geography, Chinese Academy of Sciences, Urumqi, Xinjiang, People's Republic of China

J.-Y. Jiao
State Key Laboratory of Biocontrol, Guangdong Provincial Key Laboratory of Plant Resources and Southern Marine Science and Engineering Guangdong Laboratory (Zhuhai), School of Life Sciences, Sun Yat-Sen University, Guangzhou, People's Republic of China

S. Wang
Heilongjiang Academy of Black Soil Conservation and Utilization, Harbin, People's Republic of China

W.-J. Li (✉)
State Key Laboratory of Desert and Oasis Ecology, Key Laboratory of Ecological Safety and Sustainable Development in Arid Lands, Xinjiang Institute of Ecology and Geography, Chinese Academy of Sciences, Urumqi, Xinjiang, People's Republic of China

State Key Laboratory of Biocontrol, Guangdong Provincial Key Laboratory of Plant Resources and Southern Marine Science and Engineering Guangdong Laboratory (Zhuhai), School of Life Sciences, Sun Yat-Sen University, Guangzhou, People's Republic of China
e-mail: liwenjun3@mail.sysu.edu.cn

W.-J. Li et al. (eds.), *Modern Taxonomy of Bacteria and Archaea*,
https://doi.org/10.1007/978-981-99-5720-0_6

by exploiting the biological and chemical macromolecules of cell constituents, embraces several techniques that had been used extensively in prokaryote taxonomy. Taken together, the collection of information on various constituents of the cell provided a robust approach to the generation of information used in the identification and classification of prokaryotes.

Chemotaxonomic indicators are inherent components of the microbial phenotype but are typically treated independently. Some phenotypic approaches had the drawback that some tests could not be performed on all bacterial and archaeal species. For instance, not all organisms were metabolically active in these assays, resulting in data and classification schemes that were frequently fragmented in nature. However, characteristics that include chemotaxonomic markers found in cell membranes or walls were always present to some degree. In addition to being widely employed, these features had also been shown to represent evolutionary connections and were advised for use in polyphasic research.

Since these chemical components might be identical in various species, chemotaxonomic distinctions should be used to exclude an organism from a given taxon during data interpretation, rather than similarities as the sole criterion for admission. Many groups of prokaryotes had benefited greatly from chemotaxonomic approaches since morphological and physiological features had not been able to reliably separate taxa or offer a suitable categorization. The details of the particular approaches were well-documented elsewhere, so we would not go into them here. References to sources and reviews would be used to illustrate the evolution and current state of chemotaxonomic techniques, which would be shown with instances of their application to various taxa. Possible future directions and the use of in silico approaches based on genomic data would also be explored.

6.2 History of Chemotaxonomic Biomarkers

Chemotaxonomy has a long and illustrious history. In the early 1900s, many of the chemical macromolecules that currently form the foundation of chemotaxonomic biomarkers and are commonly used to identify and classify microorganisms were first reported. Those preliminary researches showed that various types of bacteria have distinct cell membrane compositions (Salton 1994). However, it was the technical breakthroughs of the 1950s that sparked a proliferation of these approaches to microbial systematics (Kates 1964; Lechevalier and Moss, 1977; Salton 1994). Experimental studies in a controlled environment uncovered a wide variety of structural variations in these cellular biomarkers because more data was collected on various types of microorganisms. They were quickly put to use by microbiologists for both the identification and categorization of microbes (Kates 1964; Lechevalier and Moss, 1977; Salton 1994: Busse et al. 1996; Schleifer 2009).

6.3 Macromolecules of Cell Wall

Cell wall macromolecules of prokaryote microorganisms were some of the first chemotaxonomic markers to be used for prokaryote taxonomic research. However, these observations were not possible until the development of methods such as electron microscopy and chromatography, which allowed the separation and purification of cell wall material (Cummins 1956; Salton 1994).

6.3.1 Peptidoglycan

The peptidoglycan, first used by Weidel and Pelzer (1964) and might still be used in the scientific literature, was of particular significance to the classification of microorganisms. Peptidoglycan is a heteropolymer composed of glycan strands usually made up of alternating β-1,4-linked N-acetylglucosamine and N-acetyl muramic acid residues cross-linked through short peptides (Ghuysen 1968; Vollmer et al. 2008). Indeed, these taxonomic standards were still used today in the majority of taxonomic research. These studies revealed that the cell wall preparations contained three important components: sugar, amino sugars, and amino acid residues. Furthermore, it was discovered that the amino acids were limited to alanine, glutamic acid, aspartic acid, glycine isomers of diaminopimelic acid (DAP), and lysine. Later studies revealed that some species from the genus *Corynebacterium* contained ornithine and 2,4-diaminobutyric acid and that *Micrococcus radiodurans* also contained ornithine rather than DAP and lysine. Later, 2,4-diaminobutyric acid was realized as another important taxonomic biomarker (Work and Dewey 1953; Cummins 1956; Cummins and Harris 1956; Work 1957; Salton 1994). According to the difference of cross-linkages in the peptidoglycans, it would be divided into two main types (A and B) (Schleifer and Kandler 1972). In type A, the cross-linkage was achieved by linking the ω-amino group of the diamino acid at position 3 to the carboxylic group of D-alanine at position 4 of the adjacent peptide subunit, either directly or through an interpeptide bridge consisting of 1–7 amino acid residues. For the cell walls of type B, the α-carboxylic group of D-glutamic acid or threo-3-hydroxyglutamic acid at position 2 was involved in the cross-linkage (Schleife et al. 1967). This type of cross-linkage requires the presence of the diamino acid (2.4-diaminobutyric acid) in the interpeptide bridge that allows a free amino group for linkage to the carboxylic group of the adjacent peptide subunit (Schleifer and Kandler 1972; Schumann 2011) (Fig. 6.1).

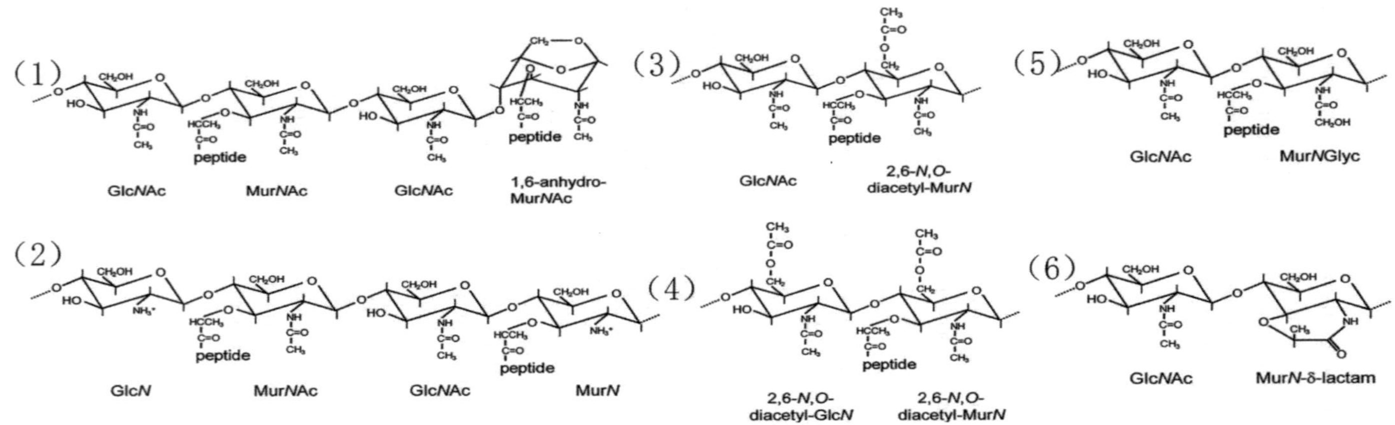

Fig. 6.1 (**a**) Structural modifications in the glycan strands of peptidoglycan. (1) Unmodified glycan strand with 1,6-anhydroMur*N*Ac-terminus; (2) N-deacetylation at Glc*N*Ac and Mur*N*Ac or both; (3) O-acetylation at Mur*N*Ac; (4) O-acetylation at Glc*N*Ac and Mur*N*Ac; (5) Glycosylation at Mur*N;* (6) Muramic acid δ-lactam. (**b**) The peptides in peptidoglycan of different species. (1) Basic structure with meso-A2pm; (2)Direct 4-3 cross-link structure; (3)I3-3 cross-link via amidated D-iAsp; (4) Basic structure with amidated iGlu; (5) Indirect 4-3 cross-link via pentaglycine bridge; (6) Indirect 4-2 cross-link via D-ornithine

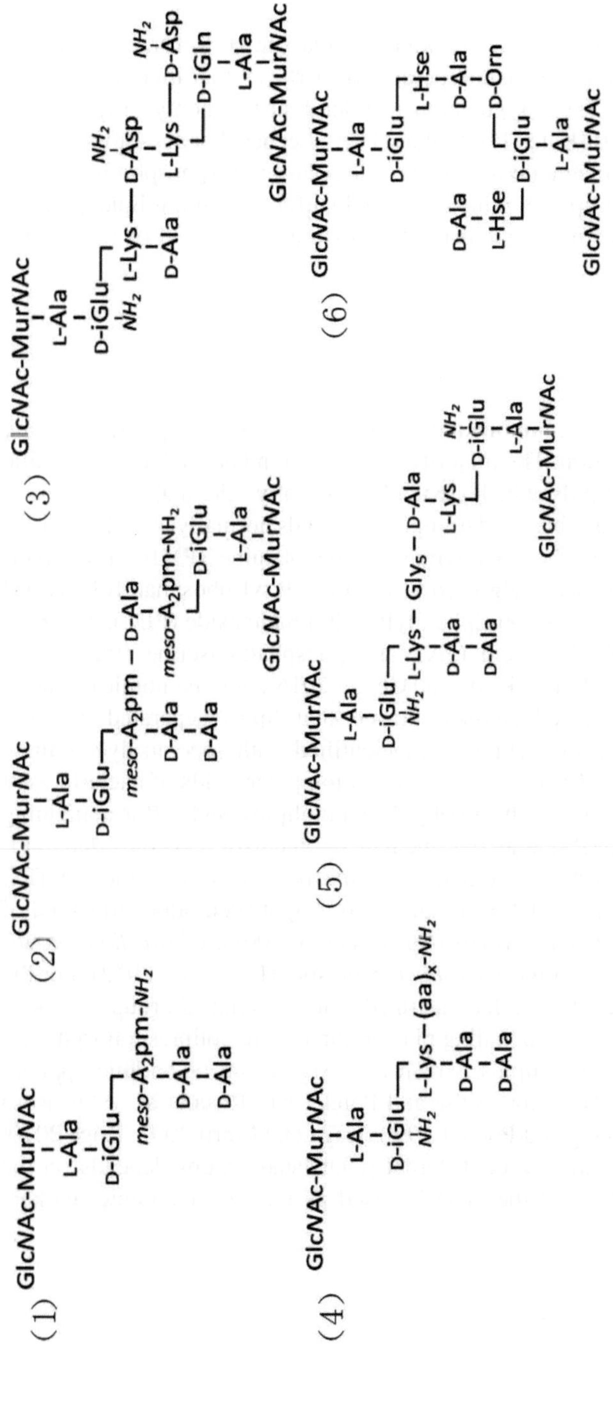

Fig. 6.1 (continued)

6.3.2 *Lipids*

Lipids have been seen as another chemotaxonomic marker. However, early studies focus more on their identification and structural properties rather than on their use as a taxonomic standard. Their application to microbial taxonomy only became more extensive in the 1970s (Kates 1964; Shaw 1970; Lechevalier and Moss, 1977). Lipids of particular use in chemotaxonomy include polar lipids (phospholipids, glycolipids and glycophospholipids, aminolipids, and sulfur-containing lipids), cellular fatty acids, and a class of terpenoid lipids known as isoprenoid or respiratory quinones.

6.3.2.1 Polar Lipids

Polar lipids were an essential component of membranes, and amphiphilic lipids are responsible for their formation. The majority of research indicates that amphiphilic lipids are glycerophospholipids, which are made up of fatty acid, a glycerol moiety, a phosphate group, and a variable head group. Polar lipids normally contained phosphatidylethanolamine (PE), phosphatidylmethylethanolamine (PME), phosphatidylglycerol (PG), diphosphatidylglycerol (DPG), lysyl-phosphatidy1glycerol (LPG), phosphatidylinositol (PI), phosphatidylinositol mannoside (PIM), phosphatidylcholine (PC), phosphatidic acid (PA), and phosphatidylserine (PS) (Kates 1964; da Costa et al. 2011b; Sohlenkamp and Geiger 2016). For taxonomic research, the vast majority of published studies showed that lipids separated on two-dimensional thin-layer chromatography and identified with specific dyes (Busse et al. 1996; da Costa et al. 2011b). The most common polar lipids of bacteria were phospholipids, glycolipids, glycophospholipids, aminolipids, and sulfur-containing lipids, but the anionic lipid (PS) as an intermediate or precursor was converted to PE by phosphatidylserine decarboxylase and could not be observed on the 2D-TLC plate (Fig. 6.2). Only some special bacterial groups might accumulate this kind of lipid (PS) as *Bdellovibrio bacteriovorus* (Nguyen et al. 2008), *Clostridium botulinum* (Evans et al. 1998), and some *Flavobacterium* spp. (Lata et al. 2012). For the archaeal groups, many studies had demonstrated that the archaeal group possesses a wide diversity of polar lipids, including phospholipids. Glycolipids, phosphoglycolipids, sulfolipids, and aminolipids, which have significant taxonomic applications (Kushwaha et al. 1981; Langworthy and Pond 1986; Rosa et al. 1986; Rosa and Gambacorta 1988; Zhang and Poulter 1992; Koga and Morii 2007; Koga 2010; Jain et al. 2014). The occurrence of different polar head groups depends on the individual archaeal family, and they can be used as unique taxonomic markers (Ulrih et al. 2009).

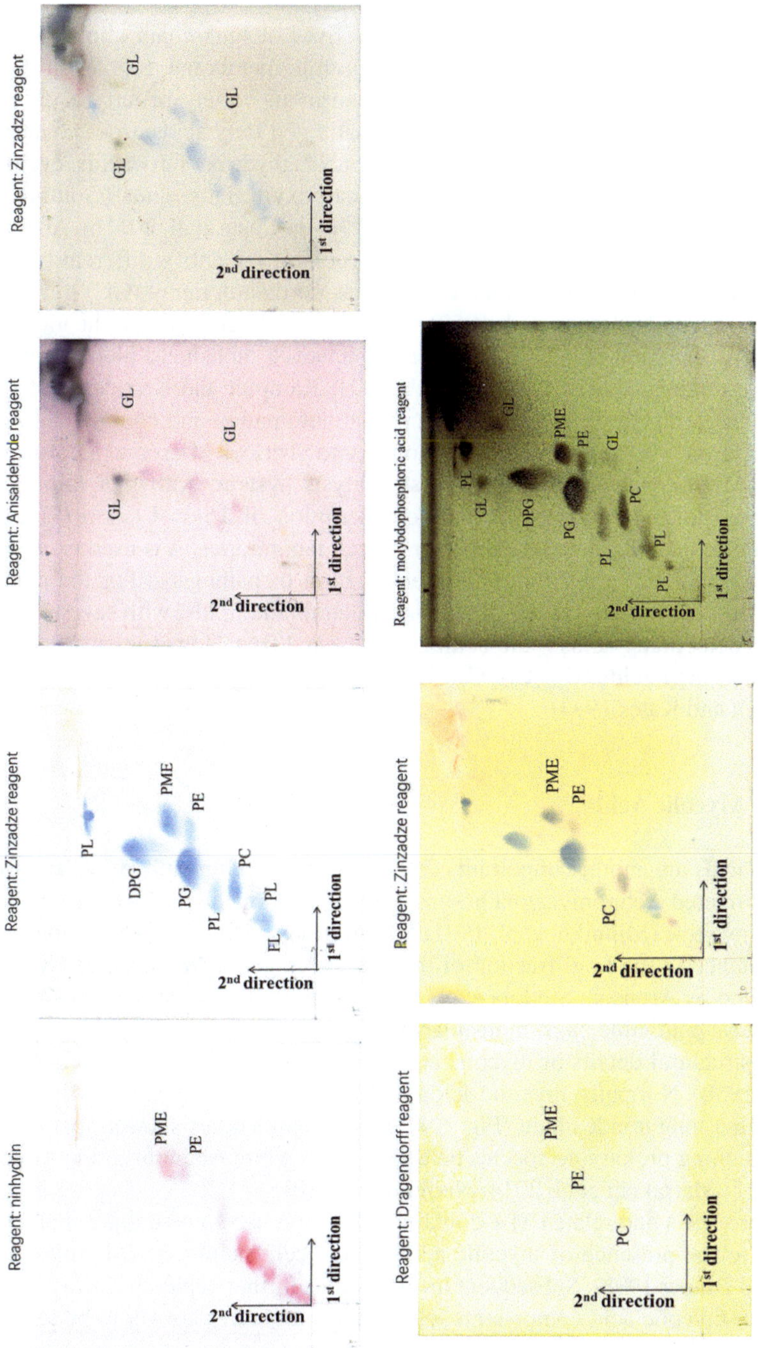

Fig. 6.2 Common polar lipids of bacteria observed on the 2D-TLC plate with different reagent (Ng et al. 2019)

6.3.2.2 Cellular Fatty Acids (CFA)

Cellular fatty acids are carboxylic acid derivatives of long-chain aliphatic molecules, with the majority residing in the cytoplasmic membrane as polar lipids and glycolipids (Kates 1964). Bacteria have an enormous variety of fatty acyl chains that comprise straight chain saturated and unsaturated fatty acids, iso- and anteiso-branched fatty acids, internally branched fatty acids, hydroxy fatty acids, cyclopropane fatty acids, ω-cyclic fatty acids, and dicarboxylic fatty acids (Goldfine and Bloch 1961; Suzuki et al. 1993; Busse et al. 1996; da Costa et al. 2011b). Abel et al. (1963) suggested that CFA analysis could successfully identify different bacterial groups after the establishment method with gas chromatography (GC) to separate organic compounds (James et al. 1952). Hence, the CFA analysis for the identification and classification of prokaryote microorganisms established a plethora of studies ensued (Bousfield et al. 1983; Kaneda 1991; Kämpfer and Kroppenstedt 1996: da Costa et al. 2011a). Later, numerous partial automation and commercialization of fatty acid analysis methods were developed, and Sherlock Microbial Identification System (MDI system) became the best analysis system, with this information included in practically every taxonomic description of a novel taxon (Fig. 6.3). Furthermore, in addition to specialized research laboratories, it is used extensively in clinical laboratories for the rapid identification of pathogens. For the archaeal groups, fatty acids were not part of archaeal membrane lipids, with several studies failing to detect fatty acids (Falb et al. 2008; Koga 2010). However, small amounts of saturated fatty acids (C_{14}, C_{16}, C_{18}) had been detected in *Halobacterium salinarum* (Pugh and Kates 1994).

6.3.2.3 Mycolic Acids

Mycolic acids are another important component of the cell membrane of mycobacteria and related Actinomycetota groups, with approximately 40%–60% content in the cell envelope (Minnikin et al. 1971; Minnikin 1982). In 1898, this kind of wax-like substance in the lipid fraction of the species of genus Mycobacteria was first investigated by Aronson. Stodola et al. (1938) proposed to name this kind of ether-soluble, unsaponifiable, high-molecular-weight hydroxy acid as "mycolic acid. The detailed structural details of mycolic acids were determined later (Asselineau and Lederer 1950). Normally, mycolic acids are high-molecular-weight β-hydroxy fatty acids with a long α side chain (Fig. 6.4). Subsequent studies showed that mycolic acid-containing prokaryote species belong to the phylum Actinomycetota (Ludwig et al. 2012; Marrakchi et al. 2014; Nouioui et al. 2018).

Mycobacteria and related MA-containing bacteria had some unique characteristics due to the presence of mycolic acids in the cell membrane (Minnikin 1982; Daffé and Draper 1998). Subsequent investigations further depict that a vast number of distinct mycolic acid components with unique structures are yet to be identified (Laneelle et al. 2011; Marrakchi et al. 2014). As a result, the structural characteristics of various types of mycolic acids made them appropriate chemotaxonomic

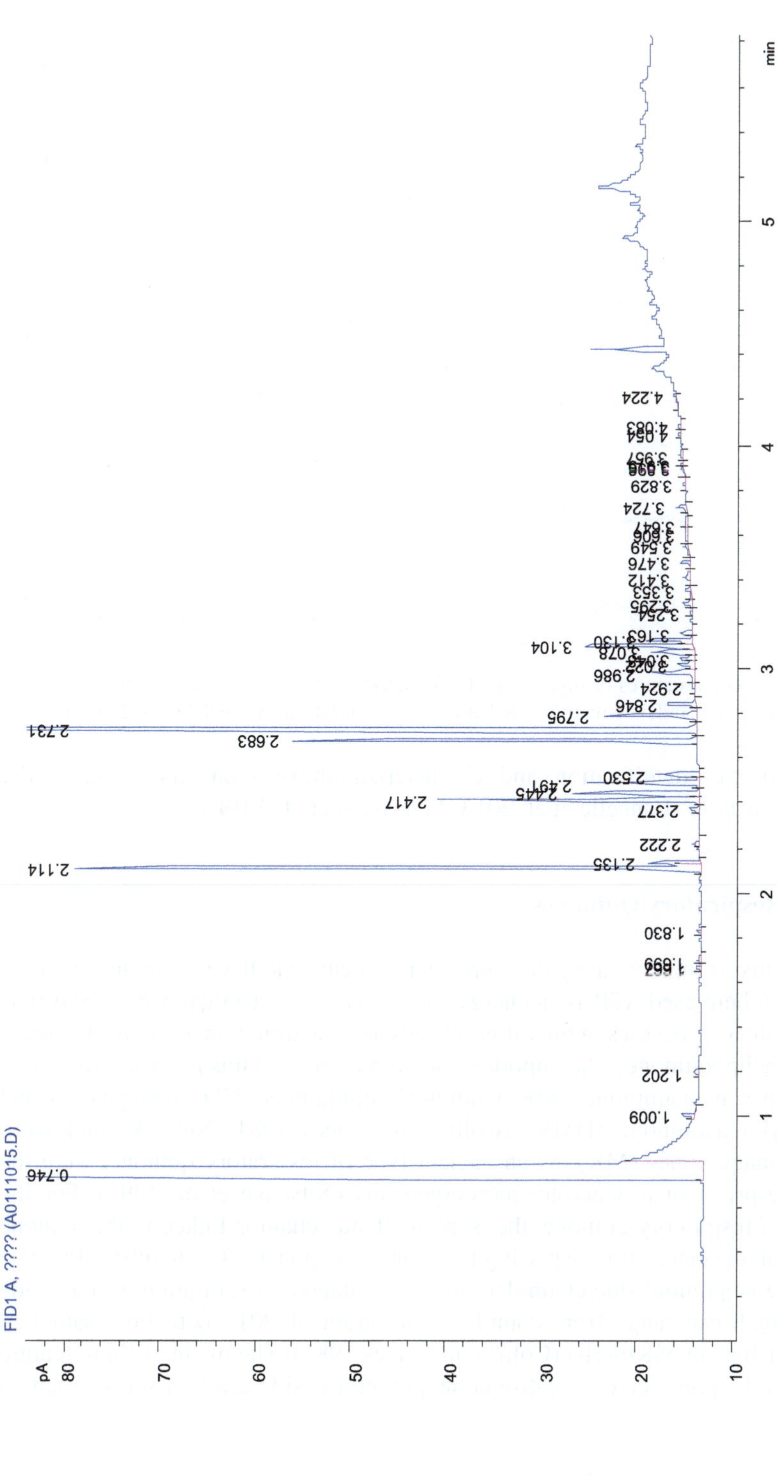

Fig. 6.3 Fatty acid profiles of strain SYSU K10001^T of the genus *Lentzea* with Microbial Identification System (MIDI) (Sherlock Version 6.1; MIDI database: TSBA6) (Fang et al. 2017)

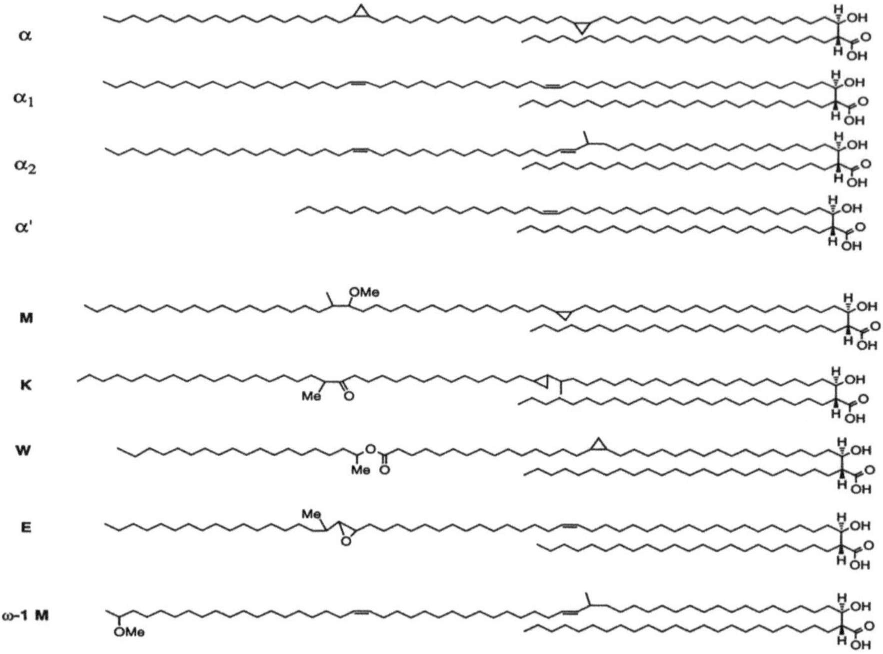

Fig. 6.4 Different structures of mycolic acids. (**a**) selection of α-mycolic acids; (**b**) a variety of oxygenated mycolic acids (M-methoxy, K-keto, W-wax ester, E-epoxy, ω-1 M – ω-1 methoxy)

markers in the identification and characterization of members of the order *Corynebacteriales* (Laneelle et al. 2013; Marrakchi et al. 2014).

6.3.2.4 Respiratory Quinones

Microorganisms inhabit vastly different environments and have different lifestyles, but most of them used ATP as an energy source. These microorganisms used different respiratory processes with different sets of electron transport mechanisms. Respiratory lipoquinones, the important electron carriers of this process, are mainly of three types, menaquinones (MK, vitamin K), ubiquinone (UQ, coenzyme Q), and 2-demethylmenaquinone (DMK) (Collins and Janes 1981; Nowicka and Kruk 2010). Menaquinones (MK), as an ancient type of respiratory quinone, were the most widespread in prokaryote microorganisms (Nitschke et al. 1995). For the structure of respiratory quinone, the isoprenoid side chain is linked to the quinone ring system, the electron-transporting functional component. The number of hydrogens in the isoprenoid side chain determines the degree of saturation of these side chains, which can range from completely unsaturated (MK-n) to fully saturated (MK-9) such as in MK-9(H_2) (Collins and Janes 1981). The distinguishing feature of UQ was the presence of a hydrophobic polyprenyl side chain of variable length

(U-n), where n is the number of isoprenyl units. Ubiquinones (UQs), another important ubiquinone quinone, are mainly distributed in some species from the phylum *Pseudomonadota* and were found in early branching bacterial phyla or methanogenic archaeal groups (Collins and Janes 1981; Zhi et al. 2014; Schütz et al. 2000; Beifuss and Tietze 2005). Naphthoquinones and benzoquinones make up the bulk of biological isoprenoid quinones, with naphthoquinones being subdivided into phylloquinones (vitamin K1) and menaquinones (vitamin K2), and benzoquinones consisting of UQs and plastoquinones. Many studies have shown that the structural variation of isoprenoid quinones in prokaryotic microorganisms is important for microbial taxonomy and phylogenetic affiliations (Page Jr et al. 1960; Bishop et al. 1962; Crane 1965; Busse et al. 1996; da Costa et al. 2011a). Consequently, the chemotaxonomic features of respiratory quinones were improved in accordance with their structural variety and make these structural differences as powerful chemotaxonomic tool.

6.3.3 Polyamines

Polyamines are non-lipophilic constituents of the cell and are a diverse group of compounds found in the majority of prokaryotes. They are not detected in some extremely halophilic archaeal species and, if present, could be found in some trace amounts (Hamana et al., 1985). They are normally derived from amino acids and changed at physiological pH. They are known to play important roles in some cellular processes. The knowledge of polyamines involved in the synthesis and stabilization of biomolecules or the regulation of the cell cycle had been well studied (Tabor and Tabor 1985; Michael 2016; Feuerstein et al. 1991; Eraso and Kaplan 2009). Furthermore, polyamines enabled bacteria to react to osmotic shock that led to an intercellular increase of inorganic cations, which was compensated by the excretion of the polycationic polyamines. Polyamines normally included the diamines 1,3-diaminopropane (DAP), 1,4-diaminobutane (putrescine, PUT) and 1,5-diaminopentane (cadaverine, CAD); triamines sym-norspermidine (NSPD), spermidine (SPD), and symhomospermidine (HSPD); the uncommon triamines aminopropylcadaverine and aminobutyl-cadaverine; and the tetraamines norspermine (TSPM), spermine (SPM), and thermospermine (TSPM) (Fig. 6.5). For chemotaxonomic purposes, the use of polyamine patterns emerged in the 1970s and had expanded to encompass a vast number of taxa (Yamamoto et al. 1979, 1983; Hamana and Matsuzaki 1993; Gvozdiak et al. 1998; Hosoya et al. 2004).

In the field of microbial systematics, the application and significance of chemotaxonomic biomarkers could not be overstated. Before the pre-molecular era with the application of sequencing methods, the taxonomic research of the particular taxa relied to a great extent on the chemotaxonomic biomarkers. Hence, some particular biomarkers could be attributed to special bacterial groups. As a result, chemotaxonomic methods were increasingly used in the descriptions of novel prokaryotic microorganisms. This had been further enhanced by molecular methods (16S rRNA

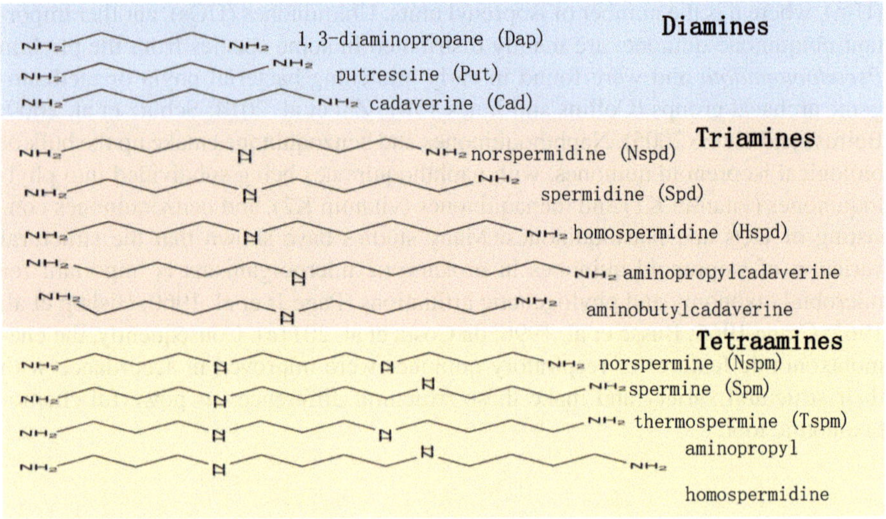

Fig. 6.5 Polyamines of the diamine, triamine, and tetraamine classes found in eukaryotes, bacteria, and archaea

gene and more recently whole genomes) and the demonstration of the stability of chemotaxonomic biomarkers concerning their presence within phylogenetic groups (Stackebrandt and Schumann 2006). Because many published taxonomic descriptions include chemotaxonomic information, the importance of chemotaxonomic biomarkers has been demonstrated by their inclusion in many published minimal standards for describing different classification levels in prokaryotes (Levy-Frebault and Portaels 1992; Freney et al. 1999; Mattarelli et al. 2014; De Lajudie et al. 2019; Arahal et al. 2007; Christensen et al. 2007; Logan et al. 2009; Mattarelli et al. 2014; Schumann et al. 2009; Oren et al. 1997).

6.4 Chemotaxonomy in the Era of Omics

It is commonly accepted that chemotaxonomic biomarkers are quite stable within different prokaryote taxa. However, with the advancements in genomes and proteomics, the applicability of these traditional chemotaxonomic markers to contemporary taxonomy is being questioned (Sutcliffe et al. 2013; Mahato et al. 2017). Traditional chemotaxonomic methods used for taxonomic descriptions were laborious and not amenable to high-throughput methods, and some methods such as the polar lipid analysis with 2D-TLC lacked resolution and information (with some unidentified lipids). More and more high-resolution methods such as NMR and MS had been used for analyzing the biomarker molecules of prokaryotes, but these had not been adopted and used in the taxonomic publication (Khuller and Goldfine 1974; Johnston et al. 2010; Garrett et al. 2012). Differences between laboratories, subjectivity in data interpretation, repeated use of reference strains/materials (which increases costs), and lack of transferable/searchable databases could influence the generation of accurate information and the exchange of related knowledge (Moore et al. 2010; Whitman 2014). These reasons generally impelled the move away from traditional chemotaxonomic methodology to genomic and phylogenomic taxonomy (Sutcliffe et al. 2013; Whitman 2014). Chemotaxonomy was considered to be antiquated compared to "omics" methods, so many would choose many multi-omics methods to study prokaryote taxonomy. This phenomenon would lead to fewer researchers to culture a wide range of microorganisms, and even fewer can perform laboratory-based chemotaxonomic investigations (Tamames and Rosselló-Móra 2012).

Many research had shown that the information on the genome in species played an increasing role in prokaryote taxonomy (Chun and Rainey 2014; Ramasamy et al. 2014; Thompson et al. 2015: Garrity 2016: Hahnke et al. 2016). The use of in silico analysis in taxonomic descriptions had gradually replaced the manual determination of some characteristics in the laboratory, with the dual objectives of accelerating the taxonomic description of taxa and improving the quality and consistency of the information supplied. Traditional taxonomic characteristics, such as DNA-DNA hybridization, are being replaced by average nucleotide identity (ANI) and digital DNA-DNA hybridization (Goris et al. 2007: Meier-Kolthoff et al. 2013: Lee et al. 2016). These advances would be contingent upon improved sequencing and algorithm analysis (Köser et al. 2012; Loman et al. 2012; Bertelli and Greub 2013), and more and more online tool as KEGG, RAST, PathoSystems, Pfam, or MetaCyc integrated knowledge-based different reference databases would encourage the in silico approach to the prediction of chemotaxonomic biomarkers (Barona-Gómez et al. 2012; Sutcliffe et al. 2013; Lawson et al. 2016; Patel et al. 2016; Mahato et al. 2017), but predictions of chemotaxonomic traits based genome information are still in trial stage (Sutcliffe 2010; Sutcliffe et al. 2013; Whitman 2014). In general, the entire phenotype of microorganisms is regulated by the genome, and a true understanding of phenotype based on genome predictions is not yet possible (Kämpfer 2014). However, some research based on multi-omics analyses has been able to identify a particular adaptability between different taxa based on genome predictions (Jensen et al. 2012), which would shed light on other important physiological growth parameters required in taxonomic descriptions.

There have been few studies on the chemotaxonomic traits of prokaryotic microorganisms based on genome information, although online databases such as KEGG, RAST, PATRIC, and Pfam have been widely used to identify genes and biosynthetic pathways associated with these biomarkers. Some examples might show the avenue to chemotaxonomy based on genome predictions. The cytoplasmic steps of the biosynthesis of bacterial peptidoglycan are catalyzed by a series of enzymes, among which some special ligases are classified as *Mur* E genes (Maruyama et al. 1988; Vollmer et al. 2008; Lugtenberg and Van Schijndel-van Dam 1972). When genomes are processed by the KEGG databases, they appear not to identify the specific *MurE* gene, presumably owing to the diversity within the sequences, as demonstrated in the PATRIC database. The KEGG database also contained pathways for glycerophospholipid metabolism that included genes for phosphatidylcholine biosynthesis (M00090 and M00091) and phosphatidylethanolamine biosynthesis (M00092 and M00093). Some studies had shown that polar lipids of different prokaryotic microorganisms were detected by the chromatography plates, which could also be predicted using the KEGG database, although some unidentified lipids could not find the corresponding genes in this database (Stamps et al. 2014; Tan et al. 2012). In addition, the pathways of fatty acid biosynthesis based on genome information were well constructed and available in the KEGG database, but many of these genes had unknown translation products and could not be assigned any EC numbers. Some enzyme-assigned EC numbers had only the function of chain elongation or the final production of single fatty acids. However, with the widespread use of automated high-throughput systems such as MIDI, the in silico method for fatty acids was not widely used in prokaryotic taxonomic studies.

The resolving capacity of chemotaxonomic features from genome information will be improved if a dedicated database related to the biochemical metabolism of these chemotaxonomic biomarkers is established. This will enable the laborious chemical experiments in the laboratory to be replaced with in silico modeling, enabling more authors to formally declare many new species that are potentially unculturable in pure forms as in the human gut microbiome (Vandamme and Sutcliffe 2021).

6.5 Conclusion

Although chemotaxonomic methods have been enormously important in the past with identification and classification schemes, it remains to be seen in what form they will be utilized in the genomic era, and the suite of methods available in the era of omics. In the future, with the knowledge of genes and related biochemical metabolism derived from genomic studies, researchers or related taxonomic journals should be encouraged to include this information in taxonomic descriptions (Vandamme and Sutcliffe 2021). The confidence in silico modeling of chemotaxonomic biomarkers could be improved by correlating information that was obtained via experimentation with information that was collected from the genomes. With the incorporation of artificial intelligence, machine learning, and neural networks, more precise phenotypic and chemotaxonomic predictions would undoubtedly be improved during prokaryotic taxonomy research. Nevertheless, a comprehensive analysis of chemotaxonomic features will be necessary for the declaration of species at a taxonomic rank above species.

References

Abel K, Deschmertzing H, Peterson JI (1963) Classification of microorganisms by analysis of chemical composition I: feasibility of utilizing gas chromatography. J Bacteriol 85(5):1039–1044. https://doi.org/10.1128/jb.85.5.1039-1044.1963

Arahal DR, Vreeland RH, Litchfield CD, Mormile MR, Tindall BJ, Oren A, Bejar V, Quesada E, Ventosa A (2007) Recommended minimal standards for describing new taxa of the family *Halomonadaceae*. Int J Syst Evol Microbiol 57:2436–2446. https://doi.org/10.1099/ijs.0.65430-0

Asselineau J, Lederer E (1950) Structure of the mycolic adds of *Mycobacteria*. Nature 166:782–783. https://doi.org/10.10387/1667S2a0

Barona-Gómez F, Cruz-Morales P, Noda-García L (2012) What can genome-scale metabolic network reconstructions do for prokaryotic systematics? Antonie Van Leeuwenhoek 101:35–43. https://doi.org/10.1007/s10482-011-9655-1

Beifuss U, Tietze M (2005) Methanophenazine and other natural biologically active phenazines. In: Muizer J (ed) Natural product synthesis II: topics in current chemistry, pp 77–113. https://doi.org/10.1007/b96889

Bertelli C, Greub G (2013) Rapid bacterial genome sequencing: methods and applications in clinical microbiology. Clin Microbiol Infect 19:803–813. https://doi.org/10.1111/1469-0691.12217

Bishop DH, Pandya KP, King HK (1962) Ubiquinone and vitamin K in bacteria. Biochem J 83:606–614. https://doi.org/10.1042/bj0830606

Bousfield IJ, Smith GL, Dando TR, Hobbs G (1983) Numerical analysis of total fatty acid profiles in the identification of coryneform, nocardioform and some other bacteria. Microbiology 129:375–394. https://doi.org/10.1099/00221287-129-2-375

Busse HJ, Denner EBM, Lubitz W (1996) Classification and identification of bacteria: current approaches to an old problem. Overview of methods used in bacterial systematics. J Biotechnol 47:3–38. https://doi.org/10.1016/0168-1656(96)01379-X

Christensen H, Kuhnert P, Busse HJ, Frederiksen WC, Bisgaard M (2007) Proposed minimal standards for the description of genera, species and subspecies of the *Pasteurellaceae*. Int J Syst Evol Microbiol 57:166–178. https://doi.org/10.1099/ijs.0.64838-0

Chun J, Rainey FA (2014) Integrating genomics into the taxonomy and systematics of the bacteria and archaea. Int J Syst Evol Microbiol 64:316–324. https://doi.org/10.1099/ijs.0.054171-0

Collins MD, Janes D (1981) Distribution of isoprenoid quinone structural types in bacteria and their taxonomic implication. Microbiol Rev 45:316–317. https://doi.org/10.1128/MMBR.45.2.316-354.1981

da Costa MS, Albuquerque L, Nobre MF, Wait R (2011a) The extraction and identification of respiratory lipoquinones of prokaryotes and their use in taxonomy. In: Rainey F, Oren A (eds) Methods in microbiology volume 38, pp 197–206. https://doi.org/10.1016/B978-0-12-387730-7.00009-7

da Costa MS, Albuquerque L, Nobre MF, Wart R (2011b) The identification of polar lipids in prokaryotes. In: Rainey F, Oren A (eds) Methods in microbiology volume 38. taxonomy of prokaryotes. Academic Press, London, pp 165–181. https://doi.org/10.1016/B978G12-387730-7.00007-3

Crane FL (1965) Distribution of ubiquinones. In: Biochemistry of quinones, pp. 183-206

Cummins CS (1956) The chemical composition of the bacterial cell wall. Int Rev Cytol 5:25–50. https://doi.org/10.1016/S0074-7696(08)62566-8

Cummins CS, Harris H (1956) The chemical composition of the cell wall in some gram-positive bacteria and its possible value as a taxonomic character. J Gen Microbiol 14:583–600. https://doi.org/10.1099/00221287-14-3-5B3

Daffé M, Draper P (1998) The envelope layers of mycobacteria with reference to their pathogenicity. Adv Microb Physiol 3:131–203. https://doi.org/10.1016/S0065-2911(08)60016-8

De Lajudie PM, Andrews M, Ardley J, Eardly B, Jumas-Bilak E, Kuzmanović N, Lassalle F, Lindström K, Mhamdi R, Martínez-Romero E, Moulin L (2019) Minimal standards for the description of new genera and species of rhizobia and agrobacteria. Int J Syst Evol Microbiol 69:1852–1863. https://doi.org/10.1099/ijsem.0.003426

Eraso JM, Kaplan S (2009) Regulation of gene expression by *Prr*A in *Rhodobacter sphaeroides* 2.4. 1: role of polyamines and DNA topology. J Bacteriol 191:4341–4352. https://doi.org/10.1128/JB.00243-09

Evans RI, McClure PJ, Gould GW, Russell NJ (1998) The effect of growth temperature on the phospholipid and fatty acyl compositions of non-proteolytic *Clostridium botulinum*. Int J Food Microbiol 40:159–167. https://doi.org/10.1016/S0168-1605(98)00029-4

Falb M, Müller K, Königsmaier L, Oberwinkler T, Horn P, von Gronau S, Gonzalez O, Pfeiffer F, Bornberg-Bauer E, Oesterhelt D (2008) Metabolism of halophilic archaea. Extremophiles 12:177–196. https://doi.org/10.1007/s00792-008-0138-x

Fang BZ, Han MX, Liu L, Zhang ZT, Liu WL, Shen JT, Wang Y, Zhang WQ, Wei DQ, Li WJ (2017) *Lentzea cavernae* sp. nov., an actinobacterium isolated from a karst cave sample, and emended description of the genus *Lentzea*. Int J Syst Evol Microbiol 67:2357–2362. https://doi.org/10.1099/ijsem.0.001958

Feuerstein BG, Williams LD, Basu HS, Marton LJ (1991) Implications and concepts of polyamine-nucleic acid interactions. J Cell Biochem 46:37–47. https://doi.org/10.1002/jcb.240460107

Freney J, Kloos WE, Hajek V, Webster JA, Bes M, Brun Y, Vemozy-Rozand C (1999) Recommended minimal standards for description of new staphylococcal species. Int J Syst Evol Microbiol 49:489–502. https://doi.org/10.1099/00207713-49-2-489

Garrett TA, O'Neill AC, Hopson ML (2012) Quantification of cardiolipin molecular species in *Escherichia coli* lipid extracts using liquid chromatography/electrospray ionization mass spectrometry. Rapid Commun Mass Spectrom 26:2267–2274. https://doi.org/10.1002/rcm.6350

Garrity GM (2016) A new genomics-driven taxonomy of bacteria and archaea: are we there yet? J Clin Microbiol 54:1956–1963. https://doi.org/10.1128/JCM.00200-16

Ghuysen JM (1968) Use of bacteriolytic enzymes in determination of wall structure and their role in cell metabolism. Bacteriol Rev 32:425–464. https://doi.org/10.1128/WMBR.32.4_Pt_2.425-464.1966

Goldfine H, Bloch K (1961) On the origin of unsaturated fatty acids in Clostridia. J Biol Chem 236:2596–2601. https://doi.org/10.1016/S0021-9258(19)61705-9

Goris J, Konstantinidis KT, Klappenbach JA, Coenye T, Vandamme P, Tiedje JM (2007) DNA-DNA hybridization values and their relationship to whole-genome sequence similarities. Int J Syst Evol Microbiol 57:81–91. https://doi.org/10.1099/ijs.0.64483-0

Gvozdiak OR, Schumann P, Griepenburg U, Auling G (1998) Polyamine profiles of Gram-positive catalase positive cocci. Syst Appl Microbiol 21:279–284. https://doi.org/10.1016/50723-2020(98)80034-9

Hahnke RL, Meier-Kolthoff JP, García-López M, Mukherjee S, Huntemann M, Ivanova NN, Woyke T, Kyrpides NC, Klenk HP, Göker M (2016) Genome-based taxonomic classification of *Bacteroidetes*. Front Microbiol 7:72. https://doi.org/10.3389/fmicb.2016.02003

Hamana K, Kamekura M, Onishi H, Akazawa T, Matsyzaki S (1985) Polyamines in photosynthetic eubacteria and extreme-halophilic archaebacteria. J Biochem 97(6):1653–1658. https://doi.org/10.1093/oxfordjournals.jbchem.a135223

Hamana K, Matsuzaki S (1993) Polyamine distribution patterns serve as a phenotypic marker in the chemotaxonomy of the Proteobacteria. Can J Microbiol 39:304–310. https://doi.org/10.1139/m93-043

Hosoya R, Hamana K, Niitsu M, Itoh T (2004) Polyamine analysis for chemotaxonomy of thermophilic eubacteria: polyamine distribution profiles within the orders *Aquificales*, *Thermotogales*, *Thermodesulfobacteriales*, *Thermales*, *Thermoanaerobacteriales*, *Clostridiales* and *Bacillales*. J Gen Appl Microbiol 50:271–287. https://doi.org/10.2323/jgam.50.271

Jain S, Caforio A, Driessen AJM (2014) Biosynthesis of archaeal membrane ether lipids. Front Microbiol 5:641. https://doi.org/10.3389/fmicb.2014.00641

James AT, Martin AJ, Smith GH (1952) Gas-liquid partition chromatography: the separation and micro-estimation of ammonia and the methylamines. Biochem J 52(2):238. https://doi.org/10.1042/bj0520238

Jensen DB, Vesth TC, Hallin PF, Pedersen AG, Ussery DW (2012) Bayesian prediction of bacterial growth temperature range based on genome sequences. BMC Genomics 13:S3. https://doi.org/10.1186/1471-2164-13-S7-S3

Johnston NC, Aygun-Sunar S, Guan Z, Ribeiro AA, Daldal F, Raetz CR, Goldfine H (2010) A phosphoethanolamine-modified glycosyl diradylglycerol in the polar lipids of *Clostridium tetani*. J Lipid Res 51:1953–1961. https://doi.org/10.1194/jlr.M004788

Kämpfer P (2014) Continuing importance of the 'phenotype' in the genomic era. In: Rainey F, Oren A (eds) Methods in microbiology volume 41. New approaches to prokaryotic systematics. Academic Press, London, pp 307–320. https://doi.org/10.1016/bs.mim.2014.07.005

Kämpfer P, Kroppenstedt RM (1996) Numerical analysis of fatty acid patterns of coryneform bacteria and related taxa. Can J Microbiol 42:989–1005. https://doi.org/10.1139/m96-128

Kaneda T (1991) Iso- and anteiso-fatty acids in bacteria: biosynthesis, function, and taxonomic significance. Microbiol Mol Biol Rev 55:288–302. https://doi.org/10.112/MMBR.55.2.288-302.1991

Kates M (1964) Bacterial lipids. In: Paoletti R, Kritchevsky D (eds) Advances in lipid research. Academic Press, New York, pp 17–90. https://doi.org/10.1016/B978-1-48319938-2.50008-X

Khuller GK, Goldfine H (1974) Phospholipids of *Clostridium butyricum*. V. Effects of growth temperature on fatty acid, alk-1-enyl ether group, and phospholipid composition. J Lipid Res 15:500–507. https://doi.org/10.1016/S0022-2275(20)36770-5

Koga Y (2010) The biosynthesis and evolution of archaeal membranes and ether phospholipids. In: Timmis KN (ed) Handbook of hydrocarbon and lipid microbiology. Springer-Verlag, Berlin, Heidelberg, pp 451–458. https://doi.org/10.1007/978-3-540-775B7-4_33

Koga Y, Morii H (2007) Biosynthesis of ether-type polar lipids in archaea and evolutionary considerations. Microbiol Mol Biol Rev 71:97–120. https://doi.org/10.1128/MMBR.00033-06

Köser CU, Ellington MJ, Cartwright EJ, Gillespie SH, Brown NM, Farrington M, Holden MT, Dougan G, Bentley SD, Parkhill J, Peacock SJ (2012) Routine use of microbial whole genome sequencing in diagnostic and public health microbiology. PLoS Pathog 8:e1002824. https://doi.org/10.1371/journal.ppat.1002824

Kushwaha SC, Kates M, Sprott GD, Smitti ICP (1981) Novel polar lipids from the methanogen Methanospirilium hungatei GP1. Biochim Biophys Acta 664:156–173. https://doi.org/10.1016/0005-2760(81)90038-2

Laneelle MA, Eynard N, Spina L, Lemassu A, Laval F, Huc E, Etienne G, Marrakchi H, Daffé M (2013) Structural elucidation and genomic scrutiny of the C_{60}–C_{100} mycolic acids of *Segniliparus rotundus*. Microbiology 159:191–203. https://doi.org/10.1099/mic.0.063479-0

Laneelle MA, Launay A, Spina L, Marrakchi H, Laval F, Eynard N, Lemassu A, Tropis M, Daffé M, Etienne G (2011) A novel mycolic acid species defines two novel genera of the *Actinbacteria, Hoyosella* and *Amycolicicoccus*. Microbiology 158:643–855. https://doi.org/10.1099/mic.0.055509-0

Langworthy TA, Pond JL (1986) Archaebacterial ether lipids and chemotaxonomy. Syst Appl Microbiol 7:253–257. https://doi.org/10.1016/S0723-2020(86)8001S-7

Lata P, Lal D, Lal R (2012) *Flavobacterium ummariense* sp. nov., isolated from hexachlorocyclohexane-contaminated soil and emended description of *Flavobacterium ceti* Vete et al. 2007. Int J Syst Evol Microbiol 62:2674–2679. https://doi.org/10.1099/ijs.0.030916-0

Lawson PA, Sankaranarayanan K, Patel NB, Busse HJ. (2016) In-silico chemotaxonomy: a tool for 21st century microbial systematics. In: Lawson PA (ed.) Microbial Systematics and Metagenomics BISMIS 2016 Abstracts Book. p. 27

Lechevalier MP, Moss CW (1977) Lipids in bacterial taxonomy - a taxonomist's view. CRC Crit Rev Microbiol 5:109–210. https://doi.org/10.3109/10408417709102311

Lee I, Kim YO, Park SC, Chun J (2016) OrthoANI: an improved algorithm and software for calculating average nucleotide identity. Int J Syst Evol Microbiol 66:1100–1103. https://doi.org/10.1099/ijsem.0.000760

Levy-Frebault VV, Portaels F (1992) Proposed minimal standards for the genus *Mycobacterium* and for description of new slowly growing *Mycobacterium* species. Int J Syst Bacteriol 42:315–323. https://doi.org/10.1099/00207713-42-2-315

Logan NA, Berge O, Bishop AH, Busse HJ, De Vos P, Fritze D, Heyndrickx M, Kämpfer P, Rabinovitch L, Salkinoja-Salonen MS, Seldin L (2009) Proposed minimal standards for describing new taxa of aerobic, endospore-forming bacteria. Int J Syst Evol Microbiol 59:2114–2121. https://doi.org/10.1099/ijs.0.013649-0

Loman NJ, Constantinidou C, Chan JZ, Halachev M, Sergeant M, Penn CW, Robinson ER, Pallen MJ (2012) High-throughput bacterial genome sequencing: an embarrassment of choice, a world of opportunity. Nat Rev Microbiol 10:599–606. https://doi.org/10.1038/nrmicro2850

Ludwig W, Euzéby J, Schumann P, Busse HJ, Trujillo ME, Kämpfer P, Whitman WB (2012) Road map of the phylum Actinobacteria. In: Berge's manual of systematic bacteriology Volume 4 The Actinobacteria. pp. 1–28. https://doi.org/10.1007/978-0-3B7-68233-4_1

Lugtenberg EJ, Van Schijndel-van Dam A (1972) Temperature-sensitive mutants of Escherichia coli K-12 with low activity of the diaminopimelic acid adding enzyme. J Bacteriol 110:41–46. https://doi.org/10.1128/jb.110.1.41-46.1972

Mahato NK, Gupta V, Singh P, Kumari R, Verma H, Tripathi C, Rani P, Sharma A, Singhvi N, Sood U, Hira P (2017) Microbial taxonomy in the era of OMICS: application of DNA sequences, computational tools and techniques. Antonie Van Leeuwenhoek 110:1357–1371. https://doi.org/10.1007/s10482-017-0928-1

Marrakchi H, Lanéelle MA, Daffé M (2014) Mycolic acids: structures, biosynthesis, and beyond. Chem Biol 21:67–85. https://doi.org/10.1016/j.chembiol.2013.11.011

Maruyama IN, Yamamoto AH, Hirota Y (1988) Determination of gene products and coding regions from the *mur*E-*mur*F region of *Escherichia coli*. J Bacteriol 170:3786–3788. https://doi.org/10.1128/jb.170.8.3786-3788.1988

Mattarelli P, Holzapfel W, Franz CM, Endo A, Felis GE, Hammes W, Pot B, Dicks L, Dellaglio F (2014) Recommended minimal standards for description of new taxa of the genera *Bifidobacterium*, *Lactobacilius* and related genera. Int J Syst Evol Microbiol 64:1434–1451. https://doi.org/10.1099/ijs.0.060046-0

Meier-Kolthoff JP, Auch AF, Klenk HP, Göker M (2013) Genome sequence-based species delimitation with confidence intervals and improved distance functions. BMC Bioinformatics 14:60. https://doi.org/10.1186/1471-2105-14-60

Michael AJ (2016) Polyamines in eukaryotes, bacteria, and archaea. J Biol Chem 291:14896–14903. https://doi.org/10.1074/jbc.R116.734780

Minnikin DE (1982) Lipids: complex lipids, their chemistry, biosynthesis and roles. In: Ratledge C, Stanford JL (eds) The biology of the mycobacteria. Academic Press, London, pp 95–184

Minnikin DE, Abdolrahimzadeh H, Baddiley J (1971) The interrelation of polar lipids in bacterial membranes. Biochim Biophys Acta 249:651–655. https://doi.org/10.1016/0005-2736(71)90148-9

Moore ER, Mihaylova SA, Vandamme P, Krichevsky MI, Dijkshoorn L (2010) Microbial systematics and taxonomy: relevance for a microbial commons. Res Microbiol 161:430–438. https://doi.org/10.1016/j.resmic.2010.05.007

Ng ZY, Fang BZ, Li WJ, Tan GY (2019) *Marinitenerispora sediminis* gen. nov., sp. nov., a member of the family *Nocardiopsaceae* isolated from marine sediment. Int J Syst Evol Microbiol 69:3031–3040. https://doi.org/10.1099/ijsem.0.003587

Nguyen NAT, Sallans L, Kaneshiro ES (2008) The major glycerophospholipids of the predatory and parasitic bacterium *Bdellovibno bacteriovorus* HID5. Lipids 43:1053–1063. https://doi.org/10.1007/S11745-008-3235-9

Nitschke W, Kramer DM, Riedel A, Liebl U (1995) From naphtho- to benzoquinones-(r)evolutionary reorganisations of electron transfer chains. In: Mathis P (ed) Photosynthesis: from light to biosphere, vol I. Kluwer Academic Publishers, Dordrecht, pp 945–950. https://doi.org/10.1007/978-94-009-0173-5_225

Nouioui I, Carro L, García-López M, Meier-Kolthoff JP, Woyke T, Kyrpides NC, Pukall R, Klenk HP, Goodfellow M, Göker M (2018) Genome-based taxonomic classification of the phylum Actinobacteria. Front Microbiol 9:2007. https://doi.org/10.3389/fmicb.2018.02007

Nowicka B, Kruk J (2010) Occurrence, biosynthesis and function of isoprenoid quinones. Biochim Biophys Acta 1797:1587–1605. https://doi.org/10.1016/j.bbabio.2010.06.007

Oren A, Ventosa A, Grant WD (1997) Proposed minimal standards for description of new taxa in the order *Halobacteriales*. Int J Syst Bacteriol 47:233–238. https://doi.org/10.1099/00207713-47-1-233

Page AC Jr, Gale P, Wallick H, Walton RB, McDaniel LE, Woodruff HB, Folkers K (1960) Coenzyme Q. XVII. Isolation of Coenzyme Q10 from bacterial fermentation. Arch Biochem Biophys 89:318–321. https://doi.org/10.1016/0003-9861(60)90062-X

Patel NB, Sankaranarayanan K, Busse HJ, Lawson PA (2016) Investigating genomic tools for polar lipid prediction. Microbial Systematics and Metagenomics BISMIS 2016 Abstracts Book. p. 41

Pugh EL, Kates M (1994) Acylation of proteins of the archaebacteria *Halobacterium cutirubrum* and *Methanobacterium thermoautotrophicum*. Biochim Biophys Acta 1196:38–44. https://doi.org/10.1016/0005-2736(94)90292-5

Ramasamy D, Mishra AK, Lagier JC, Padhmanabhan R, Rossi M, Sentausa E, Raoult D, Fournier PE (2014) A polyphasic strategy incorporating genomic data for the taxonomic description of novel bacterial species. Int J Syst Evol Microbiol 64:384–391. https://doi.org/10.1099/ijs.0.057091-0

Rosa MD, Gambacorta A (1988) The lipids of archaebacteria. Prog Lipid Res 27:153–175. https://doi.org/10.1016/0163-7827(88)90011-2

Rosa MD, Gambacorta A, Gliozzi A (1986) Structure, biosynthesis, and physicochemical properties of archaebacterial lipids. Microbiol Mol Biol Rev 50:70-B0. https://doi.org/10.1128/MMBR.50.1.70-80.1986

Salton MRJ (1994) The bacterial cell envelope-a historical perspective. In: Ghuysen JM, Hakenbeck R (eds) Bacterial cell walt. Elsevier Science, pp 1–22. https://doi.org/10.1016/SQ167-7306(08)60404-4

Schleife KH, Plapp R, Kandler O (1967) Identification of threo-3-hydroxyglutamrc add in the cell wall of *Microbacterium lacticum*. Biochem Biophys Res Commun 28:566–570. https://doi.org/10.1016/0006-291X(67)90351-8

Schleifer KH (2009) Classification of bacteria and archaea: past, present and future. Syst Appl Microbiol 32:533–542. https://doi.org/10.1016/j.syapm.2009.09.002

Schleifer KH, Kandler O (1972) Peptidoglycan types of bacterial cell walls and their taxonomic implications. Bacteriol Rev 36:407–477. https://doi.org/10.1128/MMBR.36.4.407-477.1972

Schumann P (2011) Peptidoglycan structure. In: Rainey F, Oren A (eds) Methods in microbiology volume 38. Taxonomy of prokaryotes. Academic Press, London, pp 101–129. https://doi.org/10.1016/B978-0-12-387730-7.00005-X

Schumann P, Kämpfer P, Busse HJ, Evtushenko LI (2009) Proposed minimal standards for describing new genera and species of the suborder *Mtcrococcineae*. Int J Syst Evol Microbiol 59:1823–1849. https://doi.org/10.1099/ijs.0.012971-0

Schütz M, Brugna M, Lebrun E, Baymann F, Huber R, Stetter KO, Hauska G, Toci R, Lemesle-Meunier D, Tron P, Schmidt C (2000) Early evolution of cytochrome be complexes. J Mol Biol 300:663–675. https://doi.org/10.1006/jmbi.2000.3915

Shaw N (1970) Bacterial glycolipids. Bacteriol Rev 34:365–377. https://doi.org/10.1128/MMBR.34.4.365-377.1970

Sohlenkamp C, Geiger O (2016) Bacterial membrane lipids: diversity in structures and pathways. FEMS Microbiol Rev 40:133–159. https://doi.org/10.1093/temsre/fuv008

Stackebrandt E, Schumann P (2006) Introduction to the taxonomy of *Actinobacteria*. In: Dworkin M, Fatkow S, Rosenberg E, Schleifer KH, Stackebrandt E (eds) The prokaryotes. Springer, New York, NY, pp 297–321. https://doi.org/10.1007/0-387-30743-5_16

Stamps BW, Losey NA, Lawson PA, Stevenson BS (2014) Genome sequence of *Themioanaerobaculum aquaticum* MP-01^T, the first cultivated member of *Acidobacteria* subdivision 23, isolated from a hoi spring. Genome Announc 2:eOC57Q-14. https://doi.org/10.1128/genomeA.00570-14

Stodola FH, Lesuk A, Anderson RJ (1938) The chemistry of the lipids of Tubercle bacilli. J Biol Chem 126:505–513

Sutcliffe IC, Trujillo ME, Whitman WB, Goodfellow M (2013) A call to action for the international committee on systematics of prokaryotes. Trends Microbiol 21:51–52. https://doi.org/10.1016/j.tim.2012.11.004

Sutcliffe LC (2010) A phylum level perspective on bacterial cell envelope architecture. Trends Microbiol 18:464–470. https://doi.org/10.1016/j.tim.2010.06.005

Suzuki K, Goodfellow M, O'Donnell AG (1993) Cell envelopes and classification. In: Goodfellow M, O'Donnell (eds) Handbook of new bacterial systematics. Academic Press, London, pp 195–250

Tabor CW, Tabor H (1985) Polyamines in microorganisms. Microbiol Mol Biol Rev 49:81–99. https://doi.org/10.1128/MMBR.49.1.81-99.1995

Tamames J, Rosselló-Móra R (2012) On the fitness of microbial taxonomy. Trends Microbiol 20:514–516. https://doi.org/10.1016/j.tim.2012.08.012

Tan BK, Bogdanov M, Zhao J, Dowhan W, Raetz CR, Guan Z (2012) Discovery of a cardiolipin synthase utilizing phosphatidylethanolamine and phosphatidylglycerol as substrates. Proc Natl Acad Sci U S A 109:16504–16509. https://doi.org/10.1073/pnas.1212797109

Thompson CC, Amaral GR, Campeão M, Edwards RA, Polz MF, Dutilh BE, Ussery DW, Sawabe T, Swings J, Thompson FL (2015) Microbial taxonomy in the post-genomic era: rebuilding from scratch? Arch Microbiol 197:359–370. https://doi.org/10.1007/s00203-014-1071-2

Ulrih NP, Gmajner D, Raspor P (2009) Structural and physicochemical properties of polar lipids from thermophilic Archaea. Appl Microbiol Biotechnol 84:249–260. https://doi.org/10.1007/s00253-009-2102-9

Vandamme P, Sutcliffe I (2021) Out with the old and in with the new: time to rethink twentieth century chemotaxonomic practices in bacterial taxonomy. Int J Syst Evol Microbiol 71:005127. https://doi.org/10.1099/ijsem.0.005127

Vollmer W, Blanot D, De Pedro MA (2008) Peptidoglycan structure and architecture. FEMS Microbiol Rev 32:149–167. https://doi.org/10.1111/j.1574-6976.2007.00094.x

Weidel W, Pelzer H (1964) Bagshaped macromolecules-a new outlook on bacterial cell walls. In: Nord NN (ed) Advances in enzymology—and related areas of molecular biology, vol 26. John Wiley and Sons, Inc, USA, pp 193–232. https://doi.org/WJ002/9780470122716.ch5

Whitman WB (2014) The need for change: embracing the genome. Methods Microbiol 41:1–12. https://doi.org/10.1016/bs.mim.2014.08.002

Work E (1957) Biochemistry of the bacterial cell wall. Nature 179:841–847. https://doi.org/10.1038/179841a0

Work E, Dewey DL (1953) The distribution of α, ε-diaminopimelic acid among various microorganisms. J Gen Microbiol 9:394–409. https://doi.org/10.1099/00221287-9-3-394

Yamamoto S, Shinoda S, Kawaguchi M, Wakamatsu K, Makita M (1983) Polyamine distribution in Vibrionaceae: norspermidine as a general constituent of Vibrio species. Can J Microbiol 29:724–728. https://doi.org/10.1139/m83-118

Yamamoto S, Shinoda S, Makita M (1979) Occurrence of norspermidine in some species of genera *Vibrio* and *Beneckea*. Biochem Biophys Res Commun 87:1102–1108. https://doi.org/10.1016/S0006-291X(79)80021-2

Zhang D, Poulter CD (1992) Biosynthesis of Archaebacterial ether lipids. Formation of ether linkages by prenyltransferases. J Am Chem Soc 115:1270–1277. https://doi.org/10.1021/ja00057a008

Zhi XY, Yao JC, Tang SK, Huang Y, Li HW, Li WJ (2014) The futalosine pathway played an important role in menaquinone biosynthesis during early prokaryote evolution. Genome Biol Evol 6:149–160. https://doi.org/10.1093/gbe/evu007

Chapter 7
Overall Genome Relatedness Indices: Interpretation and Defining New Taxonomic Ranks

Manik Prabhu Narsing Rao and Arinthip Thamchaipenet

7.1 Introduction

Taxonomy is derived from the Greek words *taxis* and *nomos*, which mean arrangement or order and law, respectively (Johansson and Pettersson 2002). It consists of three components: classification, nomenclature, and identification (Pitt and Barer 2012). In bacterial taxonomy, the species is the most basic taxonomic group (i.e., unit) (Gajdács 2020). Groups of species are combined to form a genus, genera into the family, families into the order, orders into class, classes into phylum, and phyla into a domain or kingdom (the highest level) and the subgroups to these main classifications (Gajdács 2020).

In 1923, Bergey's published the first edition of the *Manual of Determinative Bacteriology*, which classified bacteria into nested hierarchical classifications to denote varying degrees of relatedness (Bergey et al. 1923). This was the first modern attempt to systematically classify bacteria based on phenotypic characteristics (Hugenholtz et al. 2021). The concept that bacteria might be best classified by comparing their genomes was supported during the 1960s. Overall base compositions of

M. P. Narsing Rao
Department of Genetics, Faculty of Science, Kasetsart University, Chatuchak, Bangkok, Thailand

Omics Center for Agriculture, Bioresources, Food and Health, Kasetsart University (OmiKU), Bangkok, Thailand

A. Thamchaipenet (✉)
Department of Genetics, Faculty of Science, Kasetsart University, Chatuchak, Bangkok, Thailand

Omics Center for Agriculture, Bioresources, Food and Health, Kasetsart University (OmiKU), Bangkok, Thailand

Biodiversity Center Kasetsart University (BDCKU), Bangkok, Thailand
e-mail: arinthip.t@ku.ac.th

W.-J. Li et al. (eds.), *Modern Taxonomy of Bacteria and Archaea*,
https://doi.org/10.1007/978-981-99-5720-0_7

DNAs (mol% G + C values) were used initially (Lee et al. 1956). It was evident that bacteria with mol% G + C values that differed noticeably from one another did not belong to the same species (Rosselló-Mora and Amann 2001). A much more accurate method was required because the single values obtained from the analysis of DNA base compositions only allowed very superficial comparisons (Brenner et al. 1969). Thus, a method like DNA-DNA hybridization (DDH) was developed (Brenner et al. 1969) and remained the gold standard in bacterial taxonomy. It was noticed that phenotypically coherent microorganisms could be considered a single species if they shared DDH values above 70%. An ad hoc committee later recommended this cutoff value as an approximate threshold for circumscribing species (Wayne et al. 1987).

Woese and Fox's groundbreaking study in the 1970s established ribosomal RNA genes as standard phylogenetic markers in taxonomic studies (Woese and Fox 1977). Attempts were made to determine relationships between distantly related bacteria by cataloguing ribosomal ribonucleic acids (rRNA) (Stackebrandt et al. 1985). Currently, 98.7% 16S rRNA gene sequence similarity is recommended as the cutoff for species delineation (Chun et al. 2018). Prokaryotic species delineation has relied on a polyphasic approach (Colwell 1970) based on a consensus of all available methods: single-character tests, as well as multiple-character tests including phenotypic (morphological and other characters), chemotaxonomic (fatty acids, polar lipids, and others), and genomic data (16S rRNA gene sequence comparison, G + C content variation, and DNA-DNA hybridization) (Vandamme et al. 1996; Goodfellow 2000; Varghese et al. 2015; Young 2001). Although the 16S rRNA gene is very useful for classifying bacteria, it has poor discriminatory power for some genera and low phylogenetic resolution at the species level (Janda and Abbott 2007; Caudill and Brayton 2022). Contrarily, while DDH has been the "gold standard" for bacterial species delineation, its procedures are known to be labor-intensive, prone to experimental error, and inconsistent in terms of reproducibility (Gevers et al. 2005; Hu et al. 2022).

In the genomics era, the high-throughput sequencing method makes it simple to obtain high-quality bacterial genome sequences (Liu et al. 2019). With the rapid progress in this area, overall genome relatedness indices (OGRIs) have been developed to replace DDH (Chun and Rainey 2014). The present chapter discusses various OGRIs that have been developed, as well as their interpretation and definition of new taxonomic ranks.

7.2 Overall Genome Relatedness Indices (OGRIs)

The term "OGRI" was first coined by Chun and Rainey (2014) to describe any measurements that indicate how similar two genome sequences are. With the rapid progress in this area, many efforts have been made to develop methods to replace DDH focused on developing values that are analogous to DDH values. Various OGRI approaches have been developed, such as genome BLAST distance phylogeny

(GBDP) (Henz et al. 2004), maximal unique matches index (MUMi) (Deloger et al. 2009), average nucleotide identity (ANI) (Konstantinidis and Tiedje 2005a), and digital DNA-DNA hybridization (dDDH) (Auch et al. 2010). GBDP is a distance-type genome relatedness index that uses BLAST (Altschul et al. 1990) to obtain sets of high-scoring segment pairs between each pair of genomes, then derives a distance matrix, and finally builds a tree or a network using algorithms like UPGMA, Neighbor-Joining, BioNJ, or Neighbor-Net (Henz et al. 2004). MUMi is another distance-type index that was developed to provide higher resolution at the intraspecies level (Deloger et al. 2009). It is based on maximally unique exact matches shared by two genomes. MUMi employs the MUMmer program (Kurtz et al. 2004) to perform rapid pairwise genome comparisons with shared segments of at least 19 bp in length (Deloger et al. 2009). Goris et al. (2007) proposed the currently accepted concept of ANI, which employs the BLAST program. Richter and Rosselló-Móra (2009) discovered that the MUMmer can be used instead of BLAST. These algorithms, which use either BLAST or MUMmer to compute ANI values with directional specificity, compare two genomes and generate the calculated ANI value, which may differ depending on which genome was chosen as a query and which was chosen as a subject. An improved algorithm called OrthoANI was proposed to address this issue (Lee et al. 2016). The dDDH was developed to perform an in silico DDH and was accomplished with the software Genome-to-Genome Distance Calculator (https://ggdc.dsmz.de/ggdc.php; Meier-Kolthoff et al. 2014). Its result has been shown to have a high correlation with DDH (70% threshold) (Meier-Kolthoff et al. 2013, 2014). The tool allows users to select different parameters to calculate intergenomic distances between genomes, which are then converted into similarity values comparable to DDH (Auch et al. 2010).

7.3 Interpretation and Defining New Taxonomic Ranks

The present 16S rRNA gene sequence identity cutoff of 98.7% has been recommended for species delineation (Chun et al. 2018), which is higher than the previously accepted 97% (Stackebrandt and Goebel 1994), and necessitates assurance in the quality of the 16S gene sequences under consideration (Kim et al. 2014; Chun et al. 2018). Higher taxa boundaries were separated by steps of about 3%, so pairwise 16S identity values of smaller than 95%, 92%, 89%, 86%, and 83% were obtained, indicating affiliation with different genera, families, orders, classes, and phyla, respectively (Rosselló-Móra and Amann 2015; Konstantinidis et al. 2017). The strains showing 98.7% or higher 16S similarity will be selected for calculating OGRIs (Chun et al. 2018). Among the abovementioned OGRI, ANI and dDDH have been the most widely used and generally accepted species boundaries for ANI and dDDH value, at 95–96% (Goris et al. 2007; Richter and Rosselló-Móra 2009) and 70% (Meier-Kolthoff et al. 2013), respectively.

The ANI can be used to delineate prokaryotic species, but it is not suitable for genus or higher taxonomic ranks (Qin et al. 2014). As an alternative metric, average amino acid identity (AAI) (Konstantinidis and Tiedje 2005b) and percentage of conserved proteins (POCP) were developed (Qin et al. 2014). POCP has been used as a genomic index to delimit prokaryotic genus boundaries (Qin et al. 2014). A POCP compares the proteomes of two strains using BLAST searches. The conserved proteins (proteins from the query genome with an e-value less than 1e-5, identity greater than 40%, and an alignable region greater than 50%) between two genomes were identified by aligning all of the protein sequences from the query genome with all of the protein sequences from the other genome using the BLASTP program. A POCP of 50% was proposed to be the genus boundary (Qin et al. 2014). AAI is very similar to ANI, but instead, it analyzes amino acid sequences (Konstantinidis and Tiedje 2005b). The AAI cutoff point of 95–96% for species demarcation has been proposed and values <45% and < 45–65% for different families and genera, respectively (Konstantinidis and Tiedje 2005b; Konstantinidis et al. 2017).

The process to identify new taxa is depicted in Fig. 7.1, and some tools used to evaluate ANI, dDDH, and AAI are mentioned in Table 7.1. It should be noted that the 16S rRNA gene and AAI/ANI values proposed for defining taxonomic ranks are approximate, not absolute, thresholds that should be adjusted as needed to better capture the candidate taxa's phenotypic or ecological distinctiveness (Konstantinidis et al. 2017).

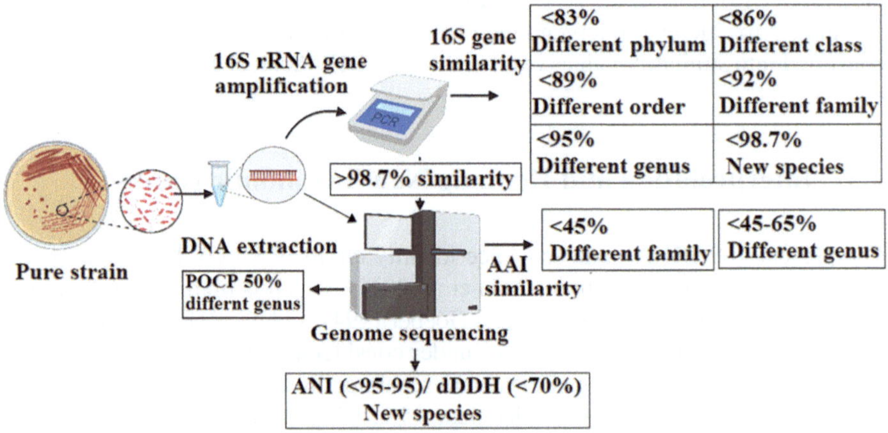

Fig. 7.1 Flowchart to identify new taxa. ANI average nucleotide identity, AAI average amino acid identity, POCP percentage of conserved proteins. The figure was created with BioRender.com

Table 7.1 Tools to identify new taxa

Tools	Function	Command line/web service	URL/references
JSpecies/JSpeciesWS	ANI	Command line/web service	http://imedea.uib-csic.es/jspecies/ (Richter and Rosselló-Móra 2009); http://jspecies.ribohost.com/ (Richter et al. 2016).
PYANI	ANI	Command line	https://github.com/widdowquinn/pyani (Pritchard et al. 2016).
OrthoANI with usearch	ANI	Command line/web service	https://www.ezbiocloud.net/tools/orthoaniu; https://www.ezbiocloud.net/tools/ani (Yoon et al. 2017).
ANI calculator	ANI	Web service	http://enve-omics.ce.gatech.edu/ani/ (Rodriguez-R and Konstantinidis 2016).
Genome-to-Genome Distance Calculator	dDDH	Web service	http://ggdc.dsmz.de/ggdc.php/ (Meier-Kolthoff et al. 2013).
AAI calculator	AAI	Web service	http://enve-omics.ce.gatech.edu/aai/ (Rodriguez-R and Konstantinidis 2016.)
EzAAI	AAI	Command line	http://leb.snu.ac.kr/ezaai (Kim et al. 2021).
Percentage of conserved proteins	POCP	Command line	https://github.com/hoelzer/pocp/blob/master/README.md (Qin et al. 2014).

ANI average nucleotide identity, AAI average amino acid identity, POCP percentage of conserved proteins, dDDH digital DNA-DNA hybridization

Acknowledgment MPNR has been awarded a postdoctorate fellowship under Kasetsart University Reinventing University Program 2022 supported by the Ministry of Higher Education, Science, Research, and Innovation and the Thailand Science Research and Innovation.

References

Altschul SF, Gish W, Miller W, Myers EW, Lipman DJ (1990) Basic local alignment search tool. J Mol Biol 215(3):403–410. https://doi.org/10.1016/s0022-2836(05)80360-2

Auch AF, von Jan M, Klenk HP, Göker M (2010) Digital DNA-DNA hybridization for microbial species delineation by means of genome-to-genome sequence comparison. Stand Genomic Sci 2(1):117–134. https://doi.org/10.4056/sigs.531120

Bergey DH, Harrison FC, Breed RS, Hammer BW, Huntoon FM (1923) Bergey's manual of determinative bacteriology, 1st edn. Williams & Wilkins Co., Baltimore

Brenner DJ, Fanning GR, Rake AV, Johnson KE (1969) Batch procedure for thermal elution of DNA from hydroxyapatite. Anal Biochem 28(1):447–459. https://doi.org/10.1016/0003-2697(69)90199-7

Caudill MT, Brayton KA (2022) The use and limitations of the 16S rRNA sequence for species classification of *Anaplasma* samples. Microorganisms 10:3. https://doi.org/10.3390/microorganisms10030605

Chun J, Oren A, Ventosa A, Christensen H, Arahal DR, da Costa MS, Rooney AP, Yi H, Xu X-W, De Meyer S, Trujillo ME (2018) Proposed minimal standards for the use of genome data for the taxonomy of prokaryotes. Int J Syst Evol Microbiol 68(1):461–466. https://doi.org/10.1099/ijsem.0.002516

Chun J, Rainey FA (2014) Integrating genomics into the taxonomy and systematics of the bacteria and archaea. Int J Syst Evol Microbiol 64(Pt 2):316–324. https://doi.org/10.1099/ijs.0.054171-0

Colwell R (1970) Polyphasic taxonomy of bacteria. In: Proceedings of the first international conference an culture collections, October 7–11, 1968, . University of Tokyo Press, pp 421–436

Deloger M, El Karoui M, Petit M-A (2009) A genomic distance based on MUM indicates discontinuity between most bacterial species and genera. J Bacteriol 191(1):91–99. https://doi.org/10.1128/JB.01202-08

Gajdács M (2020) Taxonomy and nomenclature of bacteria with clinical and scientific importance: current concepts for pharmacists and pharmaceutical scientists. Acta Pharm Hung 89(4):99–108. https://doi.org/10.33892/aph.2019.89.99-108

Gevers D, Cohan FM, Lawrence JG, Spratt BG, Coenye T, Feil EJ, Stackebrandt E, Van de Peer Y, Vandamme P, Thompson FL, Swings J (2005) Opinion: re-evaluating prokaryotic species. Nat Rev Microbiol 3(9):733–739. https://doi.org/10.1038/nrmicro1236

Goodfellow M (2000) Microbial systematics: background and uses. In: Priest FG, Goodfellow M (eds) Applied microbial systematics. Springer, Netherlands, Dordrecht, pp 1–18. https://doi.org/10.1007/978-94-011-4020-1_1

Goris J, Konstantinidis KT, Klappenbach JA, Coenye T, Vandamme P, Tiedje JM (2007) DNA-DNA hybridization values and their relationship to whole-genome sequence similarities. Int J Syst Evol Microbiol 57(Pt 1):81–91. https://doi.org/10.1099/ijs.0.64483-0

Henz SR, Huson DH, Auch AF, Nieselt-Struwe K, Schuster SC (2004) Whole-genome prokaryotic phylogeny. Bioinformatics 21(10):2329–2335. https://doi.org/10.1093/bioinformatics/bth324

Hu S, Li K, Zhang Y, Wang Y, Fu L, Xiao Y, Tang X, Gao J (2022) New insights into the threshold values of multi-locus sequence analysis, average nucleotide identity and digital DNA-DNA hybridization in delineating *Streptomyces* species. Front Microbiol 13:910277. https://doi.org/10.3389/fmicb.2022.910277

Hugenholtz P, Chuvochina M, Oren A, Parks DH, Soo RM (2021) Prokaryotic taxonomy and nomenclature in the age of big sequence data. ISME J 15(7):1879–1892. https://doi.org/10.1038/s41396-021-00941-x

Janda JM, Abbott SL (2007) 16S rRNA gene sequencing for bacterial identification in the diagnostic laboratory: pluses, perils, and pitfalls. J Clin Microbiol 45(9):2761–2764. https://doi.org/10.1128/jcm.01228-07

Johansson K-E, Pettersson B (2002) Taxonomy of mollicutes. In: Razin S, Herrmann R (eds) Molecular biology and pathogenicity of mycoplasmas. Springer US, Boston, MA, pp 1–29. https://doi.org/10.1007/0-306-47606-1_1

Kim D, Park S, Chun J (2021) Introducing EzAAI: a pipeline for high throughput calculations of prokaryotic average amino acid identity. J Microbiol 59(5):476–480. https://doi.org/10.1007/s12275-021-1154-0

Kim M, Oh HS, Park SC, Chun J (2014) Towards a taxonomic coherence between average nucleotide identity and 16S rRNA gene sequence similarity for species demarcation of prokaryotes. Int J Syst Evol Microbiol 64(Pt 2):346–351. https://doi.org/10.1099/ijs.0.059774-0

Konstantinidis KT, Rosselló-Móra R, Amann R (2017) Uncultivated microbes in need of their own taxonomy. ISME J 11(11):2399–2406. https://doi.org/10.1038/ismej.2017.113

Konstantinidis KT, Tiedje JM (2005a) Genomic insights that advance the species definition for prokaryotes. Proc Natl Acad Sci U S A 102(7):2567–2572. https://doi.org/10.1073/pnas.0409727102

Konstantinidis KT, Tiedje JM (2005b) Towards a genome-based taxonomy for prokaryotes. J Bacteriol 187(18):6258–6264. https://doi.org/10.1128/jb.187.18.6258-6264.2005

Kurtz S, Phillippy A, Delcher AL, Smoot M, Shumway M, Antonescu C, Salzberg SL (2004) Versatile and open software for comparing large genomes. Genome Biol 5(2):R12. https://doi.org/10.1186/gb-2004-5-2-r12

Lee I, Ouk Kim Y, Park S-C, Chun J (2016) OrthoANI: an improved algorithm and software for calculating average nucleotide identity. Int J Syst Evol Microbiol 66(2):1100–1103. https://doi.org/10.1099/ijsem.0.000760

Lee KY, Wahl R, Barbu E (1956) Contenu en bases puriques et pyrimidiques des acides désoxyribonucléiques des bactéries. In: Annales de l institut pasteur. vol 2. Masson editeur 120 blvd saint-germain, 75280 Paris 06, France, pp 212–224

Liu GH, Narsing Rao MP, Dong ZY, Wang JP, Che JM, Chen QQ, Sengonca C, Liu B, Li WJ (2019) Genome-based reclassification of *Bacillus okuhidensis* as a later heterotypic synonym of *Bacillus halodurans*. Int J Syst Evol Microbiol 69(11):3599–3602. https://doi.org/10.1099/ijsem.0.003666

Meier-Kolthoff JP, Auch AF, Klenk H-P, Göker M (2013) Genome sequence-based species delimitation with confidence intervals and improved distance functions. BMC Bioinformatics 14(1):60. https://doi.org/10.1186/1471-2105-14-60

Meier-Kolthoff JP, Auch AF, Klenk H-P, Göker M (2014) GBDP on the grid: a genome-based approach for species delimitation adjusted for an automated and highly parallel processing of large data sets. Hochleistungsrechnen in Baden-Wuerttemberg-Ausgewaehlte Aktivitaeten im bwGRiD 2012: Beitraege zu Anwenderprojekten und Infrastruktur im bwGRiD im Jahr 2012:83

Pitt T, Barer M (2012) Classification, identification and typing of micro-organisms. Med Microbiol:24

Pritchard L, Glover RH, Humphris S, Elphinstone JG, Toth IK (2016) Genomics and taxonomy in diagnostics for food security: soft-rotting enterobacterial plant pathogens. Anal Methods 8(1):12–24. https://doi.org/10.1039/C5AY02550H

Qin QL, Xie BB, Zhang XY, Chen XL, Zhou BC, Zhou J, Oren A, Zhang YZ (2014) A proposed genus boundary for the prokaryotes based on genomic insights. J Bacteriol 196(12):2210–2215. https://doi.org/10.1128/jb.01688-14

Richter M, Rosselló-Móra R (2009) Shifting the genomic gold standard for the prokaryotic species definition. Proc Natl Acad Sci U S A 106(45):19126–19131. https://doi.org/10.1073/pnas.0906412106

Richter M, Rosselló-Móra R, Oliver Glöckner F, Peplies J (2016) JSpeciesWS: a web server for prokaryotic species circumscription based on pairwise genome comparison. Bioinformatics 32(6):929–931. https://doi.org/10.1093/bioinformatics/btv681

Rodriguez-R LM, Konstantinidis KT (2016) The enveomics collection: a toolbox for specialized analyses of microbial genomes and metagenomes. PeerJ. Preprints

Rosselló-Mora R, Amann R (2001) The species concept for prokaryotes. FEMS Microbiol Rev 25(1):39–67. https://doi.org/10.1111/j.1574-6976.2001.tb00571.x

Rosselló-Móra R, Amann R (2015) Past and future species definitions for bacteria and archaea. Syst Appl Microbiol 38(4):209–216. https://doi.org/10.1016/j.syapm.2015.02.001

Stackebrandt E, Goebel BM (1994) Taxonomic note: a place for DNA-DNA reassociation and 16S rRNA sequence analysis in the present species definition in bacteriology. Int J Syst Evol Microbiol 44(4):846–849. https://doi.org/10.1099/00207713-44-4-846

Stackebrandt E, Ludwig W, Fox G (1985) 3 16 S ribosomal RNA oligonucleotide cataloguing. In: Methods in microbiology, vol 18. Elsevier, pp 75–107

Vandamme P, Pot B, Gillis M, de Vos P, Kersters K, Swings J (1996) Polyphasic taxonomy, a consensus approach to bacterial systematics. Microbiol Rev 60(2):407–438. https://doi.org/10.1128/mr.60.2.407-438.1996

Varghese NJ, Mukherjee S, Ivanova N, Konstantinidis KT, Mavrommatis K, Kyrpides NC, Pati A (2015) Microbial species delineation using whole genome sequences. Nucleic Acids Res 43(14):6761–6771. https://doi.org/10.1093/nar/gkv657

Wayne L, Brenner D, Colwell R, Grimont P, Kandler O, Krichevsky M, Moore L, Moore W, Murray R, Stackebrandt E (1987) Report of the ad hoc committee on reconciliation of approaches to bacterial systematics. Int J Syst Evol Microbiol 37(4):463–464

Woese CR, Fox GE (1977) Phylogenetic structure of the prokaryotic domain: the primary kingdoms. Proc Natl Acad Sci U S A 74(11):5088–5090. https://doi.org/10.1073/pnas.74.11.5088

Yoon SH, Ha SM, Lim J, Kwon S, Chun J (2017) A large-scale evaluation of algorithms to calculate average nucleotide identity. Antonie Van Leeuwenhoek 110(10):1281–1286. https://doi.org/10.1007/s10482-017-0844-4

Young JM (2001) Implications of alternative classifications and horizontal gene transfer for bacterial taxonomy. Int J Syst Evol Microbiol 51(Pt 3):945–953. https://doi.org/10.1099/00207713-51-3-945

Chapter 8
Taxonomic Approaches for Uncultivated Prokaryotes

Bhagwan Narayan Rekadwad

Abbreviations

AAI	Amino acid identity
ACGs	Axenic culture genomes
ANI	Average nucleotide identity
DNA	Deoxyribonucleic acid
eDNA	Environmental DNA
GAN	Great Automatic Nomenclator
ICNP	International Code of Nomenclature of Prokaryotes
ICSP	International Committee on Systematics of Prokaryotes
IJSEM	*International Journal of Systematic and Evolutionary Microbiology*
MAGs	Metagenome-assembled genomes
NCBI Proka RefG	NCBI Prokaryotic RefSeq Genomes
rRNA	Ribosomal ribonucleic acid
SAGs	Single-cell amplified genomes/single-amplified genomes
SWOT analysis	Strengths, weaknesses, opportunities, and threats (SWOT) analysis
WGS	Whole-genome sequencing
MLST	Multi-locus sequence typing
DDH	DNA-DNA hybridization/DNA-DNA reassociation
INSDC	International Nucleotide Sequence Database Collaboration
SRA	Sequence Read Archive

B. N. Rekadwad (✉)
Division of Microbiology and Biotechnology, Yenepoya Research Centre, Yenepoya (Deemed to be University), Mangalore, Karnataka, India

W.-J. Li et al. (eds.), *Modern Taxonomy of Bacteria and Archaea*,
https://doi.org/10.1007/978-981-99-5720-0_8

8.1 Introduction

8.1.1 A Concise Outline of Prokaryotic Taxonomy and Nomenclature over the Past 300 Years

In the age of sequencing, it has become very controversial to name and describe bacteria and archaea that have not been grown in a lab. Though modern technologies are capable of delivering more digitalized information on extant species, both cultivated and uncultivated, they still rely on Carl Linnaeus' (1735) classification system. Bacterial classification and taxonomy have been studied for over 300 years. Since the inception of Ehrenberg's classification of bacteria in 1838, the taxonomy of bacterial categorization has been explored. Bacteria were first classified into families and genera based on obvious structural differences and physiological characteristics (Winslow et al. 1917). The members of the committee came to an agreement in the final report regarding families and genera as key indicators in bacterial taxonomy. The choice was made based on existing knowledge of most natural systems used in classifying species at the time, as long as the stability of the binomial name was guaranteed by the existing committee (Winslow et al. 1920).

After three decades of continuous efforts by scientists and taxonomists, a precise system of nomenclature with the incorporation of comments by botanists and zoologists and approval by the vast majority of bacteriologists globally has been developed, albeit governed by botanical and zoological rules. The bacteriological code had a requirement for some significant microbiological provisions. "The system of bacterial nomenclature was founded on principles, rules, and recommendations." The principles are based on retroactive rules to avoid illegitimate names. The recommendations deal with subsidiary information to bring greater uniformity and cleanliness to the rules. The International Association of Microbiologists and its Judicial Commission established a Nomenclature Committee and a Judicial Commission to make provisions for rule emendation, special exceptions to rules, and interpretation of rules in doubtful cases (Buchanan et al. 1948).

Since the 1950s, various judicial commissions and committees have approved reports and corrections, and the revised Bacteriological Code came into effect in the early 1990s. Between the 1950s and the 1990s, a buried report by Krawiec (1985) mentions the use of DNA for intraspecies differentiation and the identification of loci. Furthermore, the ad hoc committee's report agreed on the common covenant that the complete DNA sequence would be the reference standard for determining phylogeny and that phylogeny should determine taxonomy. In addition, nomenclature should be consistent with and reflect genomic data (Wayne et al. 1987). Biologists have structured this world in terms of certain basic dichotomies. Traditionally, anything that was not a plant was considered an animal. Carl Woese's comments on basic dichotomies among biologists appear appropriate in this regard. As bacteria were first classified as plants and then as prokaryotes, endless debates raged about prokaryotes vs. eukaryotes. This means that prokaryotes cannot be compared to eukaryotes, though it is unclear who arrived first in the evolutionary process. As a result, bacteria must be classified based on their phylogenetic relatedness, as demonstrated by appropriate marker genes such as 16S rRNA (Woese and George 1977).

The argument was that the use of DNA sequences must be prioritized and that the time has come to support taxonomy based on whole-genome sequences, even though physically cultivated forms of bacteria are important but not required for digital insights into their characteristics. Nowadays, advances in DNA sequencing technology and various bioinformatics platforms have enabled the use of genome sequences for the taxonomic classification of bacteria, archaea (Chun et al. 2018), and viruses, and scientists are applying the aforementioned methodologies to link strains and species from other domains. An abundance of evidence suggests that uncultivables are the species that we are unable to cultivate in the laboratory due to a lack of required growth conditions. Hence, floccinaucinihilipilificated debates insisting on cultivable appear to be numerically incomprehensible.

8.1.2 Advantages and Limitations of the Prokaryotic Code

Taxonomy refers to the theory and practice of categorizing microorganisms into taxa in a systematic manner. Microbial taxonomy is currently phylogeny-guided, which means that organisms are classified according to their evolutionary relationships, which are defined by vertical genetic information transmission from mother to daughter cells (Zhu et al. 2015). Overall, the Bacteriological Code (now the Prokaryotic Code) assists us in assigning a name to new species based on existing principles, rules, and recommendations. Isolated unknown species or strains get a name following a description based on a proposed and mutually agreed-upon standard that's also timely reframed by ICNP (Oren 2019a). As a result, existing species within a genus can be classified, and genera can be arranged in an orderly fashion. Indeed, every name given to the taxon is new and unique. All of a taxon's scientific names are Latin or Latinized words treated as Latin, regardless of their origin. They are typically derived from Latin or Greek. The perfect prokaryotic code in the age of big data has been challenged by scientific minds, and some published literature has discussed the shortcomings of the prokaryotic code (Hugenholtz et al. 2021).

Over 3,400 names of bacterial and archaeal genera and 20,000 names of species have been documented in the standing nomenclature system (Parte et al. 2020). This number is small in comparison with the number of species described for plants and animals, the majority of which lacked genome-based descriptions. As a direct consequence, existing principles, rules, and recommendations are impeding the future of microbial taxonomy, particularly for uncultivated bacteria and archaea. Thousands of bacteria and archaea have been given validly published names. In contrast, nearly one-third of the bacterial and archaeal populations have not received valid names in the DNA-based modern era of taxonomy (Fig. 8.1). According to recent statements about the future of prokaryotic taxonomy, the number of cultivated species is only a few percent of the number of species that exist in nature, based on our current,

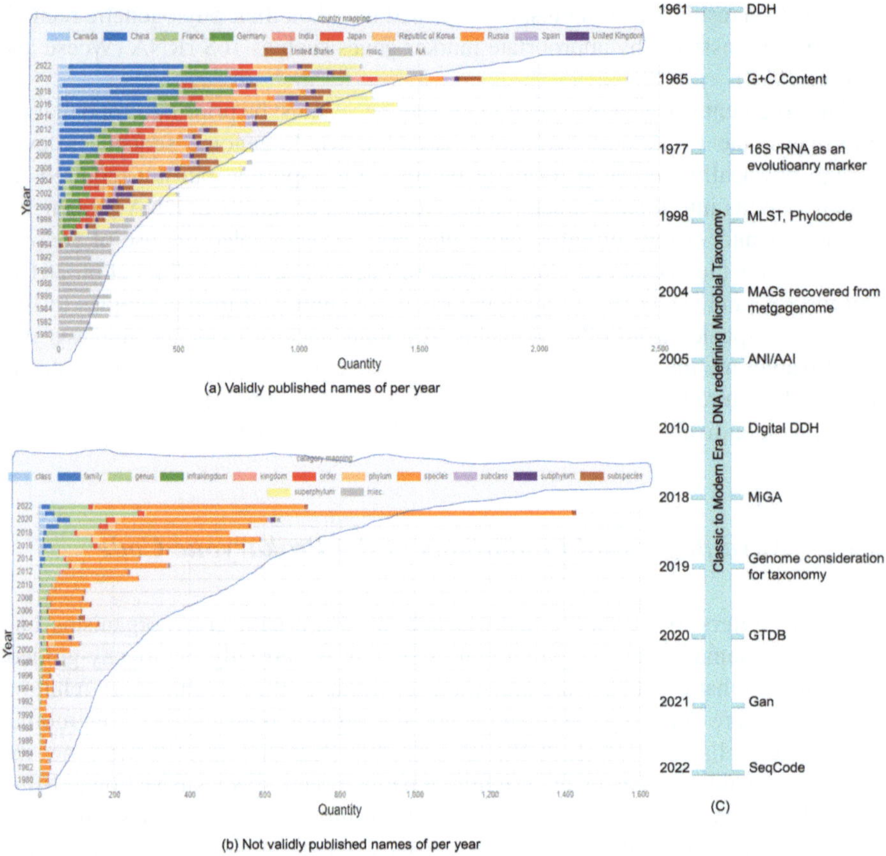

Fig. 8.1 Documented names of bacteria and archaea between years 1980 and 2022. (**a**) Validly published names per year. (**b**) Not validly published names per year (LPSN 2022). (**c**) Classical to modern taxonomy—events in the nomenclature of bacteria and archaea after use of DNA (top 1961 to 2022) as of December 24, 2022

poorly defined species concept for prokaryotes. If the existing nomenclature system is to be used in the future for the uncultivated majority, at least ten times as many names as are currently found in prokaryotic nomenclature will be required. Hence, it is very difficult to maintain the correct name of yet-to-be-cultured or uncultured (i.e., *Candidatus* taxa) identified based on genome sequences, and there is no type material available and maintained in pure culture for valid publication of names under the current rules of the Prokaryotic Code (Oren 2019b). The current Prokaryotic Code has a limitation in that genome sequences are not considered type material for naming and classifying *Candidatus* taxa that do not have a valid name.

8.2 Uncultivable Prokaryotes (Or Yet-to-Culture Prokaryotes)

An uncultivable prokaryote is a paradoxical phrase for bacteria and archaea that we have yet to achieve a cultivable form of. Inadequate procedures and ignoring natural environmental conditions are axiomatic, indicating the limitations of current growth techniques and the nutrient cocktail available to them. The inadequate knowledge about their biology for cultivation reflects the hurdles that loom in obtaining them (Stewart 2012). The actual number of prokaryotes on Earth is still a point of contention, with estimates ranging from millions to trillions. Therefore, calling them "yet-to-culture" might be an appropriate term. Currently, ICNP denotes *Candidatus* taxa as uncultivated species (Konstantinidis et al. 2017).

Uncultivated bacteria and archaea play a critical but little-understood role in the biogeochemical cycle (Anantharaman et al. 2016; Bell et al. 2018; Kato et al. 2018) and have existed for billions of years. If we look at the cultured majority, the 3,400 genera and 20,000 species (Parte et al. 2020) discovered by researchers using simple to extensive cultivation strategies around the world in the previous few centuries account for only 1 percent of all microorganisms. The remaining 99% are yet to be cultivated and are concealed in the Earth's vast environment and different habitats. Hence, we can regard the invisible majority as potentially cultivable. Again, a question arises: do the 20,000 species constitute 1% of the total number of species or less (Fig. 8.2)? It is impossible to recreate Earth's dynamic nature and environmental conditions in laboratory settings for each and every habitat; therefore, around half of the total diversity may stay uncultivable and difficult to identify indefinitely for centuries. Based on existing tests and media, a century may be required to cultivate around 5–10% of the population in the near future.

8.3 Taxonomy of Yet-to-Culture Prokaryotes

Nowadays, the development of new methodologies aids in the extraction of environmental DNA (eDNA) or genomic material from any sample, and advanced analysis technologies enable us to predict and describe existing unseen microbes (Murray and Schleifer 1994). Yet-to-culture is a distinction that limits physical experiments and descriptions as "live culture," whereas genomic insights can deduce everything equally in each species regardless of whether it is cultivable or uncultivated yet. *Candidatus* full description, like axenic culture, has never been covered by ICNP, and the lack of general benchmarks has limited *Candidatus* cataloguing. Considering a glaring discrepancy with the naming of *Candidatus* taxa that exist in nature and their poor classification according to the existing Prokaryotic Code, therefore, *Candidatus* taxa will have their own nomenclatural system, following the same principles as cultured bacteria and archaea but with their own list of legitimately published names. If widely used, this system will not only make it easier to

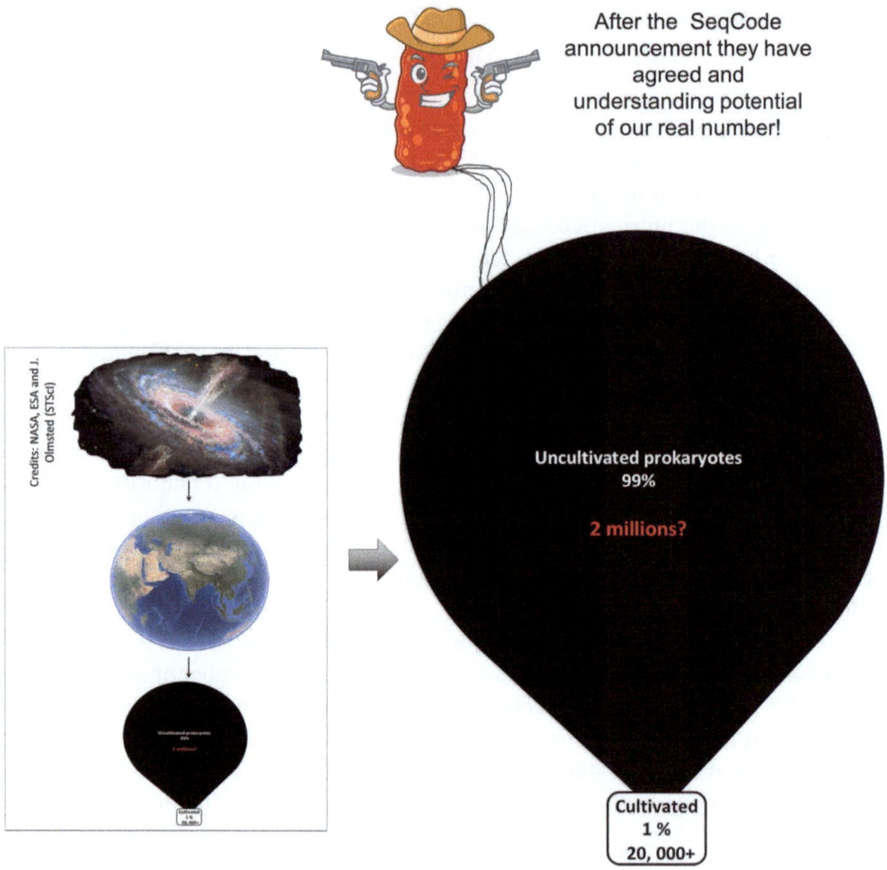

Fig. 8.2 Prokaryote numbers: cultivated vs. uncultivated—highlighting the potential of uncultivated bacteria and archaea

characterize the "uncultivated majority," but it will also give a single library of lawfully published names, eliminating synonyms and misunderstandings (Konstantinidis et al. 2017). This may be a violation of Rule 18f.

According to prokaryotic codes, new bacteria should be deposited in at least two international culture collections (Rule 30(3)(c)) before being published in the *International Journal of Systematic and Evolutionary Microbiology* (IJSEM). In the case of new novel species names published elsewhere than IJSEM, authors should submit a covering letter and the published article(s) to IJSEM. The validation requirements are the same as those for effective publication in the IJSEM. These rules have been in effect for the past two decades. In the future, there may be a significant increase in the number of fields that are not occupied by existing prokaryotic codes. Currently, the system of prokaryotic nomenclature regulated by the Prokaryote Code is based on pure cultures in laboratories (~22,000 species). The ICSP opposes any "official" prokaryotic taxonomy, and the Prokaryotic Code only deals with nomenclature that involves any international microbiological organization interfering. Thus, scientists may choose between the ICSP rules and pluralism, resulting in anarchy and the loss of an already well-ordered naming system (Oren and Garrity 2018).

8.4 New Tools and Rules for Nomenclature and Taxonomy of Prokaryotes

8.4.1 Microbial Genomes Atlas (MiGA) Webserver

The 16S RNA gene sequence has been used as a marker gene for cataloging and discriminating isolates up to the species level for three decades but not enough to differentiate some genera, e.g., *Aeromonas*. Hence, to achieve some level of accuracy, average nucleotide identity (ANI) based on whole-genome sequence (WGS) was proposed as a more accurate proxy for species and finer levels such as subspecies, strains, and pathovars (Rodriguez-R et al. 2018a). Considering the proven potential of 16S rRNA and other important proxies (such as whole genome-based aggregates of ANI and amino acid identity (AAI) and inputs from axenic culture genomes (henceforth ACG), single-cell amplified genomes (SAGs), and metagenome-assembled genomes (MAGs)), the MiGA webserver (MiGA) complements an unidentified query genomic sequence against all existing forms of sequences and genomes. Furthermore, most databases, such as the NCBI Prokaryotic RefSeq Genomes (NCBI Proka RefG), contain ACGs, SAGs, and MAGs, i.e., the Clade projects, which have been used as additional inputs in MiGA for genetic diversity, inter- and intra-evolutionary relationships among taxa, as well as identifying and calculating core-genome sets and the pangenome (Rodriguez-R et al. 2018b).

8.4.2 The Genome Taxonomy Database (GTDB)

Genomes from axenic bacteria and archaea cultures, as well as SAGs and MAGs, contributed significantly to the total genomic pool (Parks et al. 2017; Alneberg et al. 2018). GTDB is a genome-based taxonomy with rank-normalized classifications for over 150,000 bacterial and archaeal genomes from domain to genus that is phylogenetically consistent. Nearly 40% of the genomes in the GTDB, on the other hand, do not have a species designation. This constraint was overcome by using widely accepted average nucleotide identity (ANI) criteria to define species boundaries and propose species clusters that include all publicly accessible bacterial and archaeal genomes. Unlike previous ANI studies, a single representative genome has been chosen to serve as the effective nomenclatural "type" defining each species. 8792 of the 24,706 proposed species clusters are based on names that have been published.

To assign names to the growing number of genomes from uncultivated species, the remaining 15,914 species clusters were given placeholder names. This resource gives a complete domain-to-species taxonomic framework for bacterial and archaeal genomes, making it easier to study species that haven't been grown in a lab and making it easier for scientists to talk to each other (Parks et al. 2020).

8.4.3 The Great Automatic Nomenclator (GAN)

In a golden era of discovery through culturomics, genomics, and metagenomics, bacteria and archaea have an urgent and unmet need for new names for their steady taxonomic position. The time-consuming process of generating taxonomic names demands experts who have practiced classical Latin and Greek knowledge and the rules of the ICNP. The creation of millions of names is literally at your fingertips in the multi-omics era. An innovative approach has three features: en masse production of names before they are attached to taxa; combinatorial concatenation of roots from Latin and Greek, drawing on stocks of roots with pertinent meanings; and computerized automation of name formation (Pallen et al. 2021).

8.4.4 The Seqcode

Classification, nomenclature, and identification are three divisible parts of taxonomy that come into existence for naming microbes. Of these three, the nomenclature is governed by rules, but there are no rules applied for classification and identification. Naming microbe should have a sequential code to convey information about the character of organisms (Cowan 1965). Labels received for uncultivated prokaryotes (*Candidatus* taxa) such as CPR1, CPR3, and SM2F11 (Hug et al. 2016) do not express any species-specific characteristics. As cultivated species, the *Candidatus* taxa do not have priority names. As a result, providing a precise and stable name for scientific communication becomes critical for all species that exist in nature. The use of ribosomal databases, ANI/AAI aggregates, and isolated whole genomes provides appropriate species names that are derived entirely from culture investigation. The use of isolated whole genomes provides appropriate species names that are derived entirely from culture assessment. It is unethical to apply culture-based descriptions to the uncultured (Rosselló-Móra and Whitman 2019), which nullifies comments on the SeqCode in section "Threats" and "Conclusions" of SWOT analysis (Pallen 2021).

The informally called UnCode for yet-to-culture prokaryotes is termed as the SeqCode, which provides permanent names for the prokaryotes (Whitman et al. 2022). The lower taxa (species and subspecies) are exemplified by either a strain or a description in ICNP, which is applicable to the naming of all prokaryotes. SeqCode provides guiding principles, directives, and suggestions.

The Code's regulations and suggestions originate from Chap. 2's concepts. Chap. 3's rules implement the concepts. Chapter 3's recommendations augment some rules but aren't rules. They're best-practice manuals. Names against recommendations can't be rejected. Appendices may be introduced to help apply this Code but aren't legislative (Fig. 8.3).

Currently, there are two mechanisms, with the possibility of a third in the future. In the proposed procedure (blue color arrows, path 1), the preliminary registration of the name and submission of metadata into the SeqCode Directory occurs simultaneously with the preparation of the effective publication. The Registry performs data quality and name synonymy checks in conjunction with curator evaluation, resulting to the provisional acceptance of SeqCode-compliant submissions. This process maintains data accuracy and eliminates the need for errata following publishing. The time and date of priority are indicated by entering the publication's DOI into the Registry. Due to the need of the SeqCode to utilize the latest name of a taxon, the date of priority confirms this name's precedence as the only legitimate name for the taxon. The second (orange color arrows, path 2) is for already released names, such as Candidate names. The name and associated metadata are added to the Registry. Computerized checks and SeqCode curators verify compliance, and acceptance of the proposal concludes registration and establishes the time and date of precedence. At that moment, the *Candidatus* label may be eliminated. In the future, the third mechanism could be created in collaboration with one or more magazines (pink color arrows, path 3). As an integrated process for the validation of proposed names, it would involve concurrent peer review and Registry curator evaluation. The time and date of priority are established when a DOI is assigned to an accepted manuscript. For specific instances of registration via pathways 1 and 2, please consult the text and Supplementary Information. The SeqCode would be implemented on January 1, 2022, based on predecided criteria. The general considerations, principles and rules, etc. of the SeqCode can be read at SeqCode Registry website (https://disc-genomics.uibk.ac.at/seqcode/) and effective Hedland et al. 2022 (https://doi.org/10.1038/s41564-022-01214-9).

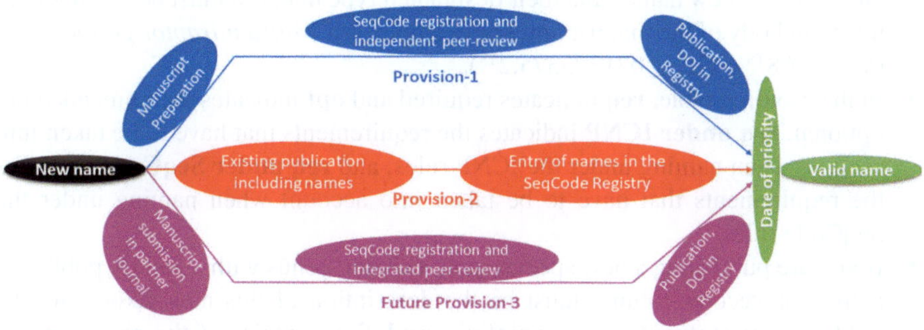

Fig. 8.3 Process for validation of name of uncultivated prokaryotes under the SeqCode (content of this figure adopted and modified, Hedlund et al. 2022)

8.5 Independent View on Adopting the SeqCode and Green Tag by the *Systematic and Applied Microbiology* (SAM)

On a wide scale, next-generation-next-level sequencing platforms allow for the sequencing and curation of ACGs, SAGs, and MAGs, which can be named under alternative code, i.e., SeqCode. The SeqCode is recognized by the *Systematic and Applied Microbiology* (SAM) (Rosselló-Móra et al. 2022). Various taxonomic databases have been developed in a short span of time, and extensive to warm discussions have arisen around the world. The development of advanced name generation algorithms and analytical tools as well as the modification of current rules to a degree leads to perfection. Microbial taxonomy is continually developing, as it was 300 years ago. According to the SeqCode's general considerations, principles, and regulations, the SeqCode will provide the taxonomically stable naming of uncultivated taxa (See Tables 8.1, 8.2, 8.3 and 8.4; mentioned tables were obtained from Copyright Clearance Center via Order Numbers 5435271373876 and 5435290224395 with permitted reuse and publication). In the future, naming axenic cultures of currently uncultivated species will take less time and effort if the SeqCode is adopted as a nomenclature scheme for uncultivated bacteria and archaea.

Note that SAM requires the inclusion of the accession number of the deposit of the genome sequence and 16S rRNA genes in one of the INSDC databases.

- All descriptions made following the nomenclatural rules of the ICNP must be given as a table in the main manuscript, as their validation in the IJSEM will require a permanent record and supplemental material is not considered (yet) to be permanent.
- The full descriptions made following the nomenclatural rules of the SeqCode can optionally be given in the supplemental material, since the permanent record is not a requirement for the SeqCode, although the registry system rules the priority and validation of the names. However, SAM requires that the name and designated type material are given in the main text. Therefore, in case of numerous taxa descriptions, SAM will consider publishing a supplementary table with all the complete protologs of the required information, but, in addition, a single table with all new names and their designated type material must be given within the main body of the manuscript. For example, see *Wolframiiraptor gerlachensis* (Wger_A8^{TS} = GCA_021323375.2^{TS}).
- In the protolog table, **req** indicates required and **opt** indicates recommended but optional. **req under ICNP** indicates the requirements that have to be taken into account when naming under the ICNP rules, and **req under SeqCode** indicates the requirements that have to be taken into account when naming under the SeqCode rules.
- If you are publishing a new species in an existing genus with a validly published name, the second column must be the description of this new taxon, and the fields ***Genus status*, *Genus etymology*, and *Type species of the genus*** can be

Table 8.1 Requirements and recommendations for describing cultivated taxa to be named under the ICNP rules. Note that SAM distinguishes between single and multiple strain descriptions, which are treated differently (Permission obtained for republication from Copyright Clearance Center (CCC) via Order number 5435271373876)

Description of cultivated taxa and naming under the ICNP rules **Name and designated type strain**
Required The taxon name and the designated type strain collection numbers need to be given in the main manuscript in the form of a protolog table. Taxon names need to be written in italics, and the designated types and culture collection numbers should be suffixed with a superscript^{T}, for example, *Salinibacter altiplanensis* (AN15^{T} = IBRC-M 11031^{T} = CECT 9105^{T}). The deposit certificates need to be submitted together with the manuscript.
Protolog
Required If taxa are named under the ICNP rules, protologs must be given in the form of a table in the manuscript. Descriptions in the supplementary material are not allowed because only permanent records are considered for validation of names. The required fields for the protolog are available in the online template. The required basic information are as follows: • Name (formed with mnemonic cues). • Etymology. • Designated type strain and two culture collection numbers. • Genome with an INSDC accession number of the designated type strain that meets minimum quality requirements. • The 16S rRNA gene accession number of the designated type strain. The gene must be almost complete (>1400 bp) if resulting from PCR amplification or complete if resulting from genome sequencing. • In the case of a new genus, designate the type species. • In the case of higher taxa, designate the type genus. • Diagnostic genomic and phenotypic properties, inferred and demonstrated physiological traits. Ideally, a phenotypic discriminative profile using relevant physiological studies is desired.
Optional The following are optional but highly recommended for more informative taxa descriptions: • Include as much metadata as possible. • As commercial biochemical tests and simple phenotyping using standard laboratory tests often do not reveal relevant properties, a more exhaustive, genome-based description of (potential) phenotypic properties is recommended instead (e.g., metabolic prediction based on functional gene annotation bioinformatics). • It will benefit the study to include ecological information, such as habitat, environmental physicochemical properties, and inferred interactions with other organisms, such as symbiosis, syntrophy, and other properties that may be relevant for the description. • Chemotaxonomic properties, such as fatty acids, polar lipids, respiratory quinones, or polyamines, unless required by the respective taxonomic subcommittees, may be regarded as optional, given that these can also be determined retrospectively. • Some fields of relevant metadata are already given in the available online template in order to guide the acquisition of such data, and these can be added as new lines in the protolog template.

(continued)

Table 8.1 (continued)

Type Material Data Quality
Required For taxonomic descriptions with nomenclature according to the ICNP rules, the type material is a living pure culture. This must be deposited in two culture collections from different countries, and the cultures must be fully available to third parties without any restriction, following the recommendation of the ICNP.
Multiple Strains ≥2 SAM recommends that authors acquire multiple non-clonal strains of the same taxon, ideally from different origins.
Required With the description of multiple strains, the following are required: • There is proof of non-clonality of the strains. For this purpose, different genome-based methods can be applied. However, for a fast-check at the initial editorial desk-based evaluation, genome fingerprinting techniques, such as PCR-RAPD profiles, should be applied, and an image should be provided in the supplementary material showing distinct fingerprints. Note that if some strains show identical fingerprints, indicating clonality, these will be considered as a single strain. If all strains of the same taxon show the same fingerprint, the description will be considered as an SSSD. • All (non-clonal) strains described in the manuscript (or a subset if >10 isolates) should be genome-sequenced in order to perform pairwise comparisons that will allow taxon circumscriptions. • The monophyletic origin is a premise for recognizing a taxon. Therefore, all strains in the study must be placed within a phylogenetic framework. SAM requires that at least a 16S rRNA phylogenetic tree is provided, and the guidelines for this are given in the text. • Genome coherence has to be evaluated using one of the OGRIs. • All strains must be phenotypically evaluated in order to reflect the intraspecific diversity.
Optional The following are optional: • In cases of large collections (>10 strains), and if well-justified, sequencing and phenotyping of only part of the collection will be allowed. • Phylogenetic inference based on additional trees beyond the 16S rRNA gene, as indicated in the text, using different gene sets (or their translated products), such as ribosomal proteins, or core genes, is encouraged in order to further support and/or demonstrate the robustness of the classification. • Chemotaxonomic markers should be determined for the whole collection, as appropriate. • To increase the interest of the study, it is highly recommended to check the abundance of the new taxon in the sample of origin by means of culture-independent methods, such as fluorescence microscopy using specific probes (FISH) and 16S rRNA gene amplicon sequencing, and/or by metagenomic read recruitments. • To increase the importance of the study, SAM highly recommends screening for the presence and relative abundance of the new taxon in the publicly available 16S rRNA gene amplicon and metagenomic datasets. Read recruitments against the genome using the publicly available metagenomes can reveal relative abundance, biogeographic distribution, and ecological patterns that increase the usefulness of the taxonomic description. For an example, see the study of the geographical distribution of *Hydrotalea lipotrueae*.
Single Strain Species Descriptions (SSSDs) If only one isolate has been obtained, SAM will consider such manuscripts for review only under exceptional cases.

(continued)

Table 8.1 (continued)

Required

To compensate for the absence of multiple strains, SAM requires increasing the interest of the study by

- Providing a high-quality genome, ideally fully closed, and comparing the gene composition with the closest relatives.
- Providing an exhaustive phenotypic description, ideally finding relevant traits in the genome and proving the expression of the capabilities.
- Providing ecological information for the isolate by reporting its relative abundance in the sample of origin by means of culture-independent methods, such as FISH, 16S rRNA gene amplicon sequencing, or metagenome read recruitments.
- Providing the biogeographical distribution by means of screening for the presence and relative abundance of the new taxon in the publicly available 16S rRNA gene amplicon and metagenomic datasets. Read recruitments against the genome using the publicly available metagenomes can reveal relative abundance, biogeographic distribution, and ecological patterns that increase the usefulness of the taxonomic description. For an example, see the study of the geographical distribution of *Hydrotalea lipotrueae*.

Optional

Any additional information that increases the value of the contribution that goes beyond SSSDs will be welcomed, such as the following:

- Evaluating the pan- and core-genomes of, for example, the genus where the strain has been classified
- Revealing taxonomically nonrelevant traits, such as prophages, mobile elements, and genomic islands that can increase the broader interest of the strain

removed. If you have more than one species, you only need to add a third, fourth, fifth, etc. column—one for each description.

- If you are publishing a new genus in addition to a new species, the second column must be the description of the genus, and then the fields ranging from ***Specific epithet* to *GC mol%*** may be filled in with a dash indicating an empty field. The genus description must be followed by the type species of the genus in a third column, and additional species will follow in the fourth, fifth, etc. columns. Should you have two or more genera, just provide the Genus, Type species, Species, etc. repeating the column series, or, alternatively, you can generate one table for each different genus.
- Please add a superscript T to the designated type strain name and strain collection numbers (e.g., AN5^{T}; IBRC-M 11031^{T} = CECT 9105^{T}) if you name under the ICNP rules.
- Please add a superscript TS to the designated type genome name and strain collection numbers (e.g., Wger_A8^{TS} = GCA_021323375.2^{TS}) if you name under the SeqCode rules.
- Empty rows (i.e., fields where no information is available) must be removed.

– Contrarily to the ICNP that requires a permanent record in order to validly publish the names, SeqCode allows the protologs to be published as supplementary material. However, SAM will only allow protologs to appear as supplementary information if the number of descriptions is large (e.g., > 10). In this latter case, the main body of the manuscript should contain a table with the new names and

Table 8.2 Requirements and recommendations for describing uncultivated taxa to be named under the SeqCode rules. The SeqCode is less restrictive than ICNP in linking a taxonomic description to a publication that is considered to be a permanent record. SeqCode requires that only new names are given in the publication proposing the name, and the full or succinct protologs, including the designated type material, can be given in the supplementary material. The SeqCode Registry is available at https://seqco.de/ and is the basis for name validation and priority establishment. SAM will follow these rules but endorse certain more restrictive requirements in order to ensure further (higher) data quality for the future. This table includes all details concerning data quality and reporting requirements, as well as recommendations for an isolate's genome, metagenome-assembled genome (MAG), or single-amplified genome (SAG) used to serve as the nomenclatural type for a species named under SeqCode, obtained largely from (Hedlund et al. 2022). (Permission obtained for republication from Copyright Clearance Center (CCC) via Order number 5435271373876)

Description of taxa and naming under the SeqCode rules
Name and designated type genome
Required The taxon name and the designated type genome, with an INSDC accession number, need to be given in the main body of the manuscript or in a table if protologs are included in the supplementary material. Exceptionally, isolate-based descriptions with a nomenclature formulated under the SeqCode rules will be recognized by SAM if the strains, due to any specific reason, cannot be deposited in two different culture collections. Taxon names need to be written in italics and the genome designated names and accession numbers suffixed with a superscript^{TS}, for example, see *Wolframiiraptor gerlachensis* (Wger_A8^{TS} = GCA_021323375.2^{TS}).
Protolog
Required The description of the taxon must be given either in the manuscript or in the supplementary material (if there are >10 protologs) depending on the volume of descriptions given in the contribution. The required fields in the protolog are given in the template in original publication of this table. The basic information required are as follows: • Name (formed with mnemonic cues) • Etymology • Interpretation of the biological properties, inferred or demonstrated physiological traits, as well as ecological information, such as habitat, environmental physicochemical properties, and inferred interactions with other organisms, such as symbiosis, syntrophy, and other properties that may be relevant for the description • Designated genome assembly (e.g., INSDC accession) and access to raw data (e.g., SRA accession) • Given registry number after submission to the SeqCode Registry system Include as much metadata as possible. Some fields of relevant metadata are already provided in the online template.
Optional Optional information can include the following: In cases where the authors wish to provide additional metadata fields, these can be added as new lines in the protolog template.

(continued)

Table 8.2 (continued)

Registry
Required Data quality will be assessed by using available automated pipelines or other approaches. Exceptions for lower data quality should be justified by the authors in the main body of the manuscript. • The assembly should be available in one of the INSDC databases. Raw data (reads) should be available in the INSDC databases (e.g., Sequence Read Archive). This is not required for names effectively published before January 1, 2023, in order to allow for existing published names (e.g., existing *Candidatus* names) and names currently undergoing peer review to be validated under SeqCode.
Optional There are several optional fields in the registry from which, if left empty/non-filled, will not lead to the invalidation of the entry. For details, check https://seqco.de/.
Type Material Data Quality
Required Genomes, MAGs, and SAGs should have minimal standards of quality to be recognized as type material. Lower standards may lead to erroneous classifications and/or confusion that will promote undesired instability in the system. The major requirements for adequate (high) quality are as follows: • Assembly quality for MAGs and SAGs: >90% complete and <5% contaminated. • For isolates as for MAGs/SAGs, read coverage ≥10x (Field et al., 2008). • Provide evidence of the discreteness of the species, taxonomic rank, and position, including the uniqueness of the species with respect to existing validly named species, as well as justifying the taxonomic rank and position based on genomic and 16S rRNA gene data. • Check for congruence between the genome-derived and 16S rRNA taxonomic assignments. • For MAGs and SAGs, compare multiple high-quality genomes representing the species from more than one sample (multiple high-quality genomic assemblies from multiple samples can support the non-chimeric nature of MAGs and provide confidence of the assembly for both MAGs and SAGs) or justify why multiple samples and/or MAGs are not available. • 16S rRNA gene >75% complete, and passes chimera checks. >80% of tRNAs present in order to further ensure high completeness of the genome (modified from Bowers et al. (2017))
Optional Genomes, MAGs, and SAGs of high quality are desirable for representing a stable nomenclatural type. Therefore, we also suggest reporting the following: High genome integrity (contig # <100; N50 >25 kb; max. contig >100 kb)

the designated type materials (i.e., the genome accession numbers), and the protologs will be given in the supplementary material.

– As with the SSSDs, SAM expects that the value and impact of a manuscript will increase by adding information that goes beyond simple genome descriptions. For MAGs and SAGs, their detection in more than a single sample is desirable, and assessing their biogeographical distribution should be evaluated. Although not compulsory, the use of alternative molecular methods to detect and quantify the abundance of MAG/SAGs in samples, such as fluorescence in situ hybridization, is recommended.

Table 8.3 Protolog table template (available online) to be used for a new genus and species. This information needs to be published as a table in the manuscript submitted to SAM, since the standard written descriptions are not necessary. This will be considered as the formal description for the name and, therefore, the effective publication of a new taxon. (Permission obtained for republication from Copyright Clearance Center (CCC) via Order number 5435271373876)

Guiding code for nomenclature **req**	**Indicate whether the rules of the ICNP or SeqCode have been followed**
Nature of the type material **req**	Indicate the **strain** or **genome sequence**
Genus name **req**	Add the generic name (e.g., *Salinibacter*)
Species name **req**	Add the full name (e.g., *Salinibacter altiplanensis*)
Genus status **req**	Select the adequate status **(gen. nov./nom. rev.)**
Genus etymology **req**	**[N.B. Genus etymology is only for new genus proposals; never for extant genera]** Add the genus etymology (e.g., Sa.li.ni.bac'ter. L. fem. pl. n. *salinae* salterns, salt-works; N.L. masc. n. *bacter* masc. equivalent of the Gr. neut. n. *bakterion* a rod; N.L. masc. n. *Salinibacter* a rod from salt-works)
Type species of the genus **req**	**[N.B. Type species of the genus is only for new genus proposals; never for extant genera]** Indicate the designated type species of the genus by adding the full specific name (e.g., *Salinibacter ruber*)
Specific epithet req	Add the specific epithet (e.g., *altiplanensis*)
Species status **req**	Select the adequate status **(sp. nov./comb. nov./nom. rev.)**
Species etymology **req**	Add the specific epithet etymology (e.g., al.ti.pla.ne'nsis, N.L. masc. adj. *altiplanensis* of the Argentinian Altiplano)
Designation of the type strain **req under ICNP**	Provide the designation of the strain as in the manuscript (e.g., $AN15^{T}$) Note that a superscript T must follow the strain designation(s)
Strain collection numbers **req under ICNP**	Provide at least two collection numbers of two international repositories (e.g., IBRC-M 11031^{T} = CECT 9105^{T}) Note that a superscript T must follow the strain designation(s)
Designated genome, MAG, or SAG **req under SeqCode**	Provide the designation of the genome, MAG, or SAG as in the manuscript (e.g., Wger_$A8^{TS}$) Note that a superscript TS must follow the genome designation(s)
Type genome, MAG, or SAG accession Nr. [INSDC databases] **req under SeqCode**	Add the genome accession number(s) with the repository identifier (e.g., GenBank = GCA_021323375.2^{TS}) Note that a superscript TS must follow the genome designation(s) if following the SeqCode rules. If following the ICNP rules, no suffix is needed.
Access to raw data (e.g., SRA accession) [opt under SeqCode]	Add the accession number to the raw data from where the MAGs were assembled and binned (required for names after January 1, 2023)
Registry number **req under SeqCode**	Add the registry number available at https://seqco.de/
Genome status **opt**	Complete/incomplete
Genome size **opt**	Add the genome size in kbp (e.g., 3580)
GC mol% **opt**	Add the GC mol% (e.g., 64.41)

(continued)

Table 8.3 (continued)

16S rRNA gene accession nr. **req**	Provide the gene accession number(s) (e.g., LT160741)
Description of the new taxon and diagnostic traits **req**	Provide the phenotypic, genotypic, and, if possible, ecological description of the new taxon emphasizing the traits used to diagnose it. This text can be as long as the normal written protologs in standard publications. In addition, add data here on chemotaxonomy, if determined
Country of origin **opt**	Add the country of origin (e.g., Argentina)
Region of origin **opt**	Add the region of origin of isolation (e.g., Salar de Antofalla)
Date of isolation (dd/mm/yyyy) **opt**	Add the isolation date
Source of isolation **opt**	Add the source of isolation (e.g., hypersaline lake)
Sampling date (dd/mm/yyyy) **opt**	Add the sampling date
Latitude (xx°xx′xx″N/S) **opt**	Add the latitude coordinates
Longitude (xx°xx′xx″E/W) **opt**	Add the longitude coordinates
Altitude (meters above sea level) **opt**	Add the altitude in meters
Number of strains in study **opt**	Add the number of strains in the study
Source of isolation of non-type strains **opt**	Add the source of isolation of the additional strains
Information related to the Nagoya Protocol **req**	Add here the permit number(s) and country of origin and/or local authority from where the permits had been obtained

8.6 The SeqCode in the Future: Unifying Prokaryote Naming

The last 60 years of microbial nomenclature saw many shifts and changes in assigned original name, and developing technologies permitted us to see new change. In order to facilitate communication on the vast genetic diversity of prokaryotes, both the SeqCode and the SeqCode Registry have been developed by Hedlund et al. (2022). With technologies like metagenomics and single-cell genomics that are getting better at putting together genomes from uncultivated prokaryotes and strong resources to help make decisions about taxonomy, the pieces are in place for prokaryotic systematics to grow over the next 10 years to cover the entire prokaryotic tree of life. This growth is likely to happen in the next 10 years. It is expected that this growth will happen over the next 10 years. This addition will help the community because it will give all prokaryotes the same way to name microorganisms. This system will be very well put together, and it will be backed up by genomic datasets that are full of data and information about that data. The level of success that the SeqCode project has will be directly related to how much the community gets involved. It is important to use the SeqCode often and in a big way. At

Table 8.4 Table template to be used for protolog descriptions of a new genus and species. This information has to be published as a table in the manuscript. The standard written descriptions are not necessary anymore. (Permission obtained for republication from Copyright Clearance Center (CCC) via Order number 5435290224395)

Genus name	Add the generic name (e.g., *Salinibacter*)
Species name	Add the full name (e.g., *Salinibacter altiplanensis*)
Genus status	Select the adequate status **(gen. nov./comb. nov./nom. rev.)**
Genus etymology	**[N.B. Genus etymology is only for new genus proposals; never for extant genera]** Add the specific etymology (e.g., Sa.li.ni.bac'ter. L. fem. pl. n. *salinae* salterns, salt-works; N.L. masc. n. *bacter* masc. equivalent of the Gr. neut. n. *bakterion* a rod; N.L. masc. n. *Salinibacter* a rod from salt-works)
Type species of the genus	**[N.B. Type species of the genus is only for new genus proposals; never for extant genera]** Indicate which is the designated type species of the genus by adding the full specific name (e.g., *Salinibacter ruber*)
Specific epithet	Add the specific epithet (e.g., *altiplanensis*)
Species status	Select the adequate status **(sp. nov. / comb. nov. / nom. rev.)**
Species etymology	Add the specific epithet etymology (e.g., al.ti.pla.ne'nsis, N.L. masc. adj. *altiplanensis* of the Argentinian Altiplano)
Description of the new taxon and diagnostic traits	Provide the phenotypic, genotypic, and, if possible, ecological description of the new taxon emphasizing on the traits that diagnose it. This text can be as long as the normal written protologs in standard publications. Also add here data on chemotaxonomy, if determined
Country of origin	Add the country of origin (e.g., Argentina)
Region of origin	Add the region of origin of isolation (e.g., Salar de Antofalla)
Date of isolation (dd/mm/yyyy)	Add the isolation date
Source of isolation	Add the source of isolation (e.g., hypersaline lake)
Sampling date (dd/mm/yyyy)	Add the sampling date
Latitude (xx°xx′xx″N/S)	Add the latitude coordinates
Longitude (xx°xx′xx″E/W)	Add the longitude coordinates
Altitude (meters above sea level)	Add the altitude in meters
16S rRNA gene accession nr.	Provide the gene accession number(s), if available (e.g., LT160741)
Genome accession number [RefSeq; EMBL; …]	Add the genome accession number(s), if available, with the repository identifier (e.g., RefSeq =NCQY00000000)
Genome status	Complete/incomplete
Genome size	Add the genome size in kbp (e.g., 3,580)
GC mol%	Add the GC mol% (e.g., 64.41)

(continued)

Table 8.4 (continued)

Number of strains in study	Add the number of strains in the study
Source of isolation of non-type strains	Add the source of isolation of the additional strains
Information related to the Nagoya Protocol	Add here the permit number(s) and country of origin and/or local authority from where the permits had been obtained
Designation of the type strain	Provide the designation of the strain as in the manuscript (e.g., AN15^T)
Strain collection numbers	Provide at least two collection numbers of two international repositories (e.g., IBRC-M 11031^T = CECT 9105^T)

- If you are publishing a new species, the second column must be the description of this new taxon, and the fields ***Genus status***, ***Genus etymology***, **and** ***Type species of the genus*** can be removed. If you have more than one species, you just need to add third, fourth, fifth, etc. columns – one for each description.
- If you are publishing a new genus in addition to a new species, the second column must be the description of the genus, and then the fields ranging from ***Specific epithet*** **to** ***GC mol%*** may be filled with a dash indicating an empty field. The genus description must be followed by the type species of the genus in a third column, and additional species will follow in the fourth, fifth, etc. columns. Should you have two or more genera, just provide the Genus, Type species, Species, etc. repeating the column series, or alternatively generate one table for each different genus.
- Please add a superscript T to the designated type strain name and strain collection numbers (e.g., AN5^T; IBRC-M 11031^T = CECT 9105^T).
- Empty rows (i.e., fields wherein no information is available) must be removed.

the same time, it is important to build a structure of governance that includes laws, committees, and membership. Also, it is important to build a governance structure that includes the SeqCode. In the next step of the SeqCode Registry's development, the focus should be on making the registration process better, increasing the number of automated inspections that are already in place, and making it work better with datasets related to informatics and biodiversity. Everything will go off without a tow bar because of this. In conclusion, a heritage of *Candidatus* names and other informal names should be recorded under the SeqCode in evaluating and bringing stability to the contemporary taxonomic framework. This would meet the goals of unifying and organizing the current taxonomic foundation (Hedlund and Whitman 2022).

References

Alneberg J, Karlsson CMG, Divne AM, Bergin C, Homa F, Lindh MV, Hugerth LW, Ettema TJG, Bertilsson S, Andersson AF, Pinhassi J (2018) Genomes from uncultivated prokaryotes: a comparison of metagenome-assembled and single-amplified genomes. Microbiome 6:173. https://doi.org/10.1186/s40168-018-0550-0

Anantharaman K, Brown C, Hug L, Sharon I, Castelle CJ, Probst AJ, Thomas BC, Singh A, Wilkins MJ, Karaoz U, Brodie EL, Williams KH, Hubbard SS, Banfield JF (2016) Thousands

of microbial genomes shed light on interconnected biogeochemical processes in an aquifer system. Nat Commun 7:13219. https://doi.org/10.1038/ncomms13219
Bell E, Lamminmäki T, Alneberg J, Andersson AF, Qian C, Xiong W, Hettich RL, Balmer L, Frutschi M, Sommer G, Bernier-Latmani R (2018) Biogeochemical cycling by a low-diversity microbial community in deep groundwater. Front Microbiol 9:2129. https://doi.org/10.3389/fmicb.2018.02129
Buchanan RE, St. John-Brooks R, Brred RS (1948) International bacteriological code of nomenclature. J Bacteriol 55:287–306. https://doi.org/10.1128/jb.55.3.287-306.1948
Chun J, Oren A, Ventosa A, Christensen H, Arahal DR, da Costa MS, Rooney AP, Yi H, Xu XW, De Meyer S, Trujillo ME (2018) Proposed minimal standards for the use of genome data for the taxonomy of prokaryotes. Int J Syst Evol Microbiol 68:461–466. https://doi.org/10.1099/ijsem.0.002516
Cowan ST (1965) Principles and practice of bacterial taxonomy—a forward look. J Gen Microbial 39:143–153. https://doi.org/10.1099/00221287-39-1-143
Hedlund BP, Chuvochina M, Hugenholtz P, Konstatinidis K, Murray A, Palmer M, Parks DH, Probst AJ, Reysenbach AL, Rodriguez-R LM, Rosselló-Móra R, Sutcliffe I, Venter SN, Whitman WB (2022) SeqCode: a nomenclatural code for prokaryotes described from sequence data. Nat Microbiol 7:1702–1708. https://doi.org/10.1038/s41564-022-01214-9
Hedlund BP, Whitman WB (2022) Towards a unifying system for the naming of cultured and uncultured prokaryotes. Nat Microbiol 7:1514–1515. https://doi.org/10.1038/s41564-022-01220-x
Hug LA, Baker BJ, Anantharaman K, Brown CT, Probst AJ, Castelle CJ, Butterfield CN, Hernsdorf AW, Amano Y, Ise K, Suzuki Y, Dudek N, Relman DA, Finstad KM, Amundson R, Thomas BC, Banfield JF (2016) A new view of the tree of life. Nat Microbiol 1:16048. https://doi.org/10.1038/nmicrobiol.2016.48
Hugenholtz P, Chuvochina M, Oren A, Parks DH, Soo RM (2021) Prokaryotic taxonomy and nomenclature in the age of big sequence data. ISME J 15:1879–1892. https://doi.org/10.1038/s41396-021-00941-x
Kato S, Sakai S, Hirai M, Tasumi E, Nishizawa M, Suzuki K, Takai K (2018) Long-term cultivation and metagenomics reveal ecophysiology of previously uncultivated thermophiles involved in biogeochemical nitrogen cycle. Microbes Environ 33:107–110. https://doi.org/10.1264/jsme2.ME17165
Konstantinidis KT, Rosselló-Móra R, Amann R (2017) Uncultivated microbes in need of their own taxonomy. ISME J 11:2399–2406. https://doi.org/10.1038/ismej.2017.113
Krawiec S (1985) Minireview: concept of a bacterial species. Int J Syst Bacteriol 35:217–220
LPSN (2022) List of prokaryotic names with standing in nomenclature, https://lpsn.dsmz.de/statistics assessed on April 30, 2022 at 14:28:59 h
Murray RG, Schleifer KH (1994) Taxonomic Notes: a proposal for recording the properties of putative Taxa of procaryotes. Int J Syst Bacteriol 44:174–176. https://doi.org/10.1099/00207713-44-1-174
Oren A (2019a) Prokaryotic nomenclature. Theory and practice of classification practice. Bergey's manual of Systematics of Archaea and bacteria. John Wiley & Sons, Inc. in association with Bergey's Manual Trust. doi: https://doi.org/10.1002/9781118960608.bm00004.pub2
Oren A (2019b) How to name new taxa of archaea and bacteria. In: Bergey's manual of systematics of archaea and bacteria. John Wiley & Sons, in association with Bergey's Manual Trust. doi: https://doi.org/10.1002/9781118960608.bm00008.pub2
Oren A, Garrity GM (2018) Uncultivated microbes—in need of their own nomenclature? ISME J 12:309–311. https://doi.org/10.1038/ismej.2017.188
Pallen MJ (2021) The status Candidatus for uncultured taxa of bacteria and archaea: SWOT analysis. Int J Syst Evol Microbiol 71:005000. https://doi.org/10.1099/ijsem.0.005000
Pallen MJ, Telatin A, Oren A (2021) The next million names for archaea and bacteria. Trends Microbiol 29:289–298. https://doi.org/10.1016/j.tim.2020.10.009

Parks DH, Chuvochina M, Chaumeil PA, Rinke C, Mussig AJ, Hugenholtz P (2020) A complete domain-to-species taxonomy for Bacteria and Archaea. Nat Biotechnol 38:1079–1086. https://doi.org/10.1038/s41587-020-0501-8

Parks DH, Rinke C, Chivochina M, Chaumeil PA, Woodcroft BJ, Evans PN, Hugenholtz P, Tyson GW (2017) Recovery of nearly 8,000 metagenome-assembled genomes substantially expands the tree of life. Nat Microbiol 2:1533–1542. https://doi.org/10.1038/s41564-017-0012-7

Parte AC, Carbasse, Meier-Kolthoff JP, Reimer LC, Göker M (2020) List of prokaryotic names with standing in nomenclature (LPSN) moves to the DSMZ. Int J Syst Evol Microbiol 70:5607–5612. https://doi.org/10.1099/ijsem.0.004332

Rodriguez-R LM, Castro JC, Kyrpides NC, Cole JR, Tiedje JM, Konstantinidis KT (2018a) How much do rRNA gene surveys underestimate extant bacterial diversity? Appl Environ Microbiol 84:e00014-18. https://doi.org/10.1128/AEM.00014-18

Rodriguez-R LM, Gunturu S, Harvey WT, Rosselló-Mora R, Tiedje JM, Cole JR, Konstantinidis KT (2018b) The Microbial Genomes Atlas (MiGA) webserver: taxonomic and gene diversity analysis of Archaea and Bacteria at the whole genome level. Nucleic Acids Res 46:W282–W288. https://doi.org/10.1093/nar/gky467

Rosselló-Móra R, Konstantinidis KT, Amann R (2022) How systematic and applied microbiology will deal with two nomenclature codes (ICNP and SeqCode) for prokaryotes, and which classification standards are recommended for new taxa descriptions. Syst Appl Microbiol. https://doi.org/10.1016/j.syapm.2022.126371

Rosselló-Móra R, Whitman WB (2019) Dialogue on the nomenclature and classification of prokaryotes. Syst Appl Microbiol 42:5–14. https://doi.org/10.1016/j.syapm.2018.07.002

Stewart EJ (2012) Growing unculturable bacteria. J Bacteriol 194:4151–4160. https://doi.org/10.1128/JB.00345-12

Wayne LG, Brenner DJ, Colwell RR, Grimont PAD, Kandler O, Krichevsky MI, Moore LH, Moore WEC, RGE M, Stackebrandt BE, Starr MP, Truper HG (1987) Report of the Ad Hoc Committee on Reconciliation of Approaches to Bacterial Systematics. Int J Syst Evol Microbiol 37:463–464. https://doi.org/10.1099/00207713-37-4-463

Whitman WB, Chuvochina M, Hedlund BP, Hugenholtz P, Konstantinidis KT, Murray AE, Palmer M, Parks DH, Probst AJ, Reysenbach AL, Rodriguez-R LM, Rossello-Mora R, Sutcliffe I, Venter SN (2022) Development of the SeqCode: a proposed nomenclatural code for uncultivated prokaryotes with DNA sequences as type. Systematic and applied microbiology 45(5):126305. https://doi.org/10.1016/j.syapm.2022.126305

Winslow CEA, Broadhurst J, Buchanan RE, Krumwiede C Jr, Rogers LA, Smith GH (1917) The families and genera of the bacteria- preliminary report of the committee of the Society of American Bacteriologists on characterization and classification of bacterial types. J Bacteriol 2:505–566. https://doi.org/10.1128/jb.2.5.505-566.1917

Winslow CEA, Broadhurst J, Buchanan RE, Krumwiede C Jr, Rogers LA, Smith GH (1920) The families and genera of the bacteria- preliminary report of the committee of the society of american bacteriologists on characterization and classification of bacterial types. J Bacteriol 2:191–229. https://doi.org/10.1128/jb.2.5.505-566.1917

Woese CR, George E (1977) Phylogenetic structure of the prokaryotic domain: the primary kingdoms. PNAS 74:5088–5090. https://doi.org/10.1073/pnas.74.11.5088

Zhu C, Delmont TO, Vogel TM, Bromberg Y (2015) Functional basis of microorganism classification. PLoS Comput Biol 11:e1004472. https://doi.org/10.1371/journal.pcbi.1004472

Chapter 9
Description and Naming of *Candidatus* Taxa

Aharon Oren

9.1 Naming Uncultivated Prokaryotes: Historical Overview

One of the conditions for valid publication of names of new species of prokaryotes is the designation of a type strain isolated in pure culture. A viable culture of that strain must be deposited in at least two publicly accessible culture collections in different countries from which subcultures must be available (Rule 18a and 30(3) of the *International Code of Nomenclature of Prokaryotes* (ICNP), the internationally recognized formal framework for naming of bacteria and archaea) (Parker et al. 2019; Oren et al. 2023). Thus, names of species, genera, etc. of prokaryotes that cannot yet be cultivated or for which no axenic cultures are available cannot be validly published under the current rules of the ICNP.

The requirement for pure cultures available from culture collections only dates from the beginning of this century. In the 1990 revision of the nomenclature code, then named the *International Code of Nomenclature of Bacteria* (ICNB, Lapage et al. 1992), Rule 18a stated: "For a species which has not so far been maintained in laboratory culture or for which a type strain does not exist, a description, preserved specimen, or illustration may serve as the type." It is interesting to note that no cultures are available for the prokaryote with the oldest validly published name – *Polyangium vitellinum* (Link 1809) – although other members of the genus have been grown in pure culture. A similar case of a genus for which cultures are available for several species but not for the type species is *Leptothrix*: the type species *Leptothrix ochracea* named by Kützing (1843) and earlier known as *Conferva ochracea* (Roth 1797) was described on the basis of morphological characteristics only. Some well-known examples of prokaryotes with conspicuous morphology

A. Oren (✉)
Department of Plant and Environmental Sciences, The Institute of Life Sciences, The Hebrew University of Jerusalem, Edmond J. Safra Campus, Jerusalem, Israel
e-mail: aharon.oren@mail.huji.ac.il

W.-J. Li et al. (eds.), *Modern Taxonomy of Bacteria and Archaea*,
https://doi.org/10.1007/978-981-99-5720-0_9

described in the early days of microbiological research are the iron oxidizing bacterium *Gallionella ferruginea* (Ehrenberg 1838), the large oval freshwater sulfur bacterium *Achromatium oxaliferum* (Schewiakoff 1893), and the large intestinal bacterium *Oscillospira guilliermondii* (Chatton and Pérard 1913). These names were included in the 1980 *Approved Lists of Bacterial Names* (Skerman et al. 1980) and are therefore validly published. The most recent example of a prokaryotic species, representing a new genus, for which no cultures are available but whose name was validly published is the giant sulfur bacterium *Thiomargarita namibiensis* (Schulz et al. 1999; Anonymous 1999).

At the meetings of the International Committee on Systematics of Bacteria (ICSB), now the International Committee on Systematics of Prokaryotes (ICSP), and its Judicial Commission held in Sydney in 1999, Rule 18a(2) of the ICNP was added, stating that as from January 1, 2001, a description, preserved (nonviable) specimen, or illustration may not serve as the type. Rule 30 was emended so that descriptions of new species or new combinations must include the designation of a type strain, so that a viable culture of that strain must be deposited in at least two publicly accessible service collections in different countries from which subcultures must be available, and so that the designations allotted to the strain by the culture collections should be quoted in the published description (De Vos and Trüper 2000; Labeda 2000). These emendations of the code are nonretroactive, so that names validly published before 2001 remain validly published also if no pure cultures can be obtained from culture collections.

As a result of these changes in the rules of the code, names of more recently discovered representatives of the genus *Thiomargarita* can no longer be validly published. Therefore, *Thiomargarita joergensenii*, *Thiomargarita nelsonii*, and *Thiomargarita magnifica* were described as *Candidatus* taxa (Salman et al. 2011; Volland et al. 2022).

9.2 The Concept of *Candidatus* Names for Uncultivated Prokaryotic Taxa

Since the early 1990s, techniques for the molecular sequence-based description of prokaryotes without the need for cultivation are widely available. Genes coding for 16S rRNA and other phylogenetically relevant markers can be retrieved from nucleic acids isolated from the environment and sequenced. It quickly became clear that the cultivated prokaryotic diversity is only a small fraction of the true diversity in nature. Organisms harboring those novel 16S rRNA genes could be visualized by fluorescence in situ hybridization. Already then it was estimated that there are at least two orders of magnitude more yet – unnamed prokaryotic species than the number of cultivated and named species.

The rules of the 1990 revision of the *International Code of Nomenclature of Bacteria* (Lapage et al. 1992) did not cater for the naming of this uncultivated majority. According to Rule 18a, a description, preserved specimen, or illustration could until the end of 2000 still serve as the type for a species which has not so far been maintained in laboratory culture. However, this is no longer possible following the above-explained emendations of the code that took effect in 2001.

We thank the establishment of the *Candidatus* concept as a provisional status for naming of putative taxa of prokaryotes that could not (yet) be cultivated in pure culture to Robert Murray (1919–2022), who, together with Karl-Heinz Schleifer and Erko Stackebrandt, published the proposal in the mid-1990s (Murray and Schleifer 1994; Murray and Stackebrandt 1995). At the time, the concept was mainly developed to cater for the nomenclature of symbionts and parasites of eukaryotic cells, because such organisms could be recognized by their molecular structures but could not be assigned to a known genus because of the lack of sufficient distinguishing characteristics (Murray and Stackebrandt 1995). However, already then, the authors understood to what extent the development of new methods and technologies could revolutionize our understanding of the true microbial diversity: "As the editors of two journals concerned with the systematics of procaryotes, we recognize that the problem will increase as the available technology gains wider use and becomes easier to apply" (Murray and Schleifer 1994).

The authors of the proposal stressed that sequence information indicating the phylogenetic position of the *Candidatus* organism should be supplemented with as much additional information as possible: the natural environment where the organism can be detected by in situ hybridization or other techniques and as many structural, metabolic, reproductive, and other phenotypic features as may be available. It was recognized that the sequences used to deduce the phylogenetic affiliation of the *Candidatus* taxon represent only a small proportion of its genome and that phylogenetically related prokaryotes can have diverse physiologies whereas physiologically similar organisms can occur in different phylogenetic lineages. A novel sequence isolated from a certain environment merely indicates that there may be a unique organism present there. The authors stressed the importance of the design of probes complementary to specific regions of the nucleic acid sequence, so that the organism from which the sequence was retrieved can be detected and identified in situ. By means of suitable phylogenetic probes, it is then possible to obtain information about the morphology of the organism and its abundance and distribution in its habitat.

The Judicial Commission of the ICSB discussed the proposals made by Murray and Schleifer (1994) at its meeting in Prague in 1994. The Judicial Commission agreed that Rule 18 of the ICNB did not provide adequate regulation for the naming of yet uncultivated prokaryotes recognized only or mainly on the basis of a gene sequence. It was recommended to restrict the term *Candidatus* to situations where it is known that an organism exists, and the term should not be used for sequences only. The preparation of the taxonomic note by Murray and Stackebrandt (1995) to introduce the *Candidatus* concept was announced at the meeting. It was stressed that the *Candidatus* status for putative taxa of prokaryotes should be implemented for describing prokaryotic entities for which more than a mere sequence is available but that could not be described in sufficient detail for valid publication of newly proposed names under the rules of the code of nomenclature (Frederiksen 1995).

The papers by Murray and Schleifer (1994) and Murray and Stackebrandt (1995) included a few examples of descriptions of *Candidatus* taxa. Based on their proposals, such a name consists of the word *Candidatus*, followed by a vernacular epithet that consists of either a genus name with a specific epithet, or only a genus name, or only a specific epithet. As currently some of their proposals are no longer in use, and as their proposals also contained errors that needed to be corrected (adopting the rules and recommendations of the ICNB/ICNP; see section "Correcting Incorrectly Formed *Candidatus* Names"), here are the original examples with explanatory comments:

"*Candidatus* intracellularis" [δ-*Proteobacteria*) Not Cultivated; Gram -; Rod; Nucleic Acid Sequence (GenBank number L15739), oligonucleotide sequence complementary to unique region of 16S rRNA 5'-GTCTTGAGGCTCCCCGAAAGGCACCTCTTAATC-3', Symbiontic (*Sus*, ileum); Mesophilic] Gebhart et al. Int. J. Syst. Bacteriol. 43: 533, 1993

This is an example of a *Candidatus* name that consists of only a specific epithet, and it probably is the only case of a name formed in this way. Soon after, the organism was grown in pure culture and its name was validly published as *Lawsonia intracellularis* McOrist et al. 1995 with type strain 1482/89, deposited as NCTC 12656; however, it is currently not found in the online catalog of the NCTC culture collection.

"*Candidatus* magnetobacterium" [(new subclass of *Proteobacteria* or new lineage) Not Cultured; Gram –; Rod; Nucleic Acid Sequence (EMBL X71838), oligonucleotide sequence complementary to unique region of 16S rRNA 5'-GCCATCCCCTCGCTTACT-3', Freshwater lake sediment; Microaerophilic; Magnetosomes, Sulfur inclusions, Mesophilic]. Spring et al., Appl. Environ. Microbiol. 59:2397, 1993

This is an example of a *Candidatus* name consisting of only a genus name, but spelled with an initial lower case letter. In the original description, published before the *Candidatus* concept was introduced, the organism was tentatively named '*Magnetobacterium bavaricum*', creating a binomial name that complies with the rules of the ICNB/ICNP. Later, proposals for two additional species of the *Candidatus* genus Magnetobacterium were published. *Candidatus* Magnetobacterium casensis Lin et al. 2014 is a malformed name as the gender of the epithet does not agree with the gender of the generic name, contravening Rule 12c(1) of the ICNP; grammatically correct is *Candidatus* Magnetobacterium casense, 'pertaining to CAS, acronym of the Chinese Academy of Sciences'. The name *Candidatus* Magnetobacterium cryptolimnobacter Zhang et al. 2021 ('a hidden rod from a lake', is correctly formed as a noun in the nominative case in apposition, following Rule 12c(2) of the ICNP.

"*Candidatus* Liberobacter asiaticum" [(α-*Proteobacteria*) Not Cultivated; Gram –; Filamentous; Nucleic Acid Sequence (GenBank number L22532), oligonucleotide sequence complementary to unique region of 16S rRNA 5'-GCGCGTATGCAATACGAGCGGCA3'; Symbiotic (*Citrus*, phloem; *Diphorina citri* (Psyllidae), hemolymph, salivary glands); Mesophilic]. Jagoueix et al., Int. J. Syst. Bacteriol. 44:386, 1994

"*Candidatus* Liberobacter africanum" [(α-*Proteobacteria*) Not Cultivated; Gram –; Filamentous; Nucleic Acid Sequence (GenBank number L22533), oligonucleotide sequence complementary to unique region of 16S rRNA 5'-GCGCGTATTTTATACGAGCGGCA-3'; Symbiotic (*Citrus*, phloem; *Trioza erytreae* (Psyllidae), hemolymph, salivary glands); Mesophilic]. Jagoueix et al., Int. J. Syst. Bacteriol. 44:386, 1994

Candidatus Liberobacter asiaticum and *Candidatus* Liberobacter africanum are malformed names: similar to the case of *Candidatus* Magnetobacterium casensis, the gender of the epithets does not agree with the gender of the generic name, contravening Rule 12c(1) of the ICNP. Liberobacter is masculine, so the correctly formed epithets are 'asiaticus' and 'africanus'. Moreover, following the guidelines of Appendix 9 of the 2008 revision of the ICNP (Parker et al. 2019), the connecting vowel after the Latin word *liber* ('free') should be *-i-* and not *-o-* which is used after Greek word elements. This recommendation already existed in the 1990 revision of the code (Lapage et al. 1992). Therefore, when the first member of the genus was cultivated (Fagen et al. 2014), the name *Liberibacter* was chosen, and the name of the type species *Liberibacter* crescens was validly published. Currently (November 2022), this is still the only cultivated representative of the genus, which further contains *Candidatus* Liberibacter asiaticus corrig., *Candidatus* Liberibacter africanus corrig. and seven additional *Candidatus* species and five *Candidatus* subspecies.

The word *Candidatus* is sometimes abbreviated in the literature as Ca., but the author of this chapter recommends full spelling. The word *Candidatus* is printed in italic type and the name that follows in roman type. Murray and Schleifer (1994) recommended using quotation marks for *Candidatus* taxa, and many authors have followed this recommendation. However, no quotation marks are used in Appendix 11 of the ICNP (Parker et al. 2019; Oren et al. 2023).

The example of *Liberibacter crescens* shows a case where a member of a genus-level taxon earlier described as *Candidatus* was brought into culture and the name was validly published under the rules of the ICNP. As the names of *Candidatus* taxa have, per definition, provisional status, the ultimate goal would be to obtain cultures of all such taxa that can be deposited in culture collections so that the names can be validly published in accordance with Rule 30 of the ICNP. The names are then deleted from the list of *Candidatus* taxa. Therefore, it is important that proposed names of new *Candidatus* taxa are correctly formed and comply with the rules of the ICNP, even if formally such names are not regulated by that code. The advantage is that the names can later be validly published without the need for corrections after the organisms are cultivated.

The *Candidatus* concept has been widely adopted by journals and databases, and many *Candidatus* names have achieved de facto standing in the scientific literature following description of such taxa in peer-reviewed publications (Pallen 2021). Notably, this includes the malformed name *Candidatus* Pelagibacter ubique, the famous SAR11 phylotype found in marine waters, being the most abundant species on Earth (Rappé et al. 2002). The Latin *ubique* is an adverb, and therefore it does not qualify as a specific epithet as it is not one of the three categories listed in Rule 12c of the ICNP. Correction to *Candidatus* Pelagibacter communis was therefore proposed (Oren 2017; Oren et al. 2020b).

Proposals to include the status *Candidatus* in the rules of the ICNP have been discussed in the past years. Further information is found in section "Toward the Establishment of a Separate Code of Nomenclature for the Uncultivated Majority of Prokaryotes." As those proposals were rejected by the ICSP, *Candidatus* names do not have standing in the nomenclature today, as also explained in Appendix 11 of the ICNP (Parker et al. 2019; Oren et al. 2023).

9.3 How Many *Candidatus* Taxa Have Been Described?

When the *Candidatus* concept was introduced in the mid-1990s, the number of proposed *Candidatus* names was very small. However, the authors predicted that the increasing application of molecular methods to the assessment of prokaryotic diversity in nature and to the study of complex symbioses makes it likely that numerous *Candidatus* examples will be recorded in the future. Therefore, it was proposed to record them in lists to be published in the IJSB/IJSEM. Authors of names of new *Candidatus* taxa were encouraged to send a reprint along with an entry in the appropriate format to the editor responsible for the preparation of such lists that should appear at appropriate intervals (annually at the outset) (Murray and Stackebrandt 1995). At its meeting in Prague in 1994, the Judicial Commission of the ICSB also recommended that a list of *Candidatus* names should be established in the IJSB (Frederiksen 1995). Implementation of the preparation of such lists was also indicated in Appendix 11 of the 2008 revision of the ICNP (Parker et al. 2019), stating that "a list in the form of a codified record of organisms of the status *Candidatus* is kept by the Judicial Commission of the ICSP in cooperation with the Editorial Board of the IJSEM and is published in that journal in appropriate intervals." However, this was not implemented at the time, and this task of the Judicial Commission and of the IJSEM Editorial Board was never formalized in the ICSB/ICSP statutes. The first lists of *Candidatus* taxa were published as late as 2020 (Oren et al. 2020a, b).

In 2020, the list editors of the IJSEM, who are responsible for the production of the monthly Notification Lists ("Notification that new names of prokaryotes, new combinations, and new taxonomic opinions have appeared in volume …, part … of the IJSEM"), the bimonthly Validation Lists ("Valid publication of new names and new combinations effectively published outside the IJSEM"), and the biannual Lists of Changes in Taxonomic Opinion ("Notification of changes in taxonomic opinion previously published outside the IJSEM"), took the initiative to publish annual curated and annotated lists of *Candidatus* taxa in the IJSEM. These lists contain the names (if necessary corrected based on the rules of the ICNP and its orthography appendix – Appendix 9), their etymologies, and references to the publications in which the names were proposed. The first of these lists was preceded by an editorial, explaining the purpose of the *Candidatus* lists. These lists have no official status and are not endorsed by the ICSP or its Judicial Commission. They are intended to comply at least in part with the guidelines of Appendix 11 of the ICNP (Parker et al. 2019) and provide an inventory of published *Candidatus* names; however, the names listed are not validly published or officially approved in any other way. The list editors of the IJSEM aim at keeping the record of published names of *Candidatus* taxa as complete as possible. Lists 2–4 therefore contain addenda to the earlier published names that were not included in the previous lists. A survey of the many papers in which new *Candidatus* taxa were published during a quarter of a century shows that the quality of the descriptions is highly variable. Many do not contain a detailed description of the taxon, and sometimes the newly proposed names were only incidentally mentioned and/or no information was given about their etymology (Oren et al. 2020a).

Most of the entries of *Candidatus* lists 1–4 were retrieved from the literature by the list editors; only occasionally did the editors receive requests from authors of new *Candidatus* names to include those in an upcoming list. To complete and if necessary correct the current lists, input from the entire community of microbiologists is always appreciated. Therefore, the list editors are looking forward to receiving requests to include names of newly published *Candidatus* taxa in future lists, as well as corrections and further suggestions from the scientific community. Comments on the sometimes far-reaching name changes proposed by the editors (see below) are also welcome as long as alternative names proposed comply with the rules and the recommendations of the ICNP and its Appendix 9.

Candidatus lists 1–4 included taxa of all ranks between subspecies and class, i.e., the taxonomic ranks covered by the rules of the ICNP until 2021. The rank of phylum was included in the code only in 2021 (Oren et al. 2021), and the names of more than 40 phyla have now been validly published (Oren and Garrity 2021a). Names of phyla are formed by adding *–ia* to the stem of the proposed type genus (Oren et al. 2021). Therefore, it is now time to also prepare lists of names of *Candidatus* phyla. More than 150 such names are already found in the literature, most of which were proposed in metagenomic sequencing studies that greatly expanded our understanding of the microbial diversity in nature. Many of the published *Candidatus* phylum names do not comply with the orthography rules of Appendix 9 of the ICNP. No subordinate taxa were defined for most of those phyla, which is problematic as the nomenclatural type of a phylum is a genus (Oren et al. 2021). Preparation of a curated list of all names of *Candidatus* phyla, including proposals for orthographic corrections where needed, will be a task for the future.

Candidatus list no. 1 (Oren et al. 2020b) contained an inventory of names of 1091 *Candidatus* taxa published between 1995 and 2018: 1 higher taxon of undefined rank, 7 classes, 1 subclass, 12 orders, 25 families, 329 genera, 706 species, and 10 subspecies. In cases where species-level names were proposed belonging to not-yet-documented *Candidatus* genera, the list editors added new *Candidatus* genera. List no. 2 (Oren and Garrity 2021b) included names published in 2019 and addenda to the first list; it contained 226 new names, including 110 species. List no. 3 (Oren and Garrity 2022) included names published in 2020 and addenda to the earlier lists; it contained 286 new names, including 160 species. List no. 4 (Oren 2022), with names published in 2021 and addenda to the earlier lists, contained 1190 new names, including 861 species. Many of the new *Candidatus* names proposed in 2021 (1 family, 159 genera, and 657 species) were published in a single paper analyzing the chicken gut microbiome (Gilroy et al. 2021).

As of November 8, 2022, the draft list of new names to be included in *Candidatus* list no. 5, with names published in 2022 and addenda to the earlier lists, contained 333 new names, including 196 species-level *Candidatus* names. These numbers include 48 new *Candidatus* genera and 101 new *Candidatus* species proposed in a paper on the microbiome of horse feces (Gilroy et al. 2022). This brings the total number of *Candidatus* taxa documented in the lists to 3126, of which 2033 are species-level taxa. These numbers do not include taxa proposed as *Candidatus* but for which pure cultures were isolated. Some of those are even available from culture collections. These are listed in separate tables in the *Candidatus* lists. They also do not include the more than 65,000 *Candidatus* names published in September 2022 to name previously unnamed taxa listed in the Genome Taxonomy Database (Pallen et al. 2022), as discussed further in section "Approaches to the Naming of Large Numbers of Uncultivated Genera and Species."

When an organism originally described as a *Candidatus* taxon is later isolated and a pure culture can be deposited in two culture collections and all other requirements of Rule 30 of the ICNP are fulfilled, the name can be validly published and the taxon loses its *Candidatus* status. Thus far, the number names of former *Candidatus* taxa that were validly published under the rules of the ICNP are small: until November 2022, names of 2 orders, 2 families, 18 genera, and 41 species have thus been "upgraded." They are listed in separate tables in the *Candidatus* lists. Examples of names of former *Candidatus* genera that are now validly published are *Aquiluna* Pitt et al. 2021, *Salinibacter* Antón et al. 2002, *Nitrosopumilus* Qin et al. 2017, and *Nitrososphaera* Stieglmeier et al. 2014. The latter genus became the type genus of the phylum *Nitrososphaerota*, formerly known as *Thaumarchaeota*. If necessary, orthographic corrections can be made at the time of valid publication of the name, as shown in the above discussed case of *Liberibacter crescens* Fagen et al. 2014 and *Candidatus* Liberibacter. Similarly, *Candidatus* Rickettsia amblyommii was upgraded to become *Rickettsia amblyommatis* Karpathy et al. 2016, and *Candidatus* Nitrosarchaeum koreensis became *Nitrosarchaeum koreense* Jung et al. 2018.

9.4 Correcting Incorrectly Formed *Candidatus* Names

As the nomenclature of *Candidatus* names is not formally covered by the rules of the ICNP and such names are not validly published under that code, there has been little "quality control" for proposed *Candidatus* names. This is in contrast to new names of cultivated bacteria and archaea: every name proposed in manuscripts submitted for publication in the IJSEM is checked by the journal's nomenclature reviewers. The list editors check the new prokaryotic names published in the journal once more when preparing the monthly Notification Lists, and they also ensure that names submitted to be included in a Validation List after having been effectively published in other journals meet the standards set by the rules of the ICNP and its orthography appendix. As there was no similar quality control mechanism for *Candidatus* names, many names proposed in the past were malformed. *Candidatus* names should be correctly formed so that, when the organism is later grown in pure culture, the name (without the *Candidatus* prefix) can be used unchanged in a Notification List or a Validation List in the IJSEM.

To alert the scientific community to the problem and to show how incorrectly formed *Candidatus* names could be corrected, Oren (2017) published "a plea for linguistic accuracy – also for *Candidatus* taxa." The list editors proposed necessary corrections in the *Candidatus* lists published annually in the IJSEM since 2020 (Oren et al. 2020b; Oren and Garrity 2021b; Oren and Garrity 2022; Oren 2022). It must be stressed once more that these lists should not be considered "Approved Lists of *Candidatus* Names" analogous to the 1980 Approved Lists of Bacterial Names (Skerman et al. 1980). The corrections made by the list editors are no more than proposals, and often alternative names can be found that comply with the rules and the recommendations of the ICNP.

The examples given below, derived from *Candidatus* lists 1–4, highlight representative cases that demonstrate how the provisions of the ICNP were often not followed when the names were proposed and show how problematic names could be corrected.

Starting January 2001, the nomenclature of prokaryotes is no longer independent of botanical and zoological nomenclature. When naming new taxa of the rank of genus or higher, due consideration is to be given to avoiding names which are regulated by the International Code of Zoological Nomenclature and the International Code of Nomenclature for algae, fungi, and plants (Principle 2 of the ICNP). Therefore, alternatives were proposed for *Candidatus* genus names with homonyms in the botanical or in the zoological nomenclature, for example:

- *Alloenteromonas* was proposed instead of *Enteromonas*, a name used in zoology.
- *Allopolytropus* was proposed instead of *Polytropus*, a name used in zoology.
- *Annandiella* was proposed instead of *Annandia*, a name used in zoology.
- *Blochmanniella* was proposed instead of *Blochmannia*, a name used in botany.
- *Navoides* was proposed instead of *Navis*, a name used in zoology.
- *Palibaumannia* (named after Paul and Linda Baumann) was proposed instead of *Baumannia*; the generic names *Baumannia* and *Baumanniella* have standing in the botanical nomenclature, and *Baumannella* is the name of an insect genus. Note, however, that it is not recommended to honor more than one person in one generic name or epithet (Appendix 9, section D (3) (b) of the 2022 revision of the ICNP; Oren et al. 2023).
- *Spencerbrownia* was proposed instead of *Brownia*; the generic names *Brownia* and *Browniella* are used in zoology.
- *Theodorhartigia* was proposed instead of *Hartigia*; the generic names *Hartigia* and *Hartigiella* are used in zoology.
- *Ruthturnera* was proposed instead of *Turnera*; the generic names *Ruthia*, *Turnera*, and *Turnerella* are used in botany and zoology.

According to Principle 3 and Rule 6 of the ICNP, the scientific names of all taxa are Latin or latinized words treated as Latin regardless of their origin. Therefore, alternative names were proposed in cases such as the following:

- *Chaera* was proposed instead of *Chaer*, named after Arnaldo Chaer, a Brazilian microbial ecologist.
- *Mesenetia* was proposed instead of *Mesenet*, named after Mesenet, goddess of childbirth in ancient Egyptian mythology.

- *Uabimicrobium* was proposed instead of *Uab*, a giant in Palauan mythology.
- *Nanopelagicus limnae* was proposed instead of *Nanopelagicus limnes*; here the epithet was formed based on Greek grammar; the proposed alternative was treated as a noun of the first declension.

Recommendation 6(3) of the ICNP states that words from languages other than Latin or Greek should be avoided as long as equivalents exist in Latin or Greek. Here are a number of examples where *Candidatus* names included words from other languages:

- *Bartonella bettongiae* was proposed to replace *Bartonella woyliei* (using the genitive form of the Latin genus name *Bettongia*, the woylie, an Australian marsupial).
- *Bartonella peramelis* was proposed to replace *Bartonella bandicootii* (using the genitive form of the Latin genus name *Parameles*, the bandicoot, another Australian marsupial).
- *Parastrichiiphilus* was proposed to replace *Benitsuchiphilus* (using the Latin genus name *Parastrichia*, a stinkbug named benitsuchi in Japanese).
- *Repentibacter* was proposed to replace *Sneabacter* (named because of its ability to sneak into the organelles of the host; the alternative is based on the Latin present participle *repens*, creeping).
- *Typhincola* was proposed to replace *Rohrkolberia*, using the Latin genus name *Typh*a, the cattail, which is called *Rohrkolben* in German (Oren 2017; Oren et al. 2020b). Later authors proposed *Candidatus* Symbiopectobacterium instead (Martinson et al. 2020), illustrating that the corrections made by the list editors are proposals only and that alternative names may be proposed. The genus name Symbiopectobacterium was later validly published (Nadal-Jimenez et al. 2022).
- *Endowatersipora glebosa* was proposed to replace *Endowatersipora palomitas* (meaning "popcorn" in Spanish). Finding a good Latin equivalent was challenging. The translation *maizae grana tosta* for "popcorn" given by the Latin authorities of the Vatican (Latinitas 2022) cannot be converted to a simple single-word epithet. Therefore, glebosa ("clumpy") was proposed as an alternative.

According to Rule 10a of the ICNP (2022 Revision; Oren et al. 2023), generic names with standing in the prokaryotic nomenclature are in the nominative case. Therefore, the following correction of a *Candidatus* genus name in the genitive case was proposed:

- Consessor ("a neighbor") was proposed to replace Consessoris ("of a neighbor").

Epulonipiscium, proposed to replace *Epulopiscium*, is a more complicated case. The Latin *piscium* ("of fishes"), being the genitive plural of *piscis*, was incorrectly translated by the authors to "of a fish." However, the ending *–um* is not only used for the genitive plural of nouns of the third declension but also as the nominative singular of neuter words of the second declension. Therefore, the list editors proposed *Candidatus Epulonipiscium* as a Neo-Latin neuter noun. "Epulo–" was corrected to "Epuloni" so that the stem of the word, found in the genitive case, is used (Latin, *epulo*, genitive *epulonis*, "a guest at a banquet").

One of the three ways to form a specific epithet is as an adjective that must agree in gender with the generic name. It is surprising how many incorrectly formed *Candidatus* names were proposed with the adjective in the wrong gender, contravening Rule 12c(1). Here is a small selection:

- *Acidiferrum panamense* was proposed to replace *Acidoferrum panamensis* (the genus is neuter; panamensis can be masculine or feminine; in addition, the connecting vowel *-o-* in the genus name *Acidoferrum* was corrected to *-i-* as the preceding word element is Latin and not Greek).
- *Azospirillum massiliense* was proposed to replace *Azospirillum massiliensis* (the genus is neuter; massiliensis can be masculine or feminine).
- *Electronema palustre* was proposed to replace *Electronema palustris* (the genus is neuter, derived from the Greek neuter noun *nema* ("thread"); palustris can be masculine or feminine).
- *Ichthyocystis hellenicus* was proposed to replace *Ichthyocystis hellenicum* (the genus is masculine; hellenicum is neuter).
- *Methanomethylicus mesodigestus* was proposed to replace *Methanomethylicus mesodigestum* (the genus is masculine; mesodigestum is neuter).
- *Roseilinea gracilis* was proposed to replace *Roseilinea gracile* (the genus is feminine; gracile is neuter).
- *Thiobarba azorica* was proposed to replace *Thiobarba azoricus* (the genus is feminine; azoricus is masculine).
- *Vermiplasma glaciale* was proposed to replace *Vermiplasma glacialis* (the genus is neuter, derived from the Greek neuter noun *plasma* ("anything formed," "image"); glacialis can be masculine or feminine).
- *Wolinella africana* was proposed to replace *Wolinella africanus* (the genus is feminine; africanus is masculine).

For specific epithets indicating geographical locations, the preferred ending is *–ensis* (masculine, feminine) or *–ense* (neuter) (see Appendix 9, section E of the ICNP). Use of this ending is not recommended in other cases. The "geographical" *–ensis/–ense* ending was not always applied where appropriate, and sometimes it was incorrectly used:

- *Neoehrlichia tanzaniensis* was proposed to replace *Neoehrlichia tanzania*.
- *Desulfobulbus rimicaridis* (of the shrimp genus *Rimicaris*) was proposed to replace *Desulfobulbus rimicarensis* (use of the "geographical" ending is not justified here).
- *Magnetococcus yuandaduensis* (of Yuandadu Park, Beijing, China) was proposed to replace *Magnetococcus yuandaducum* (probably formed as an adjective of the neuter gender while the genus name is masculine).
- *Rickettsia thierseeensis* (of the Thiersee, Germany) was proposed to replace *Rickettsia thierseensis*. Three es in a row does not look nice typographically, but here they are necessary as the *–ensis* ending follows the German word "See" ("lake").

Rule 12c(2) of ICNP, which deals with formation of specific epithets as nouns in apposition in the nominative case, is often misunderstood. It was repeatedly used in an incorrect way: the nominative case was used where the genitive case would be correct in accordance with Rule 12c(3). Here are a few examples:

- *Campylobacter infantis* ("of an infant") was proposed to replace *Campylobacter infans*; *Campylobacter infans* would mean that the bacterium is an infant.
- *Ichthyocystis spari* ("of a fish of the genus *Sparus*") was proposed to replace *Ichthyocystis sparus*; *Ichthyocystis sparus* would mean that the bacterium is a fish of the genus *Sparus*.
- *Nanosyncoccus alcis* ("of a moose") was proposed to replace *Nanosyncoccus alces*; *Nanosyncoccus alces* would mean that the bacterium is a moose, a member of the genus *Alces*.

Also the third way to form specific epithets, as a noun in the genitive case (Rule 12c(3)), occasionally caused problems, as the genitive was formed incorrectly or an adjective was used in cases where a genitive noun would be more appropriate:

- *Coxiella cheracis* ("of the crayfish genus *Cherax*") was proposed to replace *Coxiella cheraxi* (as *Cherax* is treated as a noun of the third declension and not the second declension).
- *Nitrosopumilus hexadellae* ("of the sponge genus *Hexadella*") was proposed to replace *Nitrosopumilus hexadellus*.
- *Repentibacter namystyniae* ("of the protist genus *Namysyinia*") was proposed to replace *Sneabacter namystus*. The proposed correction of *Sneabacter* to *Repentibacter* was explained above.

Sometimes, the genitive plural form of a noun is needed, as shown in the following example:

- *Myxoplasma liparidarum* ("of the snailfish family Liparidae") was proposed to replace *Myxoplasma liparidae*; here the –ae ending is the plural form of the Latin first declension, and the genitive case ("of the Liparidae") is liparidarum.

Table 5 in Appendix 9 of the ICNP shows the ways to form specific epithets from personal names as genitive nouns. These guidelines were not always correctly implemented:

- *Thiobarba childressii* (named after the biological oceanographer James J. Childress) was proposed to replace *Thiobarba childressi*.

Nouns in the ablative case are not among the allowed ways to form a specific epithet mentioned in Rule 12c. The list editors therefore proposed *Candidatus* Ovatibacter antiquus instead of *Candidatus* Ovatusbacter abovo proposed by Dirren and Posch (2016). Ovatusbacter was corrected to Ovatibacter (see below). The proposed epithet abovo ("from the egg," mythological allusion to one of the two eggs of Leda which was the primary cause of the Trojan War; expression used to indicate an ancient origin) contains "ovo," the ablative form of ovum ("egg"). The list editors proposed antiquus ("ancient") instead, but they welcome other suggestions.

Adverbs are not among the categories to be used as specific epithets:

- *Pelagibacter communis* ("common") was proposed to replace *Pelagibacter ubique* ("everywhere") (see also section "The Concept of *Candidatus* Names for Uncultivated Prokaryotic Taxa").

The list editors proposed numerous corrections in the way compound names were formed. Appendix 9 of the ICNP gives clear guidelines: following a Greek word element, *-o-* is used as a connecting vowel; following a Latin word element, the connecting vowel is *-i-*. Here are a number of examples:

- *Aquirickettsia* was proposed to replace *Aquarickettsia* (Latin *aqua*, "water").
- *Arcanibacter* was proposed to replace *Arcanobacter* (Latin *arcanus*, "secret," "hidden").
- *Cuticulibacterium* was proposed to replace *Cuticobacterium* (Latin *cuticula*, "little skin").
- *Endonucleariibacter* was proposed to replace *Endonucleariobacter* (*Nuclearia* is the name of an amoeboid genus).
- *Erobacter* was proposed to replace *Erabacter* (Greek *era*, "earth").
- *Fibrimonas* was proposed to replace *Fibromonas* (Latin *fibra*, "fiber," "filament"); note that the generic name *Fibrobacter* Montgomery et al. 1988 was validly published before the current guidelines of Appendix 9 of the ICNP took effect, although the recommendations about connecting vowels were similar already in the 1976 revision of the ICNB (Lapage et al. 1975).
- *Halochondriibacter* was proposed to replace *Halichondribacter* (*Halichondria* is the name of a sponge genus).
- *Mariniplasma* was proposed to replace *Marinoplasma* (Latin *marinus*, "of the sea," "marine"); note that the generic names *Marinobacter* Gauthier et al. 1992, *Marinobacterium* González et al. 1997, *Marinococcus* Hao et al. 1985, *Marinomonas* van Landschoot and De Ley 1984, and *Marinospirillum* Satomi et al. 1998 were validly published before the current guidelines of Appendix 9 of the ICNP took effect, although the recommendations about connecting vowels were similar in the previous revisions of the code.
- *Ovatibacter* was proposed to replace *Ovatusbacter* (Latin *ovatus*, "egg-shaped").

No connecting vowel should be added in compound names before a word element that starts with a vowel. Therefore, the list editors made the following corrections:

- *Caldarchaeum* was proposed to replace *Caldiarchaeum*.
- *Prometharchaeum* was proposed to replace *Promethoarchaeum*.
- *Syntropharchaeales* was proposed to replace *Syntrophoarchaeales*.

When forming compound names, the word stems are to be used except for the last word element. Here are examples of corrections proposed to comply with this guideline:

- *Hematobacterium* was proposed to replace *Hemobacterium* (Greek *haima*, genitive *haimatos*, stem transliterated as *haemato–*).

- *Mycoplasma haematohominis* was proposed to replace *Mycoplasma hemohominis*.
- *Epulonipiscium* was proposed to replace *Epulopiscium* (Latin, *epulo*, "a guest at a banquet," genitive *epulonis*, stem *epuloni–*).

The list editors also changed proposed *Candidatus* names that were grammatically correctly formed but were problematic for other reasons. Notable examples can be found in specific epithets of *Candidatus* taxa associated with animals. In such cases, the genitive of the animal genus name may be used as epithet of the *Candidatus*. For example, in the validly published names *Campylobacter hominis* and *Prevotella hominis*, isolated from *Homo sapiens*, the Latin *hominis* means "of a man." Naming a species *Campylobacter sapiens* would make little sense, even if the name complies with Rule 12c(1) and the other rules of the ICNP; it implies that the *Campylobacter* is wise (Latin *sapiens*, "wise"). Even worse would be the genitive *sapientis* ("of wise"); adjectives as specific epithets must be used in the nominative case (Rule 12c(1) of the 2022 revision of the ICNP, Oren et al. 2023).

- *Blochmanniella myrmotrichis* was proposed to replace *Blochmannia rufipes* (the organism was found in the ant species *Myrmothrix rufipes*, the Latin *rufipes* meaning "having red legs"; the ant may have red legs but the prokaryote associated with the ant does not; for the proposed correction of *Blochmannia* to *Blochmanniella*, see above).
- Phytoplasma citri ("of the *Citrus* tree") was proposed to replace *Phytoplasma aurantifolia*; the Neo-Latin *aurantifolia* means "having golden leaves," which is a property of the tree, not of the bacterium.
- *Profftia adelgis* was proposed to replace *Profftia virida*; the organism was found in the insect *Adelges viridis*, the Latin *viridis* meaning green; the feminine form of *viridis* is *viridis* and not *virida*, and it is the insect that is green, not the bacterium associated with it.
- *Xiphinematobacter americanus* ("American") was proposed to replace *Xiphinematobacter americani* ("of American") for a bacterium associated with the nematode *Xyphinema americanum*.

However, such corrections were not made for many of the *Candidatus* Burkholderia and *Candidatus* Paraburkholderia species listed in *Candidatus* list no. 1 (Oren et al. 2020b).

A few corrections of names of higher *Candidatus* taxa were proposed because the incorrect use of suffixes to form names of higher taxa as regulated by Rules 8 and 9 of the ICNP:

- Xenobiia was proposed to replace Xenobia as the name of the class of the genus *Xenobium*; the name of a class is formed by addition of the suffix *–ia* to the stem of the name of the type genus of the type order of the class, in this case Xenobi–.
- Nanogingivalales was proposed to replace Nanogingivales; the name of an order is formed by addition of the suffix *–ales* to the stem of the name of the type genus, in this case the *Candidatus* genus Nanogingivalis, stem Nanogingival–.

9.5 Toward the Establishment of a Separate Code of Nomenclature for the Uncultivated Majority of Prokaryotes

With the advancement of the technology to sequence DNA retrieved from the environment, genomes assembled from metagenomic data ("MAGs"), as well as single cell amplified genomes ("SAGs"), the exploration of the uncultivated majority of prokaryotes has made great progress in the last decade. By annotation of complete or nearly complete genomes, it is possible to deduce many properties of the organisms. Thus, the classification of uncultured microorganisms characterized using the new molecular techniques now enables the description of the large yet finite prokaryotic diversity in nature. Metagenomic data can therefore be used to propose new *Candidatus* taxa (Konstantinidis and Rosselló-Móra 2015).

Proposals were made to recognize genome sequences as the type material for taxonomic descriptions of prokaryotes. Emendation of the rules of the ICNP was proposed so that not only pure cultures but also genome sequences could serve as nomenclatural types. To verify the authenticity of the sequence, DNA samples could be deposited in public culture collections where feasible (Whitman 2015; Rossello-Mora et al. 2020). A formal proposal to expand the type material acceptable for valid publication of names of species, genera, and higher taxa under the rules of the ICNP to genome sequences and even gene sequences was published in 2016 (Whitman 2016). The proposal aimed at the establishment of a unified nomenclature for all prokaryotes, cultured as well as uncultured, while granting priority to *Candidatus* names based upon their date of publication in the IJSEM. To achieve this, many of the rules of the ICNP needed to be emended. For example, in Rule 15 (Table 2), the sentence "Whenever possible, the type of a species or subspecies is a designated strain" could be changed to "The type of a species or subspecies is a designated strain, sequence of genomic DNA, description, preserved specimen or illustration as described in Rule 18a." Rule 18a could be modified to state that the sequence of genomic DNA may also serve as the type material when it unambiguously identifies the species. The proposed new version of Rule 20a reads: "The nomenclatural type of a genus … is the type species or the sequence of one or more genes that unambiguously identifies the genus … A new part (c) was proposed to be added to Rule 30(3)"; "When a sequence is the type material, the accession number in a publicly available database or the sequence must be given. It is recommended that, when possible, a sample of the DNA be deposited in at least two publicly accessible service collections in different countries and the catalogue numbers be indicated" (Whitman 2016). It was further proposed that upon acceptance of sequence data as type material, all *Candidatus* names published before January 1, 2020, which are otherwise in accordance with the rules of the Code will have their priority based upon their date of publication in the IJSEM, either within a paper, a list of names of *Candidatus* taxa, or a Validation List (Whitman et al. 2019).

In addition to proposals to incorporate the nomenclature of the uncultivated majority under the rules of the ICNP, ideas for the establishment of a separate, independent nomenclatural system for uncultivated taxa were advanced, where names can be validly published and have priority while following nomenclature rules similar to those established for the cultured prokaryotes. A committee of experts, supported by an international microbiological society, should be formed to govern the new system (Konstantinidis et al. 2017). The two nomenclatural systems would run in parallel until they can be unified at a later stage.

The proposals by Konstantinidis et al. (2017) suffered from numerous misunderstandings of the way the ICNP works, especially the misconception that the ICSP governs the classification of prokaryotes (Oren and Garrity 2018). The rules of the ICNP govern nomenclature, not classification, and Principle 1(4) of the ICNP ("Nothing in this Code may be construed to restrict the freedom of taxonomic thought or action") is the key to the understanding of the nature of the code. It was also argued that pluralism will inevitably lead to chaos and to the destruction of now well-ordered nomenclature system governed by the ICNP (Oren and Garrity 2018; see also Konstantinidis et al. 2018). The question whether we can have two equivalent categories for species in the same nomenclature code, one for cultured and another for uncultured, was further discussed by Rosselló-Móra and Whitman (2019).

In 2020, an online public discussion was held of the proposals to modify the rules of the ICNP to accommodate the nomenclature of the uncultivated majority and allow the use of genome sequence data as type. More than 150 participants were included in the email thread, and collated comments summarizing the discussion were periodically circulated and posted on the ICSP website. The discussions were then summarized in an 80-page document that is available as Supplementary File 1 of Sutcliffe et al. (2020). Subsequently the ICSP members voted on the emendations to the ICNP proposed by Whitman (2016) and Whitman et al. (2019). The voting members were requested to decide on the proposals to extend the nature of the type acceptable for valid publication of a species or subspecies name to allow the use of complete or partial genome sequences as type, to articulate a general concept for what can serve as type for a species, and to allow valid publication of the name of a genus in the absence of a type species if the type is too ambiguous to circumscribe a species. All proposals were rejected by a large majority (Sutcliffe et al. 2020).

9.6 Development of the SeqCode: A Nomenclatural Code for Prokaryotes Described from Sequence Data

Following the decision of the ICNP not to enable valid publication of names of yet uncultivated prokaryotes based on gene or genome sequences as the nomenclatural types, an alternative Code of Nomenclature named the SeqCode (*Code of Nomenclature of Prokaryotes Described from Sequence Data*) was prepared to cover the nomenclature of the uncultivated majority. A consortium of scientists provided input toward the preparation of the SeqCode. Most nomenclature rules of the SeqCode are similar to those of the ICNP. Earlier published *Candidatus* names supported by DNA sequence information that meets the standards of the SeqCode would also be granted priority under the new code. It was not the intent of the initiators of the SeqCode to dissuade cultivation efforts: pure cultures or stable cocultures remain crucial, and if possible, such cultures should be made available from culture collections for the study of the physiology and growth characteristics of the living microorganisms (Murray et al. 2020). Under the SeqCode also cultivated prokaryotes can be named, including fastidious organisms that cannot be deposited into culture collections. Ultimately, if and when supported by the scientific community, the two nomenclature systems could be merged to yield a single, unified collection of validly published names.

The first draft of the SeqCode was prepared in 2020 and discussed at a series of international online workshops held in 2021 under the aegis of the International Society for Microbial Ecology (ISME). The first version of the SeqCode was launched in 2022 (Hedlund et al. 2022). It contains recommendations on minimal standards for DNA sequences. The SeqCode is shorter than the ICNP: several simplifications were implemented to make nomenclature more accessible and easier to apply (Whitman et al. 2022).

The SeqCode operates through the SeqCode Registry (2022), an online registration portal through which authors of new names can register nomenclatural types and link them to metadata, preferably prior to publication. The list curators then review and approve the names to become validly published. The SeqCode will not prevent validation of thousands of computer-generated names with little associated information; however, the initiators of the code admitted that such naming is of little value and should be discouraged as replacing alphanumeric labels with Latin names in the absence of further analyses will lead to the proliferation of confusing names: "In general, taxa should only be named when there is something to say about them" (Hedlund et al. 2022).

The yet-to-be-appointed Committee on the Systematics of Prokaryotes Described from Sequence Data (the "SeqCode Committee") will be responsible for the content of the SeqCode and administer the infrastructure required for maintenance of the SeqCode Registry. Its functions are similar to those of the ICSP for the cultivated prokaryotes. ISME will oversee and provide organizational support for the SeqCode Committee and other bodies to be appointed such as a SeqCode Reconciliation Commission with the authority to resolve disputes regarding the interpretations of its rules, similar to the function of the ICSP Judicial Commission (Hedlund et al. 2022).

The SeqCode and supplementary information are available online (SeqCode Publications 2022).

9.7 The Rules for Naming of Taxa Under the ICNP and the SeqCode Compared

Except for the ways names are validly published, the requirements of the SeqCode for naming prokaryotic taxa are similar to those of the ICNP. Here are the rules of the SeqCode that are most relevant to the naming of genera and species, comparing then with the similar provisions of the ICNP that were explained in further depth in Chap. 2.

General Consideration 5
The Code of Nomenclature of Prokaryotes Described from Sequence Data applies to the naming of all prokaryotes where the lower taxa (species and subspecies) are typified by a DNA sequence. This code is colloquially referred to as the SeqCode to distinguish it from the International Code of Nomenclature of Prokaryotes (ICNP), which applies to the naming of all prokaryotes where the lower taxa (species and subspecies) are typified by either a strain or illustration/description.

Principle 2
The nomenclature of prokaryotes is **not independent** of botanical, zoological, and viral nomenclature. When naming new taxa at the rank of genus or higher, names that are already regulated by the International Code of Zoological Nomenclature; the International Code of Nomenclature for algae, fungi, and plants; and the International Code of Virus Classification and Nomenclature must not be used.

The names formed under the SeqCode are **not independent** of the names regulated by the International Code of Nomenclature of Prokaryotes (ICNP) and formed before January 1, 2022. Before that date, legitimate names formed under the ICNP have priority. After that date, the rules of priority of the SeqCode are used to determine priority of names formed under the ICNP compared with names formed under the SeqCode.

As for genus names under the provisions of the ICNP, names of genera already in use in botany or zoology should not be used when naming new genera of prokaryotes under the SeqCode.

Principle 3
The scientific names of all taxa are Latin or Latinized words treated as Latin.

The SeqCode does not give a special status to classical Greek, as does Principle 3 of the ICNP.

Rule 8
The scientific names of all taxa must be treated as Latin and spelled only with the Latin alphabet. A species name is a binary combination of a genus name and a specific epithet; names of taxa above the rank of species are single words. Typographical signs, numbers, and additional characters cannot be used.

Principle 3 and Rule 8 of the SeqCode are very similar to Principle 3 and Rule 6 of the ICNP, so that names are treated as Latin under both codes. Alphanumeric identifiers such as, for example, SAR11 may be convenient as placeholders and can be used for communicating the underlying taxonomic or phylogenetic relationships, but are not acceptable as names under the SeqCode. Thus, the need for correctly formed Latin-based names is identical for names under the rules of the ICNP, *Candidatus* names, and names validated under the SeqCode. Recommendations for good practices for naming uncultivated taxa have been proposed by Chuvochina et al. (2019). Implementation of the SeqCode requires the same kind of nomenclature quality control as applied to names validly published under the provisions of the ICNP and now also implemented in the published lists of *Candidatus* taxa (Oren 2021).

Recommendation 9
To form new prokaryotic names, authors are advised as follows:

1. Names that are very long or difficult to pronounce should be avoided.
2. Names should differ by at least three characters from existing names of genera or species within the same genus.
3. Languages other than Latin should be avoided when Latin equivalents exist or can be constructed by combining Latin word elements. Exceptions include names derived from local items such as foods, drinks, geographic localities, and other names for which no Latin words exist.
4. Authors should not name organisms after themselves. If names are formed from personal names, they should contain only the name of one person. They may contain the untruncated family and/or first names.
5. All personal genus names should be feminine regardless of the gender identity of the person they commemorate.
6. Names should not be deliberately contentious or abusive of any person, race, religion, political belief, or ideology.
7. Names that include mnemonic cues are preferred because they promote learning and memory.

Recommendations 9.1 and 9.4 are similar to Recommendations 6.1 and 6.7 of the 2022 revision of the ICNP (Oren et al. 2023). Note that Recommendation 6.3 of the SeqCode does not give a special status to classical Greek, while the ICNP states: words from languages other than Latin or Classical Greek should be avoided if equivalents exist in Latin or Classical Greek or can be constructed by combining word elements from these two languages. Recommendation 9.5 of the SeqCode parallels Appendix 9, section D (3)(a), of the ICNP.

Rule 10
The name of a genus is a noun or adjective used as a noun, in the singular number and written with an initial capital letter.

The SeqCode does not explicitly state that names of genera may be composed in an arbitrary manner, as allowed by Rule 10a of the ICNP. Arbitrarily formed genus names are incidentally mentioned in Rule 50 of the SeqCode: "Arbitrarily formed genus names or vernacular names used as genus names take the gender assigned to them by their authors."

Rule 11
The name of a species is a binary combination consisting of the name of the genus followed by a single word known as a species epithet. The genus part of the name must begin with an initial capital letter, and the species epithet must begin with a lowercase letter.

Rule 12
A species epithet must be related to the genus name in one of three ways.

1. As an adjective. Example: *aureus* in *Staphylococcus aureus*
2. As a substantive (noun) in apposition in the nominative case. Example: *Desulfovibrio gigas*
3. As a noun in the genitive case. Example: *coli* in *Escherichia coli*

Rules 10, 11, and 12 of the SeqCode are similar to Rules 10a, 12a, and 12c of the ICNP. Thus, the ways generic names and specific epithets are formed are comparable under both codes. However, the SeqCode does not explicitly state that a specific epithet may be composted arbitrarily (ICNP Rule 12c). Curiously, the requirement of ICNP Rule 12c(1) that an adjectival specific epithet must agree in gender with the generic name (as in Latin, Classical Greek, and numerous modern languages) is only a recommendation in Rule 12 of the SeqCode.

Rule 24a
Effective publication under the SeqCode means that the name and evidence for the taxon have been published in a peer-reviewed journal or book.

Rule 24b
When a name of a new taxon is published in a work written in a language other than English, the author(s) should include a description in English in the publication.

Rule 24c
The following are not accepted as forms of effective publication.

1. Communication of new names at a meeting, minutes of a meeting, or abstracts of papers presented at meetings
2. Placing of names in listings or catalogs of collections
3. Reports in ephemeral publications, newsletters, white papers, self-published papers, or nonscientific periodicals
4. A published patent application or issued patent including the name
5. A database containing names associated with a sequence or metadata
6. Electronic material available in advance of publication (e.g., papers in press or preprints)

Rule 24a, 24b, and 24c of the SeqCode parallel Rules 25a and 25b of the ICNP.

Rule 26
A name of a new taxon, or a new combination for an existing taxon, is not validly published unless the following criteria are met:

1. The name is effectively published under the rules of the SeqCode.

2. The name is registered in the SeqCode Registry, along with mandatory data fields listed below.
3. The type of the taxon is clearly designated. In the case of species or subspecies, the type sequence is deposited according to Rule 18a and the accession number cited.
4. The taxonomic rank is designated.
5. The derivation (etymology) of a new name (and if necessary of a new combination) is given wherein one or more distinguishable roots are identified. Roots can originate from any language in use or extinct (see also Recommendation 9).

The SeqCode Registry has the same function as the IJSEM has for valid publication of names under the ICNP. Rule 26 of the SeqCode resembles Rule 27 of the ICNP, including the need to provide the etymology of the proposed names.

Note 1 When a new species or a new combination results in the proposal of a new genus, both the new genus name and the new species name or the new combination must be validly published. Publication of the new species epithet or new combination alone does not constitute valid publication of the new genus name.

Note 2 When possible, authors are recommended to include the SeqCode Registry identifier in the effective publication.

Note 3 If the information provided in the registration and the effective publication differ, the registration is considered definitive.

Note 1 to Rule 26 of the SeqCode is similar to Note 2 to Rule 27 of the ICNP.

Recommendation 26

It is recommended that the name, etymology, type information, and diagnosis of the novel taxon should be clearly identifiable in a designated section of the effective publication (i.e., the section termed the "protolog" by some microbial taxonomists). Authors are encouraged to provide additional information describing the taxon such as predicted or known physiological characteristics, ecological data, location, and additional metadata. Authors are also encouraged to submit metadata with the type sequence in one of the INSDC [International Nucleotide Sequence Database Collaboration] databases.

The word "protologue" is mentioned in the new version of Rule 27 in the 2022 revision of the ICNP (Oren et al. 2023).

Section 9. Orthography

Rule 46

All names comprise only the 26 letters of the ISO basic Latin alphabet. Diacritic signs are not to be used.

Rule 47

Any name or epithet should be written in conformity with the spelling of the word from which it is derived and in accordance with the rules of Latin grammar.

Exceptions are provided for typographic and orthographic errors and orthographic variants.

Note 1 Consult Appendix 9 of the ICNP for recommendations on forming properly Latinized names.

Note 2 In the SeqCode, an orthographic variant is a name (or epithet) applied to the same type that differs from another name only in transliteration into Latin of the same word from a language other than Latin or in its grammatical correctness. Changes in suffixes for consistency with the names of higher taxa are orthographic variants. Names transliterated from the same word and based on different types are not orthographic variants.

Typographic and Orthographic Errors

Rule 48
The original spelling of a name or epithet must be retained, except for typographical or orthographic errors.

An unintentional typographical or orthographic error later corrected by the author is to be accepted in its corrected form without affecting the status and date of valid publication. It can also be corrected by a subsequent author who may or may not mention that the spelling is corrected. However, the abbreviation "corrig." (corrigendum) may be appended to the name if an author wishes to draw attention to the correction. Succeeding authors may be unaware that the original usage was incorrect and use the spelling of the original author(s). Other succeeding authors may follow the correction of a previous author or may independently correct the spelling themselves, but in no case is the use of corrig. regarded as obligatory. None of these corrections affect the status and date of validation.

Note. The liberty of correcting a name or epithet must be used with reserve, especially if the change affects the first syllable and above all the first letter of the name or epithet.

Personal Names

Rule 49
The genitive and adjectival forms of a personal name are treated as different epithets and not as orthographic variants unless they are so similar as to cause confusion.

Gender of Names

Rule 50
The gender of genus names is governed by the following.

1. A Latin or Latinized genus name retains the gender of its language of origin. Authors must give the gender of any proposed genus name. In cases where the classical gender varies, the author has the right of choice between the alternatives.
2. Genus names that are compounds from two or more Latin words take the gender of the last component of the compound word.

3. Arbitrarily formed genus names or vernacular names used as genus names take the gender assigned to them by their authors.

Rules 48, 49, and 50 of the SeqCode parallel Rule 61, 63, and 65 of the ICNP.

9.8 How Many Names of Uncultivated Prokaryotic Taxa Will Be Needed in the Future?

At the time this chapter was written (November 2022), names of 22,805 species of prokaryotes with standing in the nomenclature had been published, or 18,825 names when excluding synonyms (LPSN 2022). This number is very small when compared to the total number of eukaryotes documented, which exceeds two million (Hobern et al. 2021; Catalogue of Life 2022). Nearly one million insect species have been described and named, and a single plant family, the *Orchidaceae* with about 28,000 species has more species than the number of validly published species names of bacteria and archaea combined.

The number of complete or draft genome sequences of uncultivated prokaryotes deposited in the Genome Taxonomy Database (GTDB) (Chaumeil et al. 2020) already exceeds the number of genomes derived from cultivated organisms (Genome Taxonomy Database (GTDB) 2022). Thanks to the ever-improving technology enabling massive sequencing of DNA retrieved from the environment and genome assembly, the numbers of genomes derived from cultures will soon be a small percentage of the total.

The question must therefore be asked, how many different species of prokaryotes exist on our planet. The estimates vary by several orders of magnitude, depending on the statistical calculation method and also on the species concept used. There is no universally agreed species concept for prokaryotes, and therefore the estimates depend on what species definition is favored by the authors. Estimates published more than two decades ago varied between several millions (e.g., $<2 \times 10^6$ in the sea, $\sim 4 \times 10^6$ in a ton of soil) (Curtis et al. 2002) to more than one billion (Dykhuizen 1998). More recent predictions diverged even more, with high numbers of hundreds of millions or billions (Wiens 2021) or even as many as a trillion (10^{12}) microbial species (Locey and Lennon 2016) to a few millions only (0.8–1.6 million based on hypervariable region V4 of the 16S rRNA gene (calculated at 97% similarity), estimated to be equivalent to 2.2–4.3 million full-length operational taxonomic units worldwide (Louca et al. 2019). The implications of these data for our views of the bacterial and archaeal diversity on Earth were further discussed by Amann and Rosselló-Móra (2016), Hugenholtz et al. (2021), Rosselló-Móra and Whitman et al. (2019), Schloss et al. (2016), Sutcliffe et al. (2021), and Yarza et al. (2014).

Naming the uncultivated majority of prokaryotes will therefore be a huge task. The task is not necessarily impossible when remembering that the botanists and zoologists have named over two million plant and animal species since Linnaeus introduced the binomial system for biological species in the middle of the eighteenth century. Still, the question should be asked whether indeed all those yet-unnamed species of prokaryotes need to be given Latin names. It may be preferable to name only species that are of sufficient interest to the scientific community, following a more in-depth characterization of their properties (Hugenholtz et al. 2021).

9.9 Approaches to the Naming of Large Numbers of Uncultivated Genera and Species

There is a general consensus that naming will remain important to provide labels for biological entities (Rosselló-Móra and Whitman 2019). For cultivated taxa, the rules of the ICNP apply, and the same rules were applied to *Candidatus* taxa (Appendix 11 of the ICNP). Similar rules based on Latin or latinized words, Latin grammar, and similar orthography rules were adopted in the SeqCode (Hedlund et al. 2022). Alphanumeric placeholder labels are hard-to-remember, and are not acceptable under either code.

In view of the large number (millions to possibly billions) of yet uncultivated prokaryotes that are waiting to be named in the future, the question must be asked how it will be possible to name all these taxa. It is not difficult to create names for a few novel organisms discovered and characterized during a research project, but generating large numbers of well-formed and meaningful names is time-consuming and tedious. Moreover, the number of experts conversant with Latin and Greek who are also familiar with the rules of the ICNP and the SeqCode and are able and willing to perform nomenclatural quality control is very small. Prokaryotic nomenclature, as currently configured, obviously cannot keep up with the future need for new well-formed names (Whitman et al. 2022).

In an attempt to provide at least a partial solution to the problem, a computer program was written that can generate large numbers (millions and more) of grammatically correct genus and species names that meet the orthography standards of the ICNP and the SeqCode. The program, written in the Python language, was named Gan: the "Great Automatic Nomenclator," paraphrasing "The Great Automatic Grammatizator," the title of a short story written in 1953 by Roald Dahl, and also referring to the Hebrew word *gan* ("garden"), reflecting the fertile productivity of the program (Pallen et al. 2021; Oren 2021; Telatin 2022). The script takes, as its input, tables of Latin, Neo-Latin, or Greek roots in a specified format. The program can generate large numbers of names by combinatorial concatenation of a small number of Latin or Greek roots, each name typically consisting of three or four word elements. By using three roots (e.g., the first root specifying a general context such as a host, a general environment, a person, or a place; the second root specifying a more specific context, such as an organ or tissue or a specific environment; and a final word element such as *–cola*, *–microbium*, *–monas* ("a form"), *–morpha*, *–plasma*, *–soma*, *–vivens*, etc.), very large numbers of correctly formed names can easily be produced. The endings listed above can be applied to any bacterium or archaeon, also when no information about their morphology exists, as is generally the case for uncultivated taxa for which a genome sequence is available and little else. If the shape of the organisms is known, endings such as *–bacillus*, *–coccus*, *–spira*, *–spirillum*, *–vibrio*, etc. can also be used. The program follows the rules specified in Appendix 9 of the ICNP for the use of connecting vowels: *-i-* following a word element of Latin origin and *-o-* when the preceding word element is of Greek origin, with no connecting vowel being added when the following word element starts with a vowel (with a few exceptions). The gender of adjectives used as specific epithet can also be adjusted to comply with Rule 12c(1) of the ICNP. Version 1.0 of the Gan program requires extensive curation of the input files and produces rather basic outputs that need some checking and editing. However, subsequent versions of the program may become more functional and user-friendly.

The power of the combinatorial approach used by the Gan program was shown in a paper describing nearly 700 new species (some cultivated, most with *Candidatus* status, identified from metagenome-assembled genomes) and more than 150 new genera from chicken feces (Gilroy et al. 2021). Word elements used in the input files included (for bird or chicken) the Latin *avis*, *gallus*, *gallina*, and *pullus* and the Greek *alektryon*, *kottos*, and *ornis*; (for intestine) the Latin *intestinum* and the Greek *enteron*; and (for excrement) the Latin *egeries*, *excrementum*, *faex*, *merda*, and *stercus* and the Greek *aphodos* and *kakke*. This enabled formation of compound *Candidatus* genus names such as *Alectryocaccobium*, *Aphodoplasma*, and *Egereicola* and specific epithets such as *avistercoris*, *excrementipullorum*, *faecavium*, *faecigallinarum*, *intestinigallinarum*, *merdigallinarum*, *ornithocaccae*, and *pullistercoris*.

The number of genus names and specific epithets names that can be created using such a combinatorial approach can be greatly increased by addition of prefixes such as *allo–*, *alteri–*, *meta–*, *neo–*, *novi–*, *paeni–*, *para–-*, and *pseudo–*, meaning "other," "beside," "new," "almost," "next to," and "false." Pallen et al. (2021) collated a list of 38 prefixes that can be used for this purpose. Use of such prepositions in prokaryotic names is not novel: *pseudo–* was first used in *Pseudomonas* Migula 1894, *neo–* in *Neorickettsia* Philip et al. 1953, and *para–* in *Paracoccus* Davis 1969. An example of the use of such a prefix in a specific epithet is *Paenibacillus allorhizo-sphaerae* ("related to but distinct from *Paenibacillus rhizosphaerae*") and *Leuconostoc* (*Weissella*) *paramesenteroides* ("beside *Leuconostoc mesenteroides*"). There even are cases where a genus name has two such prefixes, as shown by *Allopseudarcicella* ("another false *Arcicella*") and *Parapseudoflavitalea* ("beside a false *Flavitalea*"). Allowing the use of two prefixes further expands the number of names that can be generated from a small number of word elements.

The names generated by the Gan program are memorable linguistic labels as the names have a meaning ("a form from excrement," "of the excrement of chickens," etc.). This is obviously not the case for names such as *Candidatus* Cofrixana mapaxosa, *Candidatus* Didresosa oboposa, and *Candidatus* Sufrabetta lutexaria, examples of names automatically generated by Pallen and coworkers during their effort to name over 65,000 previously unnamed taxa within the Genome Taxonomy

Database (GTDB) (2022) (Pallen et al. 2022). The names were generated using a Python script that basically uses word components from all potential strings of consonants/consonant clusters and vowels found in the original Latin alphabet, with final word components that comply with phonotactic and grammatical norms of Latin. For ease of use, these final word components were restricted to the feminine gender and the first declension. The names thus consist of strings of letters with declinable Latin feminine suffixes. Still, these meaningless names are well-formed and meet the standards of the ICNP (Rules 10a and 12c, which allow named formed in an arbitrary manner), but not necessarily of the SeqCode. The etymologies were given as N.L. fem. n. .... a name/specific epithet created arbitrarily while preserving the phonotactics and morphology of Latin. The specific epithets were formed in accordance with Rule 12c(2) of the ICNP: as a noun in the nominative case in apposition. In some cases, the ending happens to resemble that of names that have a meaning in Latin, but there is no connection between genus names such as *Candidatus* Hufonitoga or *Candidatus* Hufonivita and the Latin nouns *toga* ("outer garment," "toga") or *vita* ("life").

Even though such names are devoid of etymology or meaning, the authors claimed that those newly created names "represent an important resource for the microbiology community, facilitating communication between bioinformaticians, microbiologists, and taxonomists while populating the emerging landscape of microbial taxonomic and functional discovery with accessible and memorable linguistic labels."

9.10 Concluding Comments

The system of naming uncultivated prokaryotes using the *Candidatus* concept proposed in the middle of the 1990s still works well (Oren 2021). Registration of *Candidatus* names in annual lists published in the IJSEM (Oren et al. 2020a, b; Oren and Garrity 2021b; Oren and Garrity 2022; Oren 2022) with proposed linguistic corrections where necessary (Oren 2017), as well as inclusion of *Candidatus* names in online databases, now ensures that a complete or at least a nearly complete record of *Candidatus* names published during more than a quarter of a century is readily available.

Recently, Pallen et al. (2021) published a SWOT (strengths, weaknesses, opportunities, and threats) analysis of the status *Candidatus* for uncultured taxa of prokaryotes. He concluded that increased use of *Candidatus* names provides a ready solution to the urgent challenge of naming thousands of recently recognized uncultured species and provides an alternative to the recently launched SeqCode. The key strengths of the status *Candidatus* are that it has been in long use and is specified in Appendix 11 of the ICNP. It is also widely used in the NCBI taxonomy browser (National Center for Biotechnology Information 2022), and the names are documented in the List of Prokaryotic names with Standing in the Nomenclature and the NamesforLife websites (LPSN 2022; NamesforLife 2022). If taxa are subsequently cultured, the term *Candidatus* can simply be deleted from the existing name, if it was formed in accordance with the rules of the ICNP. The most important weakness is that *Candidatus* names are not validly published and lack priority under the rules of the ICNP. The circumstances governing use of *Candidatus* status are poorly specified in the ICNP, and it is unclear when its use is allowed or denied, mandatory or merely optional. It also remains unclear whether the status *Candidatus* can be applied when viable, but impure cultures of a prokaryote are available and could be deposited in culture collections. In addition, the status of prokaryotic species that can be propagated only in association with host cells needs to be clarified. Rule 18a of the ICNP is not clear on this point. Thus, names of many species of *Chlamydia* and *Rickettsia* were validly published, but many others were described as *Candidatus* taxa. As opportunities, Pallen listed that de facto priority of a name can be established by the deposition of a genome sequence into the public domain and a description of a taxon in a peer-reviewed publication. The establishment of curated lists of *Candidatus* names in the IJSEM provides an opportunity for improving the linguistic quality of *Candidatus* names. The launch of the SeqCode that gives de jure priority to names associated with a type sequence presents a threat to the continued use of the status *Candidatus*. It ignores the current status quo, in which *Candidatus* names already have de facto priority in the literature. Time will tell whether authors, journals, editors, and online resources will prefer the SeqCode over the *Candidatus* concept as presented in Appendix 11 of the ICNP.

When large numbers of new *Candidatus* names are needed, automated high-throughput creation of names formed by combining Latin, Greek, and Neo-Latin word elements in a correct way that complies with the demands of the ICNP and the SeqCode is already possible (Pallen et al. 2021; Gilroy et al. 2021). Machine-generated names can be formed according to a highly standardized pattern, are often long, and are unlikely to arouse general interest in prokaryotic nomenclature. It is even possible to generate innumerable meaningless names consisting of more or less random strings of consonants/consonant clusters and vowels that meet the standards of the ICNP as long as they are treated as Latin, and this can easily be done by adding declinable Latin suffixes (Pallen et al. 2022).

Principle 4 of the ICNP states: "The primary purpose of giving a name to a taxon is to supply a means of referring to it rather than to indicate the characters or the history of the taxon." The SeqCode has a slightly different version of Principle 4, with the following addition: "A secondary consideration is that names should aid memorability." The future will show whether the correctly formed but meaningless names proposed for over 65,000 *Candidatus* names for genome sequences deposited in the Genome Taxonomy Database are indeed more memorable than the earlier alphanumeric designations.

Of course it remains possible to create names to boost the general interest in prokaryotic nomenclature (Oren 2020). This also applies to *Candidatus* taxa. Here is a small selection of *Candidatus* names that, in the opinion of the author, deserve to be mentioned as examples of attractive names that tell stories about the nature of the organisms or of the environments where they were found:

- *Candidatus* Desulforudis audaxviator, the epithet meaning "a courageous traveller." It is based on Jules Verne's *Voyage au centre de la Terre* (*Journey to the Center of the Earth*) and refers to its habitat: the deep subsurface.
- *Candidatus* Umbricyclops ("a round eye living in the shade"), highlighting the ovoid shape of the cells that live in shaded sediments.
- *Candidatus* Vampirococcus lugosii, a predatory bacterium, named after Bela Lugosi, who played the role of the vampire in the iconic 1931 film Dracula.

Authors who describe novel interesting but yet uncultivated taxa are therefore encouraged to search for attractive names within the framework that the rules of the ICNP and the SeqCode provide.

References

Amann R, Rosselló-Móra R (2016) After all, only millions? MBio 7:e00999–e00916

Anonymous (1999) Validation list no. 71. Validation of publication of new names and new combinations previously effectively published outside the IJSB. Int J Syst Bacteriol 49:1325–1326

Catalogue of Life (2022) Annual Release 2022. https://www.catalogueoflife.org/2022/08/30/release. Accessed 30 Oct 2022

Chatton E, Pérard C (1913) Schizophytes du caecum du cobaye. I. *Oscillospira guilliermondi* n.g., n.sp. C R Seances Soc Biol (Paris) 74:1159–1162

Chaumeil P-A, Mussig AJ, Hugenholtz P et al (2020) GTDB-Tk: a toolkit to classify genomes with the genome taxonomy database. Bioinformatics 36:1925–1927

Chuvochina M, Rinke C, Parks DH et al (2019) The importance of designating type material for uncultured taxa. Syst Appl Microbiol 42:15–21

Curtis TP, Sloan WT, Scannell JW (2002) Estimating prokaryotic diversity and its limits. Proc Natl Acad Sci U S A 99:10494–10499

De Vos P, Trüper HG (2000) Judicial Commission of the International Committee on Systematic Bacteriology IXth International (IUMS) Congress of Bacteriology and Applied Microbiology. Minutes of the meetings, 14, 15 and 18 August 1999, Sydney, Australia. Int J Syst Evol Microbiol 50:2239–2244

Dirren S, Posch T (2016) Promiscuous and specific bacterial symbiont acquisition in the amoeboid genus *Nuclearia* (Opisthokonta). FEMS Microbiol Ecol 92:fiw105

Dykhuizen DE (1998) Santa Rosalia revisited: why are there so many species of bacteria? Antonie van Leeuwenhoek 73:25–33

Ehrenberg CG (1838) Die Infusionsthierchen als vollkommene Organismen. L. Voss, Leipzig

Fagen JR, Leonard MT, Coyle JF et al (2014) *Liberibacter crescens* gen. nov., sp. nov., the first cultured member of the genus *Liberibacter*. Int J Syst Evol Microbiol 64:2461–2466

Frederiksen W (1995) Judicial Commission of the International Committee on Systematic Bacteriology. Minutes of the meetings, 2 and 6 July 1994, Prague. Czech Republic Int J Syst Bacteriol 45:195–196

Genome Taxonomy Database (GTDB) (2022). https://gtdb.ecogenomic.org. Accessed 30 Oct 2022

Gilroy R, Leng J, Ravi A et al (2022) Metagenomics investigation of the equine faecal microbiome reveals extensive taxonomic and functional diversity. PeerJ 10:e13084

Gilroy R, Ravi A, Getino M et al (2021) Extensive microbial diversity within the chicken gut microbiome revealed by metagenomics and culture. PeerJ 9:e10941

Hedlund BP, Chuvochina M, Hugenholtz P et al (2022) SeqCode: a nomenclatural code for prokaryotes described from sequence data. Nat Microbiol 7:1702–1708

Hobern D, Barik SK, Christidis L et al (2021) Towards a global list of accepted species VI: the catalogue of life checklist. Org Divers Evol 21:677–690
Hugenholtz P, Chuvochina M, Oren A et al (2021) Microbial taxonomy and nomenclature in the age of big sequence data. ISME J 15:1879–1892
Konstantinidis KT, Rosselló-Móra R (2015) Classifying the uncultivated microbial majority: a place for metagenomic data in the Candidatus proposal. Syst Appl Microbiol 38:223–230
Konstantinidis KT, Rosselló-Móra R, Amann R (2017) Uncultivated microbes in need of their own taxonomy. ISME J 11:2399–2406
Konstantinidis KT, Rosselló-Móra R, Amann R (2018) Reply to the commentary "Uncultivated microbes—in need of their own nomenclature?". ISME J 12:653–654
Kützing FT (1843) Phycologia generalis oder Anatomie, Physiologie und Systemkunde der Tange. Mit 80 farbig gedruckten Tafeln, gezeichnet und gravirt vom Verfasser. F.A. Brockhaus, Leipzig
Labeda DP (2000) International Committee on Systematic Bacteriology. IXth International (IUMS) Congress of Bacteriology and Applied Microbiology. Minutes of the meetings, 14 and 17 August 1999, Sydney, Australia. Int J Syst Evol Microbiol 50:2245–2247
Lapage SP, Sneath PHA, Lessel EF et al (eds) (1975) International code of nomenclature of bacteria and statutes of the International Committee on Systematic Bacteriology and Statutes of the Bacteriology Section of The International Union of Microbiological Societies. Bacteriological Code. 1976 Revision. Published for the International Union of Microbiological Societies by the American Society for Microbiology. Washington, D.C.
Lapage SP, Sneath PHA, Lessel EF et al (eds) (1992) International code of nomenclature of bacteria and statutes of the International Committee on Systematic Bacteriology and Statutes of the Bacteriology and Applied Microbiology Section of the International Union of Microbiological Societies. Bacteriological Code. 1990 Revision. Published for the International Union of Microbiological Societies by the American Society for Microbiology. Washington, D.C.
Latinitas. Opus Fundatum in Civitate Vaticana. Parvvm Verborvm Novatorvm Léxicvm. apparavit Cletus Pavanetto. https://www.vatican.va/roman_curia/institutions_connected/latinitas/documents/rc_latinitas_20040601_lexicon_it.html. Accessed 15 November 2022
Link HF (1809) Observationes in ordines plantarum naturales. Dissertatio prima complectens Anandrarum ordines Epiphytas, Mucedines, Gastromycos et Fungos. Herausgegeben von der Gesellschaft naturforschender Freunde zu Berlin 3:3–42
Locey KJ, Lennon JT (2016) Scaling laws predict global microbial diversity. Proc Natl Acad Sci U S A 113:5970–5975
Louca S, Mazel F, Doebeli M et al (2019) A census-based estimate of Earth's bacterial and archaeal diversity. PLoS Biol 17:e3000106
LPSN – List of prokaryotic names with standing in nomenclature (2022) lpsn.dsmz.de; bacterio.net. Accessed 30 Oct 2022
Martinson VG, Gawryluk RMR, Gowen BE et al (2020) Multiple origins of obligate nematode and insect symbionts by a clade of bacteria closely related to plant pathogens. Proc Natl Acad Sci U S A 117:31979–31986
Murray AE, Freudenstein J, Gribaldo S et al (2020) Roadmap for naming uncultivated Archaea and Bacteria. Nat Microbiol 5:987–994
Murray RG, Schleifer KH (1994) Taxonomic notes: a proposal for recording the properties of putative taxa of procaryotes. Int J Syst Bacteriol 44:174–176
Murray RG, Stackebrandt E (1995) Taxonomic note: implementation of the provisional status Candidatus for incompletely described procaryotes. Int J Syst Bacteriol 45:186–187
Nadal-Jimenez P, Siozios S, Halliday N et al (2022) *Symbiopectobacterium purcellii*, gen. nov., sp. nov., isolated from the leafhopper *Empoasca decipiens*. Int J Syst Evol Microbiol 72:005440
NamesforLife (2022). https://www.namesforlife.com. Accessed 16 November 2022
National Center for Biotechnology Information *(*2022*)*. https://www.ncbi.nlm.nih.gov/taxonomy. *Accessed 16 November 2022*
Oren A (2017) A plea for linguistic accuracy—also for *Candidatus* taxa. Int J Syst Evol Microbiol 67:1085–1094

Oren A (2020) Prokaryotic names: the bold and the beautiful. FEMS Microbiol Lett 367:fnaa096
Oren A (2021) Nomenclature of prokaryotic *Candidatus* taxa: establishing order in the current chaos. New Microbes New Infect 44:100932
Oren A (2022) Lists of names of prokaryotic *Candidatus* taxa. Candidatus List no. 4. Int J Syst Evol Microbiol 72:005545
Oren A, Arahal DR, Göker M et al (2023) International code of nomenclature of prokaryotes. Prokaryotic Code (2022 Revision). Int J Syst Evol Microbiol 73:005585
Oren A, Arahal DR, Rosselló-Móra R et al (2021) Emendation of rules 5b, 8, 15, and 22 of the international code of nomenclature of prokaryotes to include the rank of phylum. Int J Syst Evol Microbiol 71:004851
Oren A, Garrity GM (2018) Uncultivated microbes—in need of their own nomenclature? ISME J 12:309–311
Oren A, Garrity GM (2021a) Valid publication of the names of forty-two phyla of prokaryotes. Int J Syst Evol Microbiol 71:005056
Oren A, Garrity GM (2021b) Lists of names of prokaryotic *Candidatus* taxa. Candidatus List no. 2. Int J Syst Evol Microbiol 71:004671
Oren A, Garrity GM (2022) Lists of names of prokaryotic *Candidatus* taxa. Candidatus list no. 3. Int J Syst Evol Microbiol 72:005186
Oren A, Garrity GM, Parker CT et al (2020b) Lists of names of prokaryotic *Candidatus* taxa. Candidatus List no. 1. Int J Syst Evol Microbiol 70:3956–4042
Oren A, Garrity GM, Trujillo ME (2020a) Registration of names of prokaryotic Candidatus taxa in the IJSEM. Int J Syst Evol Microbiol 70:3955
Pallen MJ (2021) The status *Candidatus* for uncultured taxa of *bacteria* and *archaea*: SWOT analysis. Int J Syst Evol Microbiol 71:005000
Pallen MJ, Rodriguez-R LM, Alikhan NF (2022) Naming the unnamed: over 65,000 *Candidatus* names for unnamed *archaea* and *bacteria* in the genome taxonomy database. Int J Syst Evol Microbiol 72:005482
Pallen MJ, Telatin A, Oren A (2021) The next million bacterial names. Trends Microbiol 29:289–298
Parker CT, Tindall BJ, Garrity GM (eds) (2019) International code of nomenclature of prokaryotes. Prokaryotic Code (2008 revision). Int J Syst Evol Microbiol 69:S1–S111
Rappé MS, Connon SA, Vergin KL et al (2002) Cultivation of the ubiquitous SAR11 marine bacterioplankton clade. Nature 418:630–633
Rossello-Mora R, Konstantinidis KT, Sutcliffe I et al (2020) Opinion: response to concerns about the use of DNA sequences as types in the nomenclature of prokaryotes. Syst Appl Microbiol 43:126070
Rosselló-Móra R, Whitman WB (2019) Dialogue on the nomenclature and classification of prokaryotes. Syst Appl Microbiol 42:5–14
Roth AW (1797) Catalecta botanica quibus plantae novae et minus cognitae describuntur atque illustrantur, Fasc. 1. Bibliopolio I.G. Mulleriano, Lipsiae
Salman V, Amann R, Girnth AC et al (2011) A single-cell sequencing approach to the classification of large, vacuolated sulfur bacteria. Syst Appl Microbiol 34:243–259
Schewiakoff W (1893) Über einen neuen bacterienähnlichen Organismus des Süsswassers. Habilitationsschrift, Universität Heidelberg. C. Winter, Heidelberg
Schloss PD, Girard RA, Martin T et al (2016) Status of the archaeal and bacterial census: an update. MBio 7:e00201–e00216
Schulz HN, Brinkhoff T, Ferdelman TG et al (1999) Dense populations of a giant sulfur bacterium in Namibian shelf sediments. Science 284:493–495
SeqCode Publications (2022). https://disc-genomics.uibk.ac.at/seqcode/page/publications. Accessed 30 Oct 2022
SeqCode Registry (2022). https://seqco.de. Accessed 30 Oct 2022
Skerman VBD, Sneath PHA, McGowan V (1980) Approved lists of bacterial names. Int J Syst Bacteriol 30:225–420

Sutcliffe IC, Dijkshoorn L, Whitman WB, on behalf of the ICSP Executive Board (2020) Minutes of the International Committee on Systematics of Prokaryotes online discussion on the proposed use of gene sequences as type for naming of prokaryotes, and outcome of vote. Int J Syst Evol Microbiol 70:4416–4417

Sutcliffe IC, Rosselló-Móra R, Trujillo ME (2021) Addressing the sublime scale of the microbial world: reconciling an appreciation of microbial diversity with the need to describe species. New Microbes New Infect 43:100931

Telatin A (2022) Gan: The Great Automatic Nomenclator. https://github.com/telatin/gan. Accessed 30 Oct 2022

Volland J-M, Gonzalez-Rizzo S, Gros O et al (2022) A centimeter-long bacterium with DNA contained in metabolically active, membrane-bound organelles. Science 376:1453–1458

Whitman WB (2015) Genome sequences as the type material for taxonomic descriptions of prokaryotes. Syst Appl Microbiol 38:217–222

Whitman WB (2016) Modest proposals to expand the type material for naming of prokaryotes. Int J Syst Evol Microbiol 66:2108–2112

Whitman WB, Chuvochina M, Hedlund BP et al (2022) Development of the SeqCode: a proposed nomenclatural code for uncultivated prokaryotes with DNA sequences as type. Syst Appl Microbiol 45:126305

Whitman WB, Sutcliffe IC, Rossello-Mora R (2019) Proposal for changes in the International Code of Nomenclature of Prokaryotes: granting priority to *Candidatus* names. Int J Syst Evol Microbiol 69:2174–2175

Wiens JJ (2021) Vast (but avoidable) underestimation of global biodiversity. PLoS Biol 19:e3001192

Yarza P, Yilmaz P, Pruesse E et al (2014) Uniting the classification of cultured and uncultured bacteria and archaea using 16S rRNA gene sequences. Nat Rev Microbiol 12:635–645

Chapter 10
Applications of Bioactive Compounds from Novel Microbial Taxa

Murad Muhammad, Bhagwan Narayan Rekadwad, Tayyiba Habib, Lei Dong, Wael N. Hozzein, and Wen-Jun Li

10.1 Introduction

According to Carstairs-McCarthy (2010), the prefix "bio" and the suffix "active" are derivatives of the ancient Greek and Latin words for "life" and "energetic," respectively. Bioactive compounds are synthesized plants or found in food material from plants (NIH-NCI). Bioactive chemicals are substances produced and used by

M. Muhammad · T. Habib
State key laboratory of Desert and Oasis and Ecology, Xinjiang Institute of Ecology and Geography, Chinese Academy of Sciences, Urumqi, People's Republic of China

B. N. Rekadwad
Division of Microbiology and Biotechnology, Yenepoya Research Centre, Yenepoya Deemed to be University, Mangalore, Karnataka, India

L. Dong
State Key Laboratory of Biocontrol, Guangdong Provincial Key Laboratory of Plant Resources and Southern Marine Science and Engineering Guangdong Laboratory (Zhuhai), School of Life Sciences, Sun Yat-Sen University, Guangzhou, People's Republic of China

W. N. Hozzein
Botany and Microbiology Department, Faculty of Science, Beni- Suef University, Beni- Suef, Egypt

W.-J. Li (✉)
State key laboratory of Desert and Oasis and Ecology, Xinjiang Institute of Ecology and Geography, Chinese Academy of Sciences, Urumqi, People's Republic of China

State Key Laboratory of Biocontrol, Guangdong Provincial Key Laboratory of Plant Resources and Southern Marine Science and Engineering Guangdong Laboratory (Zhuhai), School of Life Sciences, Sun Yat-Sen University, Guangzhou, People's Republic of China
e-mail: liwenjun3@mail.sysu.edu.cn

W.-J. Li et al. (eds.), *Modern Taxonomy of Bacteria and Archaea*,
https://doi.org/10.1007/978-981-99-5720-0_10

organisms that have a measurable impact on their physiological functions (Xu et al. (2017).

A type of chemical compound in a minute quantity that may assist in promoting health, inhibiting disease, or becoming a cure for an ailment is termed a bioactive compound.

These bioactive substances may have an impact on all life forms. The type of the bioactive chemical, the molecules it is coupled with, the cumulative dose, the exposure pathways, and the bioavailability of the system are only a few of the variables that determine whether a bioactive chemical has a positive or negative impact on a system. Exposure to numerous bioactive compounds may have a variety of impacts, including healing, immune system stimulation, cancer prevention, delayed aging, lowered fever, and others (Panossian et al. 2021). Nowadays, it has been reported that many animals (e.g., limulus amebocyte lysate (LAL) from horseshoe crab (*Limulus polyphemus*) to test endotoxins (Krisfalusi-Gannon et al. 2018)) and microorganisms are capable of producing bioactive compounds (astaxanthin, *Paracoccus carotinifaciens*; canthaxanthin, *Lactobacillus pluvalis*; phycocyanin, *Pseudomonas* sp.; violacein, *Chromobacterium violaceum*; xanthomonadin, *Xanthomonas oryzae*; β-carotene, *Dunaliella salina*; riboflavin, *Ashbya gossypii*; etc.) of medicinal as well as food application (Rani et al. 2021). In this context, novel bacterial taxa are of particular interest for the discovery of new bioactive compounds. For instance, *Streptomyces* and *Micromonospora* were extensively studied for its ability to produce a wide range of bioactive compounds (Bérdy 2012; Baltz 2016; Genilloud 2017). The discovery of bioactive compounds from novel bacterial taxa is important as it can lead to the identification of new classes of compounds with novel structures and bioactivities (Hamed et al. 2021). Recent studies show that novel bacterial taxa isolated from scarcely explored environments of the Comau Fjord in Northern Patagonia (Chile) were belonging to *Arthrobacter*, *Corynebacterium*, *Brevibacterium*, and *Kocuria*. Isolates belonging to the genera *Arthrobacter*, *Brachybacterium*, *Curtobacterium*, *Rhodococcus*, and *Streptomyces* showed significant antimicrobial activity against Gram-negative *Pseudomonas aeruginosa*, *Escherichia coli*, and *Salmonella enterica*, as well as Gram-positive *Staphylococcus aureus* and *Listeria monocytogenes*. Isolated non-mycelial strains of *Brachybacterium*, *Curtobacterium*, and *Rhodococcus* are a good source of bioactive chemicals because their antimicrobial activity are rarely documented. Furthermore, each isolated strain has at least one gene coding NRPS, PKS I, or PKS II biosynthetic gene cluster (Undabarrena et al. 2016). This chapter aims to provide an overview (Fig. 10.1) of the latest advances in the identification and applications of bioactive compounds from novel bacterial taxa and drug development. This chapter will discuss the strategies for the isolation and identification of novel bacterial taxa, as well as the bioactive compounds that have been identified from these bacteria and their potential applications.

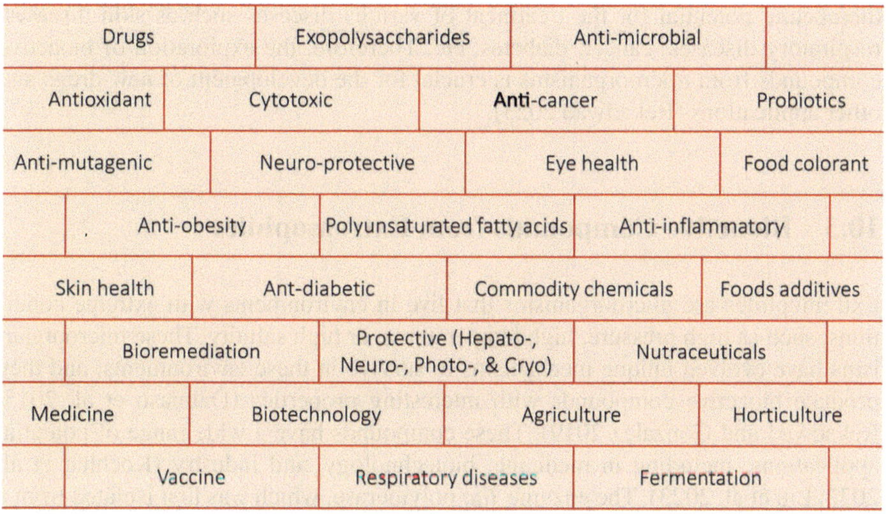

Fig. 10.1 The wall of microbial applications

10.2 Importance of Bioactive Compounds from Microorganisms

Bioactive compounds from microorganisms are particularly important due to their unique structures and diverse biological activities (Chen and Wu 2020). Microorganisms, such as bacteria and fungi, have been shown to produce a wide range of bioactive compounds, including antibiotics, antifungals, anticancer agents, immunomodulators, and enzymes with industrial applications (Undabarrena et al. 2016). Historically, natural biomolecules (Dias et al. 2012) and microbial products have been used for the development of drugs, agricultural agents, and industrial products individually (Demain and Sanchez 2009; Abdel-Razek et al. 2020) and synergistically as well (Karbowiak et al. 2023). One of the main advantages of bioactive compounds from microorganisms is their potential to target specific pathogens with high efficacy and specificity (Karthikeyan et al. 2022). For example, some bacterial compounds have been shown to selectively target certain bacterial species, viz., *K. pneumoniae*, *E. coli*, *S. enterica serovar Typhimurium*, and *B. subtilis* (Mondhe et al. 2014), including antibiotic-resistant strains (Hernando-Amado et al. 2023), without harming the beneficial bacteria in the host's microbiome. Moreover, the unique structures of bioactive compounds from microorganisms offer the potential for the development of novel drugs with improved pharmacokinetics, reduced toxicity, and increased potency (Penesyan et al. 2010; Katz and Hover 2018; Atanasov et al. 2021; Li et al. 2023). Furthermore, the discovery of bioactive compounds from previously unexplored bacterial taxa has the potential to lead to the identification of novel compounds with unique clusters, e.g., TRAP cluster, and bioactivities (Rekadwad et al. 2021; Sanjana et al. 2023), which may have therapeutic potential for the treatment of various diseases such as skin diseases, respiratory diseases, cancer, diabetes, etc. Therefore, the exploration of bioactive compounds from microorganisms is crucial for the development of new drugs and other applications (Rekadwad 2023).

10.3 Bioactive Compounds from Extremophiles

Extremophiles are microorganisms that live in environments with extreme conditions, such as high pressure, high temperatures, or high salinity. These microorganisms have evolved unique mechanisms to survive in these environments, and they produce bioactive compounds with interesting properties (Dalmaso et al. 2015; Rekadwad and Gonzalez 2019). These compounds have a wide range of potential applications, including in medicine, biotechnology, and industry (Kochhar et al. 2022; Liu et al. 2023). The enzyme *Taq* polymerase, which was first isolated from a bacterium known as *Thermus aquaticus* that resides in hot springs, is one example of a bioactive substance produced by thermophilic bacterium (Chien et al. 1976). Taq polymerase is a DNA polymerase with the ability to withstand high temperatures, making it a crucial tool for the polymerase chain reaction method for amplication of DNA/RNA into multiple copies (Ishino and Ishino 2014; Raddadi et al. 2015) (Fig. 10.2).

Similarly, a member of the microbial rhodopsin family, halorhodopsin is a seven-transmembrane retinylidene protein, which is produced by a type of archaea called *Halobacterium salinarum* (Oesterhelt 1995, 1998). Halorhodopsin is a light-driven chloride ion pump, and it has potential applications in optogenetics, a field of neuroscience that uses light to control the activity of neurons (Zhang et al. 2007). Many extremophilic bacteria are also a potential source of bioactive compounds (see Table 10.1) with antimicrobial properties, e.g., thermophilic *Bacillus licheniformis* produces an antibiotic called bacitracin that is effective against a range of Gram-positive bacteria (Luo et al. 2016). Hence, the evaluation of existing extremophiles for novel biomolecules or the discovery of novel bioactive compounds from novel bacteria and archaea becomes an exciting area of research with many potential applications in various fields.

10.4 Recent Reports on Bioactive Compounds from Novel Microbial Taxa

The scientific community has been able to identify and characterize previously unknown bacterial species from a variety of environments in recent years, thanks to advancements in genomic sequencing and bioinformatics. These environments include soil, water, marine, and deep-sea sediments, to name just a few. These

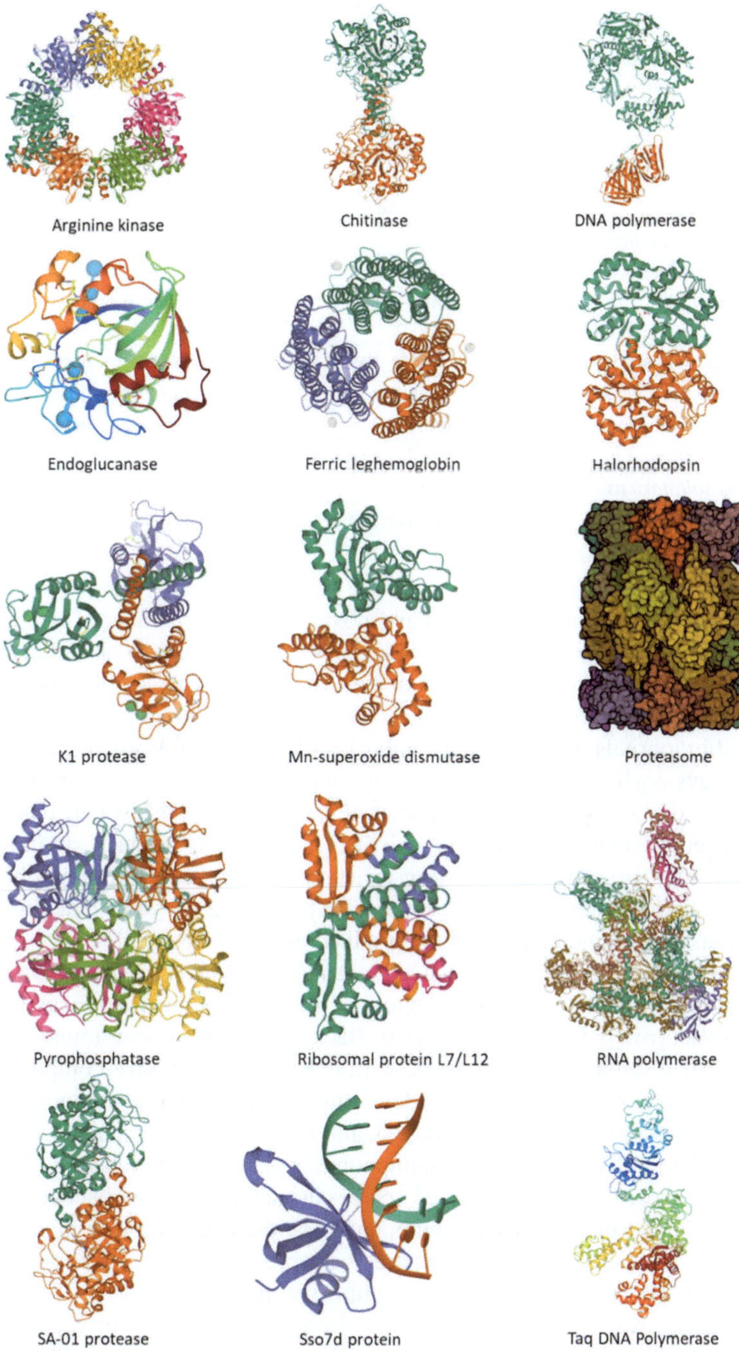

Fig. 10.2 Structure of various bioactive compounds adopted from RCSB PDB – US data center for the global Protein Data Bank (RCSB.org)

Table 10.1 Bioactive compounds from extremophiles

Extremophile	Bioactive compound	Potential applications
Pyrococcus furiosus	Endoglucanase	Biofuels, bioremediation
Sulfolobus acidocaldarius	Chitinase	Biotechnology, bioremediation
Thermus thermophilus	Arginine kinase	
Aeropyrum pernix	K1 protease	Enzyme technology, biotechnology
Aquifex aeolicus	Pyrophosphatase	
Methanopyrus kandleri	Proteasome	
Thermus scotoductus	SA-01 protease	
Acidithiobacillus ferrooxidans	Ferric leghemoglobin	Industrial uses, bioremediation
Deinococcus radiodurans	Mn-superoxide dismutase	Medicine, radiation protection
Thermococcus kodakarensis	RNA polymerase	Molecular biology, biotechnology
Thermococcus sibiricus	DNA polymerase	Molecular biology, diagnostics
Thermus aquaticus	Taq DNA polymerase	
Sulfolobus solfataricus	Sso7d protein	Molecular biology, nanotechnology
Haloarcula marismortui	Ribosomal protein L7/L12	Protein engineering, structural biology

recently found bacteria frequently have novel metabolic pathways and produce novel bioactive compounds, both of which have the potential to be utilized in the creation of new pharmaceuticals, pesticides, and other types of bioproducts. For instance, the majority of the research on novel bacteria and archaea that has been published in journals that have been authorized by Bergey's Manual Trust (https://www.bergeys.org/) and the Microbiology Society (https://www.microbiologyresearch.org/) describes several new bacterial species from soil, water, plants, animals, extreme environments, and pristine niches that produce antimicrobial compounds that are active against drug-resistant pathogens. In addition, it has been discovered that certain bacteria are capable of producing enzymes that could have applications in bioremediation and the production of bioenergy.

Plants have been found to have the highest number of bioactive compounds, with over 200,000 tentative compounds reported, and species such as *Taxus* spp. and *Catharanthus roseus* frequently reported. Bacteria have also been found to have a high number of bioactive compounds, with over 30,000 tentative compounds frequently reported. Animals have a moderate number of bioactive compounds, with over 10,000 tentative compounds reported, and various marine organisms frequently reported. The number of bioactive compounds isolated from archaea is relatively low, and the most frequently reported species are unknown. The discovery of new bacterial taxa and their bioactive compounds could change the way medicine and farming are done, and research in this area have presented many exciting discoveries (see Table 10.2) in past few years (unpublished literature).

A North Sea water sample yielded heterotrophic bacteria that thrive on complex marine media. Serial dilution and spread plating were used to isolate. Phylogenetic analysis of almost complete 16S rRNA gene sequences showed that strain HEL-45^{T} was 97.4% similar to Sulfitobacter mediterraneus and 96.5% to *Staleya guttiformis*.

Table 10.2 Number of bioactive compounds isolated from different kingdoms and their tentative numbers and reported species

Kingdom	Number of bioactive compounds	Tentative number	Most frequently reported species
Plants	High	Over 200,000	*Taxus* spp., *Catharanthus roseus*, *Hypericum perforatum*, *Curcuma longa*, *Salvia miltiorrhiza*, *Ginkgo biloba*, *Echinacea purpurea*, *Panax ginseng*, *Camellia sinensis*, *Zingiber officinale*, etc.
Animals	Moderate	Over 10,000	Various marine organisms, e.g., sponges, inhabitant of mangroves, horseshoe crab, corals, *Conus* spp., *Lonomia* spp., *Dendrobatidae* spp., *Siphonochelus australis*, *Plectranthus* spp., *Centruroides* spp., *Terebridae* spp., *Chelydra serpentina*, *Hymenoptera* spp., *Phyllomedusa* spp., etc.
Bacteria	High	Over 30,000	*Streptomyces* spp., *Bacillus* spp., *Pseudomonas* spp., *Micromonospora* spp., *Nocardiopsis* spp., *Actinomadura* spp., *Arthrobacter* spp., *Salinispora* spp., *Vibrio* spp., *Mycobacterium* spp., etc.
Archaea	Low	Unknown	Unknown

Gram-negative, nonmotile, obligate aerobe HEL-45 T requires sodium and 1–7% sea salts to grow, lacks bacteriochlorophyll, produces storage granules, and has been phylogenetically identified as *Oceanibulbus indolifex* gen. nov., sp. nov. The strain HEL-45^T exhibits the ability to synthesize indole and various indole derivatives, including 3,3-bis-(indol-3-yl)-propane-1,2-diol, 3-indole-carbaldehyde, and indole-3-carboxylic-thiomethylester. Furthermore, it was discovered that bioactive compounds were produced by the bacterium. The study identified three distinct cyclic dipeptides, specifically cyclo-(Phe,Pro), cyclo-(Tyr,Pro), and cyclo-(Leu,Pro). These cyclic dipeptides are identified for their limited antiviral, antibiotic, and antitumor properties (Wagner-Döbler et al. 2004). A polyphasic approach was used to examine an actinomycete strain SBHS Strp1^T that was isolated from a brown hypermagnesian ultramafic soil. It possesses morphological and chemical characteristics common to *Amycolatopsis* strains. It shared a closer phylogenetic relationship with *Amycolatopsis kentuckyensis*, *Amycolatopsis tolypomycina*, *Amycolatopsis lexingtonensis*, *Amycolatopsis pretoriensis*, and *Amycolatopsis rifamycinica*. Based on the combined genotypic and phenotypic information of the organisms, *Amycolatopsis plumensis* sp. nov. was named as the novel species. The genus *Amycolatopsis* is a source of novel bioactive compounds but is grossly underspecified and capable of growth in presence of various antibiotics and chemical inhibitors (Saintpierre-Bonaccio et al. 2005).

Strains of the Gram-negative bacterium *Pseudomonas* are often found in soil and water, and they are known for their ability to adapt to different environments and are frequently reported as pathogenic for several reasons. Their capacity to acquire antibiotic resistance is one factor. The ability of *Pseudomonas* species to develop resistance to numerous antibiotics makes them challenging to treat and raises the risk of infections getting worse (Lister et al. 2009). In addition, certain *Pseudomonas* species generate virulence factors, such as toxins and enzymes, that contribute to their pathogenicity. *Pseudomonas aeruginosa*, for instance, produces pyocyanin, a toxin that damages cells and tissues, and elastase, an enzyme that degrades elastin in lung tissue, thereby contributing to the onset of pneumonia. Moreover, they are capable of causing opportunistic infections. *Pseudomonas* species are opportunistic pathogens, meaning they can infect individuals with compromised immune systems, such as those with cancer, HIV/AIDS, or cystic fibrosis. These infections can be as mild as a skin infection or as severe as pneumonia or sepsis (Oliver et al. 2015). In contrast, there are some *Pseudomonas* strains that produce small-molecule compounds that have inhibitory activities other pathogenic bacteria. For instance, the bacterium *Pseudoalteromonas flavipulchra* JG1 has been found to produce a protein known as PfaP, as well as various small-molecule compounds that exhibit inhibitory effects against *Vibrio anguillarum*. The protein known as PfaP is comprised of 694 amino acids and has a molecular mass of 77.0 kDa. The PfaP exhibited a sequence identity of 58% with the l-lysine oxidase AlpP found in *Pseudoalteromonas tunicata* D2 and a sequence identity of 54% with the marinocine antimicrobial protein present in *Marinomonas mediterranea* MMB-1. The present study reports the identification of five small molecules (compounds 1–5) that exhibit antibacterial activity. These molecules have been identified as p-hydroxybenzoic acid, N-hydroxybenzoisoxazolone, 6-bromoindolyl-3-acetic acid, 2′-deoxyadenosine, and trans-cinnamic acid. The compound containing bromine exhibited antibacterial properties against Gram-positive and Gram-negative bacteria (Yu et al. 2012). Similarly, an genomic investigation evaluates the genomes of 194 *Pseudomonas* type strains, discerning the abundance of different classes of biosynthetic gene clusters and the coding attributes of genes responsible for antimicrobial peptides. These findings hold potential for exploitation in the fields of biotechnology and pharmaceuticals (Saati-Santamaría et al. 2022).

Bacteria possessing the phylum *Actinobacteria* represent a significant branch of a family tree that contains a great deal of genetic variation (Goodfellow et al. 2012). Actinobacteria are a group of bacteria that have a Gram-positive cell wall structure and a relatively high proportion of guanine and cytosine nucleotide bases in their genetic material. In addition, actinobacteria have a Gram-positive cell wall structure, which is anaother important feature. These microorganisms are widely considered to be free-living and cosmopolitan due to their ability to successfully adapt to a wide variety of ecosystems, including those found in aquatic and terrestrial environments (Qin et al. 2011; Dhakal et al. 2017; Goodfellow et al. 2018; Zamora-Quintero et al. 2022). In order to address the issue of increasing antimicrobial resistance, various alternative extremophiles have been discovered. The thermophilic actinobacteria are able to produce their own distinct antimicrobial agents. These agents have the potential to combat severe infections such as invasive aspergillosis, which is caused by fungus *Aspergillus fumigatus* and represents a significant risk to human life. A total of 36 different thermophilic *Actinobacteria* strains from samples were taken from "windrows." Molecular techniques were used to identify these strains, and it was discovered that they are able to survive and even thrive at temperatures of 50 °C. Some of these strains – DJT 15, DJT 32, and DJT 36 – were identified as *Streptomyces thermoviolaceus* subsp. *apingens*, *Saccharomonospora viridis*, and *Saccharomonospora glauca*. The aforementioned strains, DJT 32 and 36, exhibited inhibitory activity against the fungal pathogen *Aspergillus fumigatus* (Halket et al. 2020). In a similar way, genomic analysis of *Streptomyces* species using antiSMASH showed that biosynthetic gene clusters, such as non-ribosomal peptide synthase (NRPS) and 3-polyketide synthase (PKS) gene clusters, are responsible for the production of microbial antibiotics (Borba et al. 2023).

10.5 Genomics Analysis of Actinomycete *Micromonospora*

Some actinomycetal species are uncommon and slow-growing, meaning that they are rarely isolated from environmental or clinical samples and have a slower growth rate than other actinomycetes. These species are typically more difficult to study and exploit for biotechnological purposes, but they can be a rich source of novel compounds with potentially valuable biological activities. These organisms have been found in a variety of places, like soil, freshwater, and the ocean. They are thought to produce a number of bioactive compounds that could be used in medicine, such as antibiotics, antifungals, antitumor agents, and immunosuppressants. In this context, *Micromonospora* species can be a representative example of actinomycete that is rare and grows slowly producing potential bioactive compound (unpublished data).

Biosynthetic gene clusters in genome of actinomycete strain *Micromonospora* sp. L5 (CP002399) downloaded from NCBI repository can be identified using antiSMASH (Blin et al. 2021). A total 19 of biosynthetic gene clusters in *Micromonospora* sp. L5 were found, i.e., region 1–19 (see Fig. 10.3). A rare actinomycete, *Micromonospora* sp. L5 is able synthesize following biochemical: isorenieratene, phosphonoglycans, ebelactone, loseolamycin A1/loseolamycin A2, tetrachlorizine, lymphostin/neolymphostinol B/lymphostinol/neolymphostin B, peucechelin, komodoquinone B, largimiycin, largimycin, tautomycetin, lobosamide A/lobosamide B/lobosamide C, formicamycins A-M, nocathiacin, crochelin A, naphthyridinomycin, tiancimycin, and quinolidomicin A (Unpublished analysis).

10.6 Issues Need to Be Addressed for Discovery of Bioactive Compounds from Novel Taxa

The discovery of bioactive compounds from novel taxa involves identifying and characterizing new compounds that have the potential to be used in pharmaceuticals, health, medicine, agriculture, and other industries (Cragg and Newman 2013).

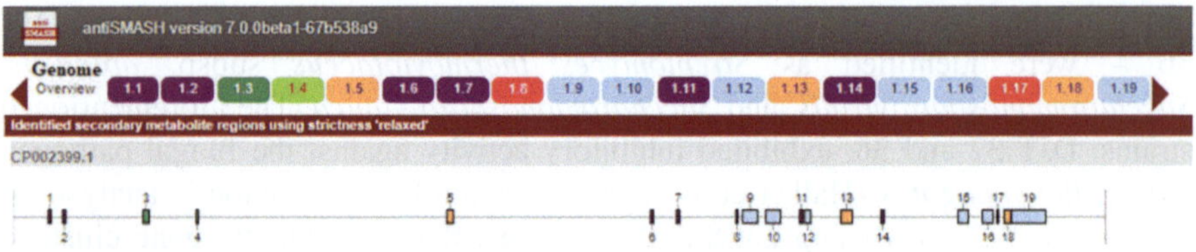

Region	Type	From	To	Most similar known cluster
Region 1	terpene	221,533	242,010	isorenieratene
Region 2	terpene	313,898	334,259	phosphonoglycans
Region 3	NRPS	822,679	865,922	ebelactone
Region 4	NAGGN	1,159,977	1,174,789	
Region 5	T3PKS	2,761,821	2,802,873	loseolamycin A1/loseolamycin A2
Region 6	terpene	4,058,457	4,077,583	tetrachlorizine
Region 7	terpene	4,222,972	4,243,981	lymphostin/neolymphostinol B/lymphostinol/neolymphostin B
Region 8	NRPS-independent-siderophore	4,603,155	4,614,958	peucechelin
Region 9	T2PKS, oligosaccharide, other	4,644,202	4,739,578	komodoquinone B
Region 10	transAT-PKS, NRPS, NRPS-like, PKS-like	4,791,253	4,886,022	largimiycin 38/largimycin 39
Region 11	betalactone	5,007,285	5,029,399	tautomycetin
Region 12	oligosaccharide, terpene	5,040,049	5,079,051	lobosamide A/lobosamide B/lobosamide C
Region 13	T2PKS	5,269,338	5,341,912	formicamycins A-M
Region 14	terpene	5,524,214	5,545,416	nocathiacin
Region 15	NRPS, T1PKS, NRPS-like	6,014,732	6,080,342	crochelin A
Region 16	PKS-like, NRPS, T1PKS, NRPS-like	6,169,329	6,237,189	naphthyridinomycin
Region 17	NRPS-independent-siderophore	6,261,797	6,273,581	FW0622
Region 18	T1PKS	6,308,998	6,352,726	tiancimycin
Region 19	T1PKS, NRPS-like, 2dos, NRPS	6,358,138	6,579,562	quinolidomicin A

Fig. 10.3 antiSMASH: detection of biosynthetic gene clusters in *Micromonospora* sp. L5 (NCBI Accession No.: CP002399)

The Biodiversity Act's restrictions on access to novel taxa are just one of several problems that serve as obstacles to the discovery of bioactive chemicals from novel taxa (NBA 2022) and other laws of the country, critical steps and throughthput technology requiremet for identification of compunds, requirement of sophisticated analytical techniques such as nuclear magnetic resonance (NMR) spectroscopy, mass spectrometry, and X-ray crystallography, and if a bioactive compound is identified, it may not be feasible to produce it at a large scale for commercial use (Harvey et al. 2015). Factors such as yield, cost, and environmental impact must be considered when developing production methods (Newman and Cragg 2016). Additionally, issues related to intellectual property related to novel taxa and biomolecules must be addressed. There are often complex legal and ethical considerations when it comes to patenting and commercializing bioactive compounds from novel taxa, particularly if they are sourced from indigenous communities or other groups that may have a claim to the intellectual property (D'Souza and Sadananda 2018; Dhar et al. 2019).

10.7 Future Perspectives

The potential utilization of bioactive compounds derived from emerging microbial taxa is promising. The growing fascination with natural products and their potential health advantages has led researchers to investigate novel microbial sources for bioactive compounds that exhibit diverse applications, such as their utilization as pharmaceuticals, food additives, and agricultural products. The identification and characterization of new microbial taxa that produce bioactive compounds will be facilitated by developments in genomics and bioinformatics tools. In addition, the advancement of novel extraction technologies pertaining to the manufacturing, retrieval, and refinement of said compounds will expedite their commercialization and augment their accessibility for diverse applications. It is anticipated that the utilization of bioactive compounds derived from emerging microbial taxa will experience an increase in the foreseeable future, thereby making a substantial contribution to diverse sectors of the economy. In addition, the growing need for environmentally sustainable products will serve as a catalyst for the utilization of bioactive compounds derived from emerging microbial taxa across multiple sectors. Bioactive compounds have the potential to serve as natural preservatives within the food industry, thereby decreasing the reliance on synthetic preservatives.

10.8 Conclusion

In conclusion, the discovery and characterization of bioactive compounds from novel bacterial taxa is an important area of research that has the potential to lead to the development of new drugs and agricultural products. Researchers have identified a wide range of bioactive compounds produced by bacterial strains belonging to various genera, including *Streptomyces*, *Micromonospora*, *Arthrobacter*, *Brachybacterium*, *Curtobacterium*, and *Rhodococcus*. These compounds have been shown to have potent antimicrobial, anticancer, and immunomodulatory properties, among others. The development of sophisticated analytical techniques has facilitated the identification and characterization of these compounds, although issues related to yield, cost, and environmental impact must be considered when developing production methods. In addition, there are complex legal and ethical considerations related to intellectual property when it comes to patenting and commercializing bioactive compounds from novel taxa. Thus, the potential applications of bioactive compounds from novel bacterial taxa are diverse and promising, and continued research in this area will likely lead to new discoveries with medical and commercial applications.

Acknowledgments This research was financially supported by the Third Xinjiang Scientific Expedition Program (Grant No. 2022xjkk1200), National Natural Science Foundation of China (Nos: 32000005 and 32061143043). BNR acknowledges part of funds received for research from Yenepoya (Deemed to be University) (Project No.: YU/SeedGrant/104-2021).

References

Abdel-Razek AS, El-Naggar ME, Allam A, Morsy OM, Othman SI (2020) Microbial natural products in drug discovery. PRO 8(4):470

Atanasov AG, Zotchev SB, Dirsch VM, the International Natural Product Sciences Taskforce, Supuran CT (2021) Natural products in drug discovery: advances and opportunities. Nat Rev Drug Discov 20:200–216

Baltz RH (2016) *Streptomyces* and *Saccharopolyspora* hosts for heterologous expression of secondary metabolite gene clusters. J Ind Microbiol Biotechnol 43(2–3):247–267

Bérdy J (2012) Thoughts and facts about antibiotics: where we are now and where we are heading. J Antibiot 65(8):385–395

Blin K, Shaw S, Kloosterman AM, Charlop-Powers Z, van Wezel GP, Medema MH, Weber T (2021) antiSMASH 6.0: improving cluster detection and comparison capabilities. Nucleic Acids Res 49(W1):W29–W35

Borba MP, Witusk JP, Cunha DM, de Lima-Morales D, Martins AF, Van Der Sand S (2023) Whole-genome sequencing-based characterization of *Streptomyces* sp. 6(4): focus on natural product. Access Microbiol 5(3):acmi000466.v3

Carstairs-McCarthy A (2010) The evolution of morphology. Oxford University Press

Chen J, and Wu Q (2020) Recent advances in the discovery, biosynthesis and biotechnological production of bioactive compounds produced by microorganisms. In Bioactive Compounds (pp. 1–32). Springer, Cham

Chien A, Edgar DB, Trela JM (1976) Deoxyribonucleic acid polymerase from the extreme thermophile Thermus aquaticus. J Bacteriol 127(3):1550–1557

Cragg GM, Newman DJ (2013) Natural products: a continuing source of novel drug leads. Biochim Biophys Acta 1830(6):3670–3695

Dalmaso GZ, Ferreira D, Vermelho AB (2015) Marine extremophiles: a source of hydrolases for biotechnological applications. Mar Drugs 13(4):1925–1965

Demain AL, Sanchez S (2009) Microbial drug discovery: 80 years of progress. J Antibiot (Tokyo) 62(1):5–16

Dhakal D, Pokhrel AR, Shrestha B, Sohng JK (2017) Marine rare actinobacteria: isolation, characterization, and strategies for harnessing bioactive compounds. Front Microbiol 8:1106

Dhar N, Razdan VK, Mahajan G (2019) Intellectual property issues in natural product drug discovery from traditional knowledge. Nat Prod Rep 36(4):578–592

Dias DA, Urban S, Roessner U (2012) A historical overview of natural products in drug discovery. Meta 2(2):303–336

D'Souza C, Sadananda MA (2018) Intellectual property rights and traditional knowledge: challenges and opportunities in the pharmaceutical industry. J Intellect Prop Rights 23(2):96–104

Genilloud O (2017) Actinomycetes: still a source of novel antibiotics. Nat Prod Rep 34(10):1203–1232

Goodfellow M, Kämpfer P, Busse H-J, Trujillo ME, Suzuki K, Ludwig W, Whitman WB (2012) Road map of the phylum actinobacteria. Bergey's manual of systematic bacteriology. Springer, New York, pp 1–28

Goodfellow M, Nouioui I, Sanderson R, Xie F, Bull AT (2018) Rare taxa and dark microbial matter: novel bioactive actinobacteria abound in Atacama Desert soils. Antonie Van Leeuwenhoek 111:1315–1332

Halket G, Herron P, Munro C (2020) Isolation of thermophilic Actinobacteria from compost and identification of bioactive compounds with antimicrobial properties. Access Microbiol 2:7A. https://doi.org/10.1099/acmi.ac2020.po0835

Hamed MBA, Liu Y, Liu M, Khalil AA, Abdelmohsen UR, Li W (2021) Streptomyces: a potential source for new antibiotics. Antibiotics 10(6):633

Harvey AL, Edrada-Ebel R, Quinn RJ (2015) The re-emergence of natural products for drug discovery in the genomics era. Nat Rev Drug Discov 14(2):111–129

Hernando-Amado S, Laborda P, Martínez JL (2023) Tackling antibiotic resistance by inducing transient and robust collateral sensitivity. Nat Commun 14:1723

Ishino S, Ishino Y (2014) DNA polymerases as useful reagents for biotechnology—the history of developmental research in the field. Front Microbiol 5:465

Karbowiak M, Szymański P, Zielińska D (2023) Synergistic effect of combination of various microbial hurdles in the biopreservation of meat and meat products-systematic review. Foods 12(7):1430

Karthikeyan A, Joseph A, Nair BG (2022) Promising bioactive compounds from the marine environment and their potential effects on various diseases. J Genet Eng Biotechnol 20(1):14

Katz M, Hover B (2018) Discovery of novel microbial natural products. Curr Opin Chem Biol 45:1–8

Kochhar N, Kavya IK, Shrivastava S, Ghosh A, Rawat VS, Sodhi KK, Kumar M (2022) Perspectives on the microorganism of extreme environments and their applications. Curr Res Microb Sci 3:100134

Krisfalusi-Gannon J, Ali W, Dellinger K, Robertson L, Brady TE, Goddard MKM, Tinker-Kulberg R, Kepley CL, Dellinger AL (2018) The role of horseshoe crabs in the biomedical industry and recent trends impacting species sustainability. Front Mar Sci 5:185

Li SR, Tan YM, Zhang L, Zhou CH (2023) Comprehensive insights into medicinal research on imidazole-based supramolecular complexes. Pharmaceutics 15(5):1348

Lister PD, Wolter DJ, Hanson ND (2009) Antibacterial-resistant Pseudomonas aeruginosa: clinical impact and complex regulation of chromosomally encoded resistance mechanisms. Clin Microbiol Rev 22(4):582–610

Liu Y, Zhang N, Ma J, Zhou Y, Wei Q, Tian C, Fang Y, Zhong R, Chen G, Zhang S (2023) Advances in cold-adapted enzymes derived from microorganisms. Front Microbiol 14:1152847

Luo H, Hallen-Adams HE, Walton JD (2016) Processing of the antimicrobial peptide bacitracin in *Bacillus licheniformis* involves a novel metalloprotease. J Bacteriol 198(8):1222–1231

Mondhe M, Chessher A, Goh S, Good L, Stach JE (2014) Species-selective killing of bacteria by antimicrobial peptide-PNAs. PLoS One 9(2):e89082

NBA (2022) National Biodiversity Act- The biological diversity act, India

Newman DJ, Cragg GM (2016) Natural products as sources of new drugs from 1981 to 2014. J Nat Prod 79(3):629–661

Oesterhelt D (1995) Structure and function of halorhodopsin. Israel J Chem 35:475–494

Oesterhelt D (1998) The structure and mechanism of the family of retinal proteins from halophilic archaea. Curr Opin Struct Biol 8:489–500

Oliver A, Mulet X, López-Causapé C, Juan C (2015) The increasing threat of Pseudomonas aeruginosa high-risk clones. Drug Resist Updat 21-22:41–59

Panossian AG, Efferth T, Shikov AN, Pozharitskaya ON, Kuchta K, Mukherjee PK, Banerjee S, Heinrich M, Wu W, Guo D (2021) Evolution of the adaptogenic concept from traditional use to medical systems: pharmacology of stress-and aging-related diseases. Med Res Rev 41:630–703

Penesyan A, Kjelleberg S, Egan S (2010) Development of novel drugs from marine surface associated microorganisms. Mar Drugs 8(3):438–459

Qin S, Xing K, Jiang JH, Xu LH, Li WJ (2011) Biodiversity, bioactive natural products and biotechnological potential of plant-associated endophytic actinobacteria. Appl Microbiol Biotechnol 89:457–473

Raddadi N, Cherif A, Daffonchio D, Neifar M, Fava F (2015) Biotechnological applications of extremophiles, extremozymes and extremolytes. Appl Microbiol Biotechnol 99(19):7907–7913

Rani A, Saini KC, Bast F, Mehariya S, Bhatia SK, Lavecchia R, Zuorro A (2021) Microorganisms: a potential source of bioactive molecules for antioxidant applications. Molecules 26(4):1142

Rekadwad B, Gonzalez JM (2019) Multidisciplinary involvement and potential of thermophiles. Folia Microbiol 64:389–406

Rekadwad BN (2023) Microbial systematics: biomolecules and natural macromolecules applications. CRC Press, (Taylor & Francis), Boca Raton

Rekadwad BN, Li WJ, Rekha PD (2021) The diversity of unique 1,4,5,6-Tetrahydro-2-methyl-4--pyrimidinecarboxylic acid coding common genes and universal stress protein in Ectoine TRAP cluster (UspA) in 32 *Halomonas* species. BMC Res Notes 14(1):296

Saati-Santamaría Z, Selem-Mojica N, Peral-Aranega E, Rivas R, García-Fraile P (2022) Unveiling the genomic potential of *Pseudomonas* type strains for discovering new natural products. Microb Genom 8(2):000758

Saintpierre-Bonaccio D, Amir H, Pineau R, Tan GYA, Goodfellow M (2005) Amycolatopsis plumensis sp. nov., a novel bioactive actinomycete isolated from a New-Caledonian brown hypermagnesian ultramafic soil. Int J Syst Evol Microbiol 55(Pt 5):2057–2061

Sanjana NS, Rekadwad BN, Rekha PD, Arun AB, Suryavanshi MV (2023) Current scenario of application of anti-infective compounds from microbial origin perspective. In: Rekadwad (ed) Microbial systematics: biomolecules and natural macromolecules applications. CRC Press, (Taylor & Francis), Boca Raton, pp 1–13

Undabarrena A, Beltrametti F, Claverías FP, González M, Moore ERB, Seeger M, Cámara B (2016) Exploring the diversity and antimicrobial potential of marine actinobacteria from the Comau Fjord in Northern Patagonia, Chile. Front Microbiol 7:1135

Wagner-Döbler I, Rheims H, Felske A, El-Ghezal A, Flade-Schröder D, Laatsch H, Lang S, Pukall R, Tindall BJ (2004) *Oceanibulbus indolifex* gen. nov., sp. nov., a North Sea alphaproteobacterium that produces bioactive metabolites. Int J Syst Evol Microbiol 54(Pt 4):1177–1184

Xu N, Wei L, Liu J (2017) Biotechnological advances and perspectives of gamma-aminobutyric acid production. World J Microbiol Biotechnol 33:1–11

Yu M, Wang J, Tang K, Shi X, Wang S, Zhu WM, Zhang XH (2012) Purification and characterization of antibacterial compounds of *Pseudoalteromonas flavipulchra* JG1. Microbiology 158(Pt 3):835–842

Zamora-Quintero AY, Torres-Beltrán M, Guillén Matus DG, Oroz-Parra I, Millán-Aguiñaga N (2022) Rare actinobacteria isolated from the hypersaline Ojo de Liebre Lagoon as a source of novel bioactive compounds with biotechnological potential. Microbiology 168(2):001144

Zhang F, Wang LP, Brauner M, Liewald JF, Kay K, Watzke N, Wood PG, Bamberg E, Nagel G, Gottschalk A, Deisseroth K (2007) Multimodal fast optical interrogation of neural circuitry. Nature 446(7136):633–639

Chapter 11
Exploring Microbial Dark Matter and the Status of Bacterial and Archaeal Taxonomy: Challenges and Opportunities in the Future

Wen-Jun Li, Bhagwan Narayan Rekadwad, Jian-Yu Jiao, and Nimaichand Salam

11.1 Exploration of Microbial Dark Matter Through the Lenses of Next-Generation Sequencing

There is no question about our planet harboring a vast majority of diverse organisms and microorganisms. It has been frequently said that more than 1% of actual bacterial and archaeal taxa have been cultured in axenic form, described and validly published (Yarza et al. 2014). Indeed, this may require attention based on "scaling laws." Scaling rules have the strongest and most conclusive connections in biology. Information on both micro- and macro-biology should be available using a global compilation (i.e., approximately 35,000 sites and approximately 5.6×10^6 species). As a result, global-scale compilation exposes commonness and rarity connections that correctly anticipate the abundance of dominant species over a scale of 30 orders of magnitude to all Earth's microorganisms. On the basis of this scaling equation and the log normal model of biodiversity, it was calculated that there might be one

W.-J. Li (✉)
State Key Laboratory of Biocontrol, Guangdong Provincial Key Laboratory of Plant Resources and Southern Marine Science and Engineering Guangdong Laboratory (Zhuhai), School of Life Sciences, Sun Yat-Sen University, Guangzhou, People's Republic of China

State Key Laboratory of Desert and Oasis Ecology, Xinjiang Institute of Ecology and Geography, Chinese Academy of Sciences, Urumqi, People's Republic of China
e-mail: liwenjun3@mail.sysu.edu.cn

B. N. Rekadwad
Division of Microbiology and Biotechnology, Yenepoya Research Centre, Yenepoya (Deemed to be University), Mangalore, Karnataka, India

J.-Y. Jiao · N. Salam
State Key Laboratory of Biocontrol, Guangdong Provincial Key Laboratory of Plant Resources and Southern Marine Science and Engineering Guangdong Laboratory (Zhuhai), School of Life Sciences, Sun Yat-Sen University, Guangzhou, People's Republic of China

W.-J. Li et al. (eds.), *Modern Taxonomy of Bacteria and Archaea*,
https://doi.org/10.1007/978-981-99-5720-0_11

trillion different microbial species on Earth. Hence, while moving forward to the conclusions, microbiologist needs to do considerable efforts to ensure sampling and retrieving quality sequences using culture-independent technology (Loceya and Lennon 2016) rather than just assessing microbial diversity of samples. As of today, more than 25,000 taxa are retrieved through axenic culture worldwide and represented by classified sequences. Less than 107 species have been recorded overall by the Earth Microbiome Project (EMP), 29% of which have only been found twice. Finding the potentially 99.999% of microbial species that are still unidentified will be considerably aided by strong correlations and requires a more comprehensive investigation of commonness and rarity (Loceya and Lennon 2016). Therefore, our knowledge of the microbial world is limited by our limited understanding of the biology of the primary branches of the tree of life, i.e., bacteria, archaea, eukaryotes, and candidate phyla radiation (CPR) (Hug et al. 2016). In reality, the need for molecules for cure sparked microbiology research; this has primarily focused on cultivation of axenic cultures up to the present day. There are some projects that have been initiated, including EMP, MDM, and GEBA, that have tried to elucidate the existing structure of microbial life in this world.

11.2 Microbial Dark Matter and "Reverse Axenic" Cultivation

The existence of MDM can be investigated by studying their genetic make-up – gene to genome level – through genomic research, which can aid in understanding their function and possible applications in biotechnology and medicine. The genomic investigation of MDM frequently employs methods like DNA sequencing and metagenomics. This enables scientists to discover new strains of bacteria and archaea and to gain a better understanding of the complexity and diversity of the microbial world. This method is scientifically and technically different from traditional microbiology research that allows cultivation, characterization, and identification. In contrast, "reverse axenic" cultivation of microorganisms refers to the identification of microorganisms using sequence data generated by any culture-independent method followed by isolation in axenic culture by applying an appropriate cultivation strategy. This is useful for exploring MDM in the most effective ways and fulfilling the needs of society in various aspects.

In the realm of microbiology, genomic analysis of MDM directly from complex environmental samples has been a considerable problem due non-surety of the complete sequence. This is mainly due to the complexity of the samples and the difficulty of extracting and purifying the DNA of the current bacterial and archaeal taxa from samples. Recent advances in DNA sequencing methods, like next-generation sequencing (NGS) and single-cell sequencing, have substantially enhanced the accuracy and range to study MDM. NGS enables the quick and high-throughput sequencing of enormous amounts of DNA, hence enabling the investigation of complex microbial communities in a single experiment. Single-cell sequencing, on the other hand, enables the isolation and sequencing of single/individual cells for the study of uncultivated microorganisms such as TM6 (McLean et al. 2013), TM7 (Marcy et al. 2007) and OP9 lineage (Dodsworth et al. 2013). In addition, advancements in computing and bioinformatics have made it possible to evaluate and comprehend the vast quantities of data generated by these sequencing technologies. This includes the creation of novel methodologies for data assembling, annotation, and functional analysis that enable researchers to glean relevant insights from the MDM genomic data (Rinke et al. 2013). These developments in DNA sequencing and computation have made it possible to conduct genomic analysis of MDM directly from complicated environmental samples and to acquire a deeper understanding of the diversity and function of uncultivated microorganisms in various ecosystems.

The development of DNA sequencing, computing technologies such as CONCOCT (Alneberg et al. 2014), and curated databases reduced the challenge of genomic exploration from MDM data directly obtained from complex samples. Newer computational tools can be used for obtaining single-amplified genomes (SAGs) and extraction of metagenome-assembled genomes (MAGs) through single-cell genomic and metagenomic approaches, respectively. This has emerged as one of the most efficient techniques for enabling taxa-level genomic analysis of intricate microbial ecosystems without the requirement for axenic culturing, which sheds light on MDM composition (Anantharaman et al. 2016; Jiao et al. 2021a).

SAGs and MAGs are two important components of today's sequencing. SAGs are typically amplified genomic DNA from individual cells using whole-genome amplification (WGA) technologies, whereas MAGs are genomes recovered from metagenomic data sets using computational binning tools (see Fig. 11.1). Genomes sequenced through SAG and MAG technologies should have high quality, which enables scientists to predict the reliability for use and the metabolic potential of MDM based on marker genes (9). However, even subpar MAGs and SAGs can help researchers understand three fundamental questions in science: who are they? Where are they? What they can do? With this knowledge, microbiologists can easily create studies to investigate in situ physiology and create specialized cultivation techniques (Jiao et al. 2021a). Currently, statistics obtained from IMG indicate (IMG/GOLD 14) that genome sequencing projects show a sudden fall in 2020 and a slight rise in late 2022 (Fig. 11.2a). Similarly, considering the source of isolation of bacteria, i.e., samples used for isolation of bacteria by relevance, indicates that more than 55% of the bacteria have been isolated from medical samples. This shows another shady root, which means no such research has been conducted in environmental microbiology (11.6% isolates from environmental niches). Relevance is not good in the agriculture sector (9.2%) and other areas. This could be due to an ongoing pandemic situation caused by a catastrophic situation (Fig. 11.2b). Moreover, huge studies have been conducted on bacteria (see Fig. 11.2c), considering opinions in previous statements (Fig. 11.2a), which are mainly contributed by the medical sector, followed by metagenomic. In the last 25 years, a few studies on archaea, which is less common than viruses, have been conducted.

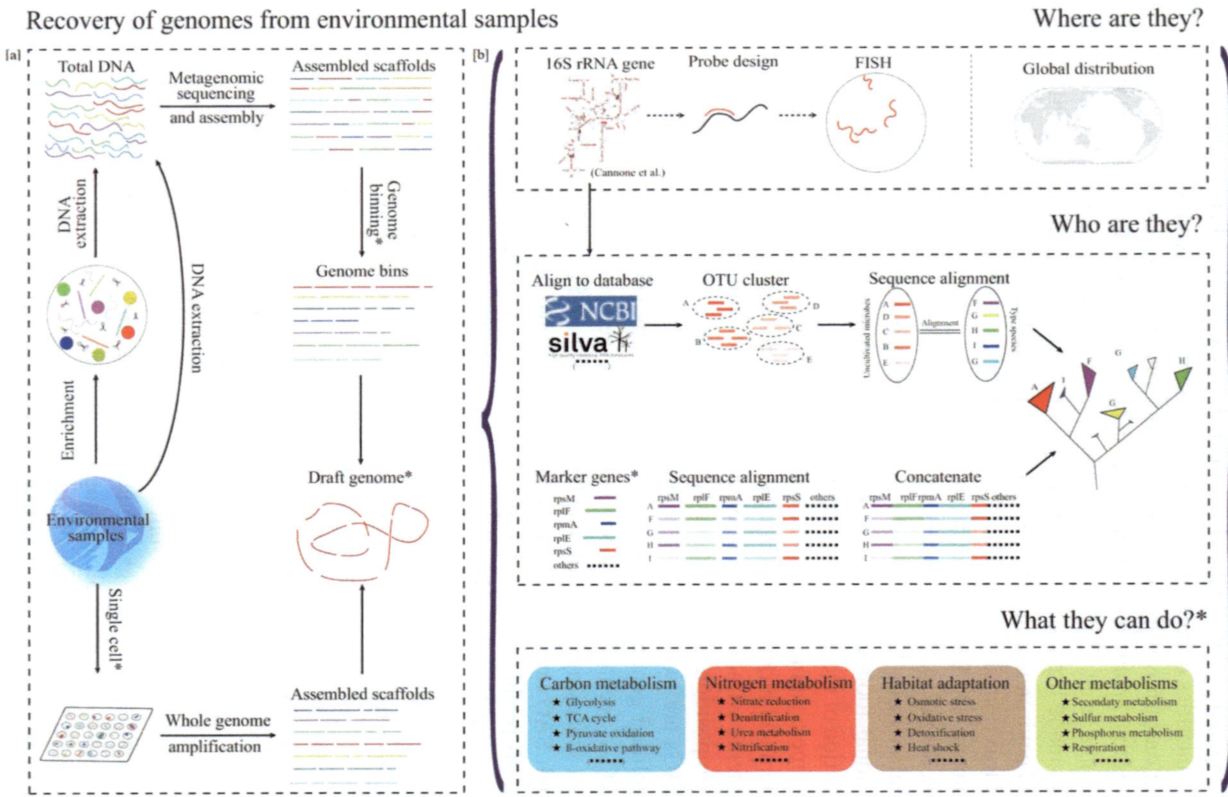

Fig. 11.1 Research plan for resolving basic and scientific problems in MDM investigations: (**a**) standard pipeline for making SAGs or MAGs. Specialized equipment, such as flow cytometry, microfluidics, or micromanipulators, are needed for single-cell isolation, and clean rooms are also necessary for downstream handling. There are other genome-binning tools, including MetaBAT (Kang et al. 2015), MaxBin (Wu et al. 2014), GroopM (Imelfort et al. 2014), MetaWatt (Strous et al. 2012), and CONCOCT (Alneberg et al. 2014). The Genome Criteria Consortium recommended SAG and MAG information as basic standards for draught genome quality in order to improve communication and more thorough comparative genomic investigations of microbial diversity. (**b**) A flowchart that shows how genomic data can be incorporated, along with downstream analysis based on MDM genomes. The phylogenetic marker genes are adaptable enough to take into account changes over time. Multi-omics, FISH, MAR, NanoSIMS, Raman, and other techniques used in combination would offer strong support for microbial function and a framework for developing tailored cultivation techniques.

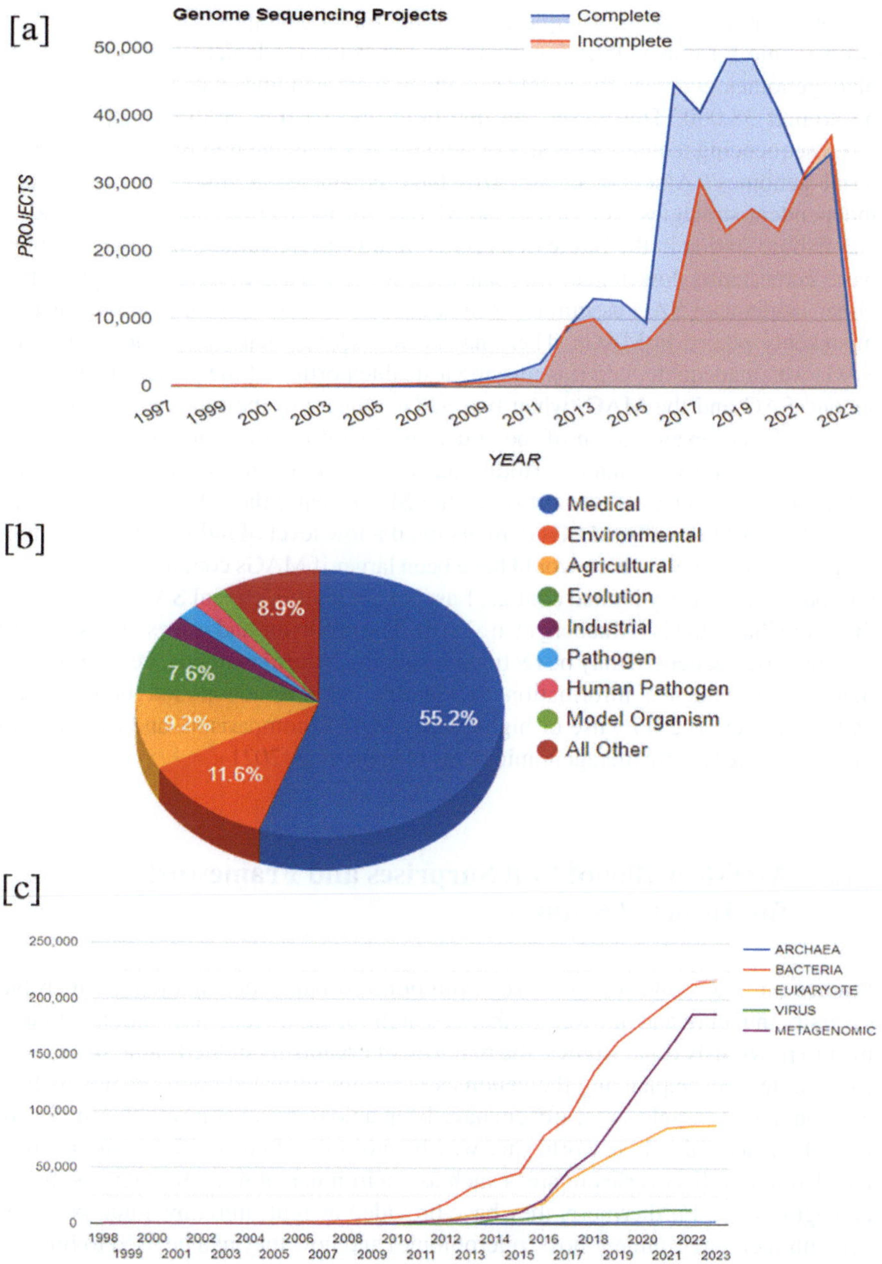

Fig. 11.2 The number of genomic projects: (**a**) represents the number of genome sequencing projects from 1997–2023 for comparison, (**b**) relevance of sequencing projects, and (**c**) genome sequencing projects by domain. Statistics were captured from IMG/GOLD14 on January 23, 2023 (https://gold.jgi.doe.gov/statistics)

On the other hand, the completeness of microbial genomes is a question if there is no axenic form available. Figure 11.2a, for example, depicts over 48,000 complete genomes, whereas the number of incomplete genomes has been estimated to be around 38,000. This raises the question: are existing culture-independent or DNA sequencing technologies and strategies competent enough for recovering bacterial genomes? Alneberg et al. 2018 have conducted a study on two culture-independent strategies, i.e., SAGs and MAGs, for recovering bacterial genomes at an offshore station in the Baltic Sea Proper. In this investigation, the SAGs obtained were consistently lower than the matching MAGs. This could be brought on by either inadequate SAG assembly or metagenome contigs that the binning method mistakenly placed in MAGs. There has been evidence that the smaller of the two SAGs was inadequate, i.e., it is missing a sizable portion of the genes shared by the second SAG and the MAG, when two SAGs aligned to the same single MAG, as seen by deeper examination of the situation. Therefore, their findings back up the previously noted explanation. Additionally, combining the parts covered by the SAGs would cover a larger portion of the MAG than either of the two SAGs did separately. Additionally, MAGs demonstrated a low level of redundancy (measured as duplicated SAGs), which would have been larger if MAGs contained a significant number of incorrectly binned contigs. Last but not least, identical SAGs also are less thorough than MAGs (Alneberg et al. 2018). The problems and issues of low-quality and low-size sequences may be rectified by quality control methods, i.e., approaches that utilize sequence information to identify and exclude nontarget sequences (Maruyama et al. 2017), use of high-quality MAGs, comparison, and examination of in situ coverage by metagenomic reads (Meziti et al. 2021).

11.3 MDM: A Bag of Full Surprises and Framework for Targeted Studies

Studies on SAGs and MAGs have revolutionized our understanding of microbial diversity in nature and allowed further research for significant new findings. In contrast to previously done studies, the majority of organisms studied came from medical samples. By sequencing the genomes of yet-uncultivated microorganisms from environmental samples, researchers have been able to uncover a wealth of information about the microbial world that was heretofore unknown. These studies have enabled major discoveries in areas such as environmental microbiology, evolution, and agriculture. Furthermore, they have provided insights into how microbes interact with their environment/host (microbiome) and how they can be used to improve human health. In the last 10 years, research on uncultivated bacteria and archaea using next-generation sequencing technology has made great strides. Researchers have been able to identify and study SAGs and MAGs of uncultivated or *Candidatus* phyla in unprecedented detail, uncovering new insights into their roles in the environment and their potential applications in various sectors as mentioned before. By metagenomics analysis of uncultivated bacteria and archaea from environmental samples, we were able to obtain a number of MAGs or SAGs representing major lineages from environmental samples or discoveries. These genomes came from all around the world (Table 11.1).

The following are examples of some seminal studies in the fields of systematic and evolutionary microbiology: Hug et al. (2016) used over a thousand MDM genomes and public genomic data (Eloe-Fadrosh et al. 2016) to reconstruct the tree of life and to describe a hyper-diverse clade of MDM called the candidate phyla radiation (CPR), which subdivides the domain Bacteria. A new version of the tree of life replicates the expected groups of organisms at the majority of taxonomic levels and is, for the most part, coherent with the tree that was produced by using the conventional SSU rRNA (16S rRNA) gene sequencing. The values of support for clades are strong at the species through class levels (>85%), with support ranging from moderate to too strong for phyla (>75% in most cases); however, the branching

Table 11.1 Comprehensive genome analysis of *Candidatus* Prometheoarchaeum syntrophicum MK-D1

Assembly details	
Total genome length	4,427,796 bp
Contig N50 size	4,427,796
Annotated genome features	
Features	**Number of genes**
Protein coding sequences (CDS)	4163
Repeat regions	43
tRNA	36
rRNA	2
Protein features	
Hypothetical proteins	3375
Proteins with functional assignments	788
Proteins with EC number assignments	406
Proteins with GO assignments	361
Proteins with pathway assignments	322
Proteins with PATRIC genus-specific family (PLfam) assignments	0
Proteins with PATRIC cross-genus family (PGfam) assignments	496
Specialty genes	
Antibiotic resistance	12
Antimicrobial resistance (AMR) genes	
AMR mechanism	**Genes**
Antibiotic activation enzyme	KatG
Antibiotic target in susceptible species	Ddl, folP, gyrA, gyrB, Iso-tRNA, rho
Protein altering cell wall charge conferring antibiotic resistance	GdpD, PgsA

order of the deepest branches cannot be determined with certainty. As a direct result of our decision to prioritize taxon sampling above the number of genes utilized in tree construction, we have a lower level of support for the placement of deep branches. The Eukarya, which comprises fungi, plants, animals, and protists, is a branch within the Archaea, more precisely within the archaeal TACK superphylum, and is a sibling to the Lokiarchaeota. This was a recent hypothesis that was put forward. It is interesting to note that this placement is not apparent in the SSU rRNA tree, which possesses the three-domain topology that was presented by Woese and coworkers in the year 1990. There are two competing ideas for the genesis of Eukarya: the two-domain Eocyte tree and the three-domain tree. Additional research to resolve these and other deep linkages will be reinforced with the availability of genomes for a broader range of organisms. The ribosomal protein tree has several significant advantages over the SSU rRNA gene tree, the most important of which are that it resolves the deeper radiations to a greater degree and that it includes organisms whose SSU rRNA gene sequences are either insufficient or nonexistent. It has been demonstrated that ribosomal proteins exhibit compositional biases throughout all three domains. These biases are driven by lifestyles that are either thermophilic, mesophilic, or halophilic, as well as by an ancient genetic code. If there were more genome sequences for non-extremophile Archaea in the DPANN lineages, these compositional biases might be easier to understand. Moreover, the incorporation of eukaryotes with the existing marker gene collection is difficult due to the overall lack of observable similarity. To generate a solid phylogeny that incorporates all forms of life, additional methodological advances are necessary (Hug et al. 2016; Zhu et al. 2019).

On a similar note, because it is hard to grow microorganisms, there isn't much phylogenetic diversity in the microbial genomes that are currently available. These challenges are somewhat addressed by Parks et al. (2017). They report on the reconstruction of genomes from more than 1500 metagenomes found in the environment and the gastrointestinal tracts of nonhumans, i.e., animals. They did this so they could focus on metagenomes that were likely to have microbial populations from lineages that had not been studied enough. Nearly half of the genomes are 90% complete with 5% contamination, and 50% of the genomes meet preset standard filtering parameters. They were able to recover 7903 genomes from 64,295 MAGs, with 7280 belonging to Bacteria and only 623 to Archaea. On the other hand, the percentage of recovered good-quality genomes is very low, i.e., 12.29%. This indicates either our wet lab protocols for DNA extraction and sequencing require more attention and standardization or we do not have enough analysis power, i.e., powerful algorithms, to analyze the obtained data. However, this 12.29% recovery expanded the diversity of the prokaryotic phylogeny by more than 30% and included 20 new phyla of Archaea and Bacteria (Parks et al. 2017). Due to the limitations of cultivation methods and the nonavailability of native growth and environment, studies on MAGs and SAGs offer the best solution and bridge for understanding the origins and complexity of microscopic cellular life (Spang et al. 2015). The evidence at hand supports theories that the first eukaryotic cell originated from the union of an alphaproteobacterial (mitochondrial) endosymbiont and an archaeal host cell. The host cell belongs to the archaeal phylum Lokiarchaeota, which has many eukaryotic characteristics. Uncertainty surrounds the origin of the complexity that distinguishes eukaryotic cells. In phylogenomic investigations, the uncultivated Archaea in the "Asgard" superphylum are grouped with eukaryotes, and their DNAs are richer for proteins that were previously thought to be unique to eukaryotes. Notably, multiple homologs of eukaryote membrane-trafficking machinery parts, including TRAPP and Sec23/24 domains, are encoded in thorarchaeal genomes. Additionally, thorarchaeal coat proteins that are involved in vesicle production have characteristics with their eukaryotic counterparts (Zaremba-Niedzwiedzka et al. 2017). The closest archaeal ancestors of eukaryotes are in the newly found superphylum Asgard. The archaeal ancestor of eukaryotes may be a sister group to all other archaea or part of the Asgard superphylum (i.e., a fight between two-domain vs. three-domain in the tree of life). It is difficult to say something that may be inappropriate due to the lack of additional data supporting the origin and complexity of Asgard archaeal genomes. The phylogenomic analysis platform for Asgard comparative genomics suggests that the revealed genomic composition of Asgard archaea may be a chosen way of the organism for the survival of the fittest or dynamic evolution, or it may be due to a eukaryotic ancestor within archaea. This means that Asgard archaea represent one of biology's greatest mysteries (Liu et al. 2021).

More understanding of Asgard archaea and Asgard archaea-like organisms becomes a challenge to existing theories on the origin of life, last universal common ancestor (LUCA), cell first theory, protein first theory, the RNA world, and the universal tree of life (ToL). They also make complex glitches in understanding the scientific community about the prevailing biological concepts concerning viruses. This may lead to a question of how eukaryogenesis raised superorganisms (Bell 2022). Based on the decade-long research followed by the isolation of the Asgard archaeon "*Candidatus* Prometheoarchaeum syntrophicum," the entangle-engulf-endogenize model has been proposed for eukaryogenesis in the Asgard archaea (McKay et al. 2019; Imachi et al. 2020).

11.3.1 Comprehensive Genome Analysis of Candidatus *Prometheoarchaeum Syntrophicum* MK-D1

Candidatus Prometheoarchaeum syntrophicum MK-D1 is the first Asgard Archaea strain cultivated in the laboratory and has an axenic culture. The first Asgard archaeon to be cultivated coexists metabolically with hydrogen-scavenging bacteria. Its complete genome study verifies the Asgard Archaea's existence (López-García and Moreira 2020). On the basis of genome annotation statistics of strain MK-D1 in PATRIC (Fig. 11.3), this genome seems to be of poor quality (Wattam et al. 2017).

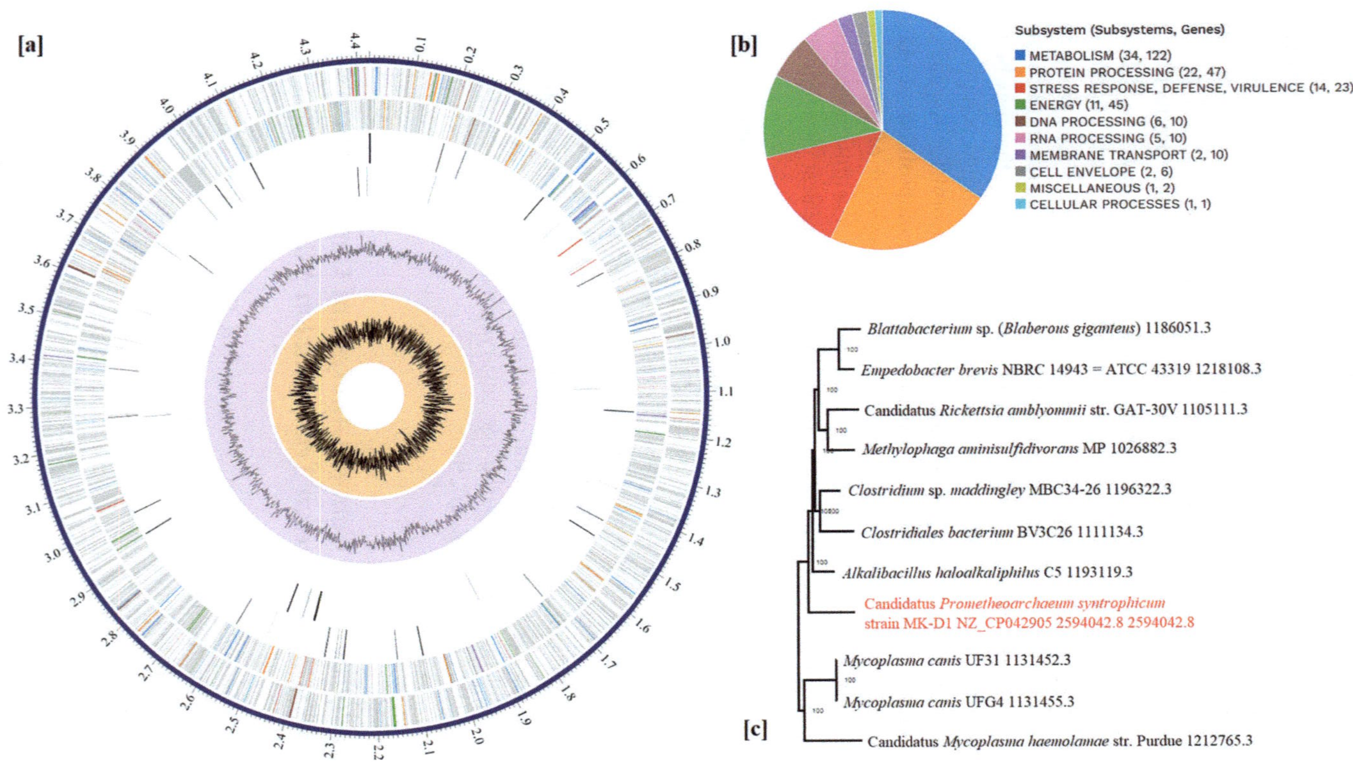

Fig. 11.3 Comprehensive genome analysis of *Candidatus* Prometheoarchaeum syntrophicum MK-D1 (NCBI Accession No.: NZ_CP042905): the annotation statistics and comparison. (**a**) Genome annotation using BV-BRC version 3.28.9 (previous name PATRIC), (**b**) subsystem analysis (subsystems, genes), and (**c**) phylogenetic analysis with fast bootstrapping

The comprehensive genome analysis shows that genome of *Candidatus*PrometheoarchaeumsyntrophicumMK-D1 has a total length of 4,427,796 bp and an average of 31.17% (G + C) content (Table 11.1). The RAST tool kit (RASTtk) (Brettin et al. 2015) was used to annotate the MK-D1 genome, and it was given the genomic identifier 2594042.8. Using genetic code 11, the MK-D1 genome, which belongs to the superkingdom Archaea, was annotated. The taxonomy of this genome is cellular organisms > Archaea > Asgard group > *Candi datus*Lokiarchaeota > *CandidatusPrometheoarchaeum* > *Candidatus* Prometheoarchaeumsyntrophicum. The MK-D1 genome consists of 2 rRNA genes, 36 tRNA genes, and 4163 CDS. The MK-D1 genome annotation discloses the presence of too many hypothetical proteins (total CDS 3375) as compared to functional proteins (total nos. 788). It also contains proteins with mapped KEGG pathways (total no. 322) (Kanehisa et al. 2016), gene ontology (total no. 361) (Ashburner et al. 2000), and a total of 406 proteins with Enzyme Commission (EC) numbers (Schomburg et al. 2004). PATRIC annotation suggests that MD-K1 has 496 proteins that belong to the cross-genus protein families (PGFams) and is devoid of proteins belonging to the genus-specific protein families (PLFams) (Davis et al. 2016). The distribution of the genome annotations is shown graphically in a circular map (Fig. 11.3a) from the outer ring of contigs to the innermost ring of GC skew. The second and third vertical lines indicate forward and reverse strands of CDS, followed by vertical lines of RNA genes, AMR genes, virulence factors, GC content, and GC skew. Colors indicated on the CDS forward and reverse strand help to identify respective subsystems (see Fig. 11.3b) and specific genes (Overbeek et al. 2005). Annotated genes are identified as drug targets (Zhu et al. 2009; Law et al. 2014), antibiotic resistance genes (McArthur et al. 2013), virulence factors (Mao et al. 2015; Chen et al. 2016), and transporters (Saier Jr et al. 2015) (see Table 11.1). A k-mer-based AMR gene detection method allowed to assigning of identities to representative sequence variants of AMR gene (Wattam et al. 2017). A phylogenetic analysis of MK-D1 was also performed (Fig. 11.3b), and the closest reference genomes were compared and identified by Mash/MinHash (Ondov et al. 2016) in BV-BRC version 3.28.9. To ascertain the evolutionary position of this genome, PGFams (Davis et al. 2016) were chosen from these genomes. The nucleotides for each of these sequences were mapped to the protein alignment after the protein sequences from these families were aligned with MUSCLE (Edgar 2004). Fast bootstrapping (Stamatakis 2014) was used to produce the support values in the tree after the joint set of amino acid and nucleotide alignments were concatenated into a data matrix. RaxML (Stamatakis et al. 2008; Stamatakis 2014) was then used to analyze the matrix (Fig. 11.3c).

Studies like MK-D1, which is an Asgard Achaea whose genome was recovered from MDM and grown from samples that were first used for MDM analysis, are changing how we think about the way life started and how organisms change over time. Surprisingly, the metabolic adaptability of candidate phyla (Kantor et al. 2013) in MDM is another noteworthy potential that attracts scientific attention. For instance, dissimilatory sulfate reduction (Wrighton et al. 2012; 2013; Seitz et al. 2016) and anaerobic methane (CH_4) oxidation in Korarchaeota MAGs indicate that these two pathways are generally expressed by distinct syntrophy partners contained in a single strain (McKay et al. 2019). MDM's functional adaptability reveals their "colorful" side, surprising us and providing us with a greater perception of the world we lived within. In addition, genomic analysis of MDM has revealed new microbial functions which could be applied to biomedical, bioenergy, and biotechnology advancements, as well as bioremediation.

From a health perspective, those microorganisms adopted human habitat, and part of human and animal microbiomes have started to gain antimicrobials that reminded us about the need for novel and effective antibiotics. In search of novel antibiotics, Kim and his team decided to target the hidden diversity of uncultivated microorganisms. Lewis found soil actinomycetes capable of producing Lassomycinare, a potent inhibitor of Mycobacteria, by making the chaperone ClpC1 that breaks down ATP. Teixobactin, which is made by a -proteobacterium, is almost completely resistant-free. It binds to immutable lipid II and lipid III targets. These examples imply that microorganisms evolved natural compounds that address resistance and tolerance, and mining uncultured bacteria could revive antibiotic research (Lewis 2018). The identification of some new antibiotic compounds, such as tetarimycin A (Kallifidas et al. 2012), was accomplished through the expression of biosynthetic gene clusters from metagenomic DNA rather than through the use of conventional cell culture methods. Additional discoveries along these lines are anticipated to assist in alleviating the current antibiotic crisis (Wright 2017). From a cow rumen metagenome, Hess et al. described 27,755 potential carbohydrate-active enzyme genes, and they functionally cloned genes and evaluated the heterologous production of 90 enzymes in *E. coli* (Hess et al. 2011). These enzymes have the potential to be utilized in the processing of lignocellulosic agricultural biomass waste for either the biofuels or specialty chemical industries. Finally, stable-isotope probing and metagenomic sequencing were used to identify MDM that took part in the breakdown of hydrocarbons. This finding suggests that MDM plays a crucial role in the natural breakdown of petroleum in the wake of the Deepwater Horizon oil spill. It also suggests that MDM could potentially be stimulated for use in bioremediation strategies (Dojka et al. 1998) to remove oil spills (Dombrowski et al. 2016).

11.4 Challenges and Opportunities

MDM science has moved from a period in which data were scarce to one in which there is an abundance of data as a result of the ease with which DNA sequencing can now be performed (Konstantinidis et al. 2017). The quantity of genetic data stored in public databases will continue to expand, and an ever-increasing number of MDM will be found and investigated. The question of how to derive the greatest benefit from massive amounts of data will continue to be at the forefront of study in the field of microbiology; however, the time has come to start thinking about what we should be doing in the post-genomic era. The ability to communicate more effectively is the primary obstacle to overcome. Due to the fluid nature of the taxonomy, communication regarding MDM can be challenging in the current scenario. The most recent version of the International Code of Nomenclature of Prokaryotes (ICNP), which is responsible for regulating the naming of bacteria and archaea, does not cover uncultivated microorganisms, which makes it difficult for scientists to communicate with one another (Konstantinidis et al. 2017). For example, the names "Ca. Bathosphaeria" and "Tepidiformia," which were addressed by using MAGs (Mehrshad et al. 2018) and axenic cultures, respectively, have been attributed to the same genus within an earlier uncultivated class that is derived from the phylum *Choroflexi* (Kochetkova et al. 2020). Both of these names were derived from the word "Bathosphaeria." Previous to that, other perspectives referred to this class as TK10, TK17, or OLB14, respectively. There is little questioning that inconsistency and synonymies in the nomenclature of microorganisms contribute to misunderstanding within the science community and even beyond (e.g., agriculture, medicine, biotechnology, and existing rules of nomenclature). Therefore, there is an immediate need for a set of standards that may be used to name, assemble, and transmit information regarding MDM. To accomplish this goal, it has been suggested that DNA be considered as nomenclature type for taxonomic classifications of prokaryotes instead of or in addition to pure cultures (Whitman 2015), and it has been advised that MDM communities have an independent nomenclatural system.

The development of a robust nomenclature and taxonomy for the prokaryotes, even those that have never been cultivated, was the motivation for the recent proposal of a road plan for naming yet-to-be-uncultivated Archaea and Bacteria. Nevertheless, participation from the research community is still necessitated by these guidelines (Murray et al. 2020). As of today, we know that existing nutrient media do not support the cultivation of most prokaryotes and they can't be grown in pure cultures. So they can't be named according to the rules and suggestions of the International Code of Nomenclature of Prokaryotes (ICNP). However, Hedland et al. proposed a new system for naming uncultivated prokaryote genome sequences or *Candidatus* through a robust nomenclature system for the prokaryotes named the SeqCode. This code makes it possible to publish the names of prokaryotes based on the sequences of an isolate genome, a genome put together from a metagenome, or a single-amplified genome. Other than that, the way names are made and the rules for priority are the same as with the ICNP. It works through the SeqCode Registry (https://seqco.de), which is a registration portal where names and nomenclatural types can be registered, validated, and linked to metadata. They talk about the two ways that SeqCode lets you register and validate names, including *Candidatus* names, and give examples for both – cultured and uncultured. There are suggestions about the minimum standards for DNA sequences. So, the SeqCode gives a consistent and objective way to name all prokaryotes, whether or not they can be grown in a laboratory (Hedlund et al. 2022).

The second and biggest dilemma is that it is still hard to figure out how things work by looking at genome-wide information. Metabolic predictions are based on genome-wide information from SAGs and MAGs of MDM and are more of a possibility than a strong proof. Over 50% of the genes in MDM genomes can't be given any of the functions which makes it very hard to figure out what MDM does. For example, Vavourakis et al. pointed out that the crucial genes of the Wood-Ljungdahl (WL) pathway were found in new clades, even in the phylum *Actinomycetota* (earlier referred to as *Actinobacteria*), which are usually thought of as aerobic chemoheterotrophs (Vavourakis et al. 2018). We still don't know if MDM can use the WL pathway to fix the carbon dioxide (CO_2) and, if so, what are the electron donors for CO_2 fixation. Only recently the function and evolution of the WL pathway in the phylum *Actinomycetota* were highlighted by Jiao et al. (2021b) to be either present in LCA or acquired independently through multiple horizontal gene transfer events. In the post-genomic epoch, this area of study will need to be moved forward by studying how MDM works. As cultivation has always had a big impact on functional research, more and more people are trying to use culture-dependent, culture-independent approaches and advanced bioinformatics pipelines/tools to bridge the gap between MDM and their function. Metatranscriptomics, metabolomics, and metaproteomics are all ways to find out how genes and biochemical pathways are expressed and to study how metabolites change over time. Along with these omics approaches, other tools like fluorescence in situ hybridization (FISH), microautoradiography (MAR), nanometer-scale secondary-ion mass spectrometry (NanoSIMS), and even Raman techniques can show how specific MDM species help with assimilation (Wagner et al. 2006; Neufeld and Murrell 2007; Behrens et al. 2008). Additionally, a useful technique for determining the function of one or more target genes is the biosynthesis and heterologous production of distinct functional MDM genes (Lloyd et al. 2013).

The cultivation of MDM is still challenging, despite the success we have made utilizing methods that are independent of cultivation (Xian et al. 2020), which brings us to the third challenge. The postulates of Koch provide an outline of the processes that must be taken to prove a causal association between a particular bacterium and a disease. The postulates include isolating the microbe from the diseased host, demonstrating that the microbe is present in all cases of the illness and producing the disease when introducing the microbe to a host that is susceptible to it, and isolating the very same infectious agent from the diseased host that was experimentally infected with the disease. Koch was successful in establishing a causal relationship between a particular microbe and a disease. It is because of his efforts that the germ theory of illness was established. This conceptual framework continues to be an important guiding principle in microbiology (Koch 1882). The scientific community relies heavily on pure cultures because they make it possible to ascribe the physiological characteristics of purified organisms unambiguously and because they provide essential resources. Despite the gains that have been made in cultivation-independent methods, isolating resistant MDM remains intrinsically difficult. Fortunately, the insights gained through SAGs and MAGs, multi-omics strategies, and different function-based labeling techniques, including MAR and nanoSIMs, can provide the foundation required for testing hypotheses to relate species to functions and for rationally designing tailored cultivation tactics. For example, Chen et al. (2018) used 39 SAGs/MAGs of Micrarchaeota and Parvarchaeota to deduce their metabolic potential. They subsequently constructed low-oxygen enrichment procedures to corroborate the organisms' hypothesized microaerobic/anaerobic lifestyle (Chen et al. 2018). A small percentage of the diverse microbiological community developed on artificial media is one of the oldest microbiological mysteries. Therefore, the "uncultivable" microbial majority becomes the Earth's largest unknown biological reservoir for new biochemicals. This new biochemical reservoir may provide original growth conditions to microorganisms and natural growth ingredients. This was proven by an isolation chip (ichip) that possesses several hundred small diffusion chambers. The ichip recovers microbial species manyfold more than ordinary cultivation methods and recovers phylogenetically new taxa. The novel approach allows basic and applied research on a wide range of previously inaccessible microbes (Nichols et al. 2010). It is important to note that the design of growth methods frequently ignores the interactions between microorganisms, which makes it difficult to grow MDM with syntrophic partners. Therefore, networks that are backed by data, such as co-occurrence networks derived from data sets of 16S rRNA genes and functional networks derived from multi-omics data (Xian et al. 2020; Zamkovaya et al. 2021; Salam et al. 2021), would offer a great deal of potential for guiding the design of culture strategies. In general, the imaginative use of fresh methods to cultivate and isolate unique microbes still presents us with limitless opportunities for new discoveries.

11.5 The Prospects for MDM Research in the Future

There is an incredible variety of microorganisms that live on Earth, the vast majority of which have not been grown, and the ecological roles that they play are just dimly understood. We identified three significant issues with MDM in the section under "Challenges and Opportunities" above. These challenges are to determine where they are, who they really are, and what they are capable of doing. To advance MDM research in the future, we propose three different strategies. First, large-scale investigations of microbial community dynamics (MDM) that integrate various omics approaches (i.e., metagenomics, metatranscriptomics, metaproteomics, metabolomics, and culturomics) are still required to investigate a range of natural environments, especially microbiomes of extreme habitats, which open our subconscious on the taxonomic and functional diversity of microbial life. This is because extreme habitats contain organisms that have adapted to survive in harsh conditions. Second, we must work to improve the linkages between taxa and the functions that they perform in their natural environments. Integration of omics, FISH, MAR, NanoSIMS, Raman, and heterologous expression tools, which offer insights into the mechanisms of species, may now be used to study this phenomenon, which can now be examined. Third, we need to design and investigate new methods of culture that are directed by information derived from several omics and network approaches to put hypotheses to the test and obtain enrichments or pure cultures. We also highlight the need to establish MDM research centers to tackle these challenges in a variety of ways, including developing testing facilities and providing technical assistance, as well as training the next generation of scientists and providing academic communication platforms. To gain a deeper understanding of MDM, it is necessary to combine a wide variety of academic fields and to have knowledge that spans multiple dimensions. The investigation of MDM is like to opening a magic box, which may or may not contain a treasure chest full of surprises. The science of MDM is currently in its infancy, and future examination of MDM will likely yield unexpected and intriguing answers concerning the evolution of MDM and the functions that MDM plays in nature. However, as a scientist, I can tell you that the great things exhibited by the "box" are not the conclusive result of MDM research; rather, they serve as a starting point for future investigation of ecological and evolutionary phenomena.

Funding This study was supported by the National Natural Science Foundation of China (nos. 92251302, 32370011, 32200007, 31972856, 91951205, and 32061143043), Key-Area Research and Development Program of Guangdong Province (2022B0202110001), and the National Key Research and Development Program of China (2021YFD1500300). BNR duly acknowledges funding support from Yenepoya (Deemed to be University) (project no.: YU/SeedGrant/104/2021).

Author Contributions WJL and JYJ designed this work, BNR worked on the manuscript and explored ideas, NS edited the manuscript, and all coauthors contributed to writing of the manuscript.

References

Alneberg J, Bjarnason BS, de Bruijn I, Schirmer M, Quick J, Ijaz UZ, Lahti L, Loman NJ, Andersson AF, Quince C (2014) Binning metagenomic contigs by coverage and composition. Nat Methods 11(11):1144–1146. https://doi.org/10.1038/nmeth.3103

Alneberg J, Karlsson CMG, Divne AM, Bergin C, Homa F, Lindh MV, Hugerth LW, Ettema TJG, Bertilsson S, Andersson AF, Pinhassi J (2018) Genomes from uncultivated prokaryotes: a comparison of metagenome-assembled and single-amplified genomes. Microbiome 6:173. https://doi.org/10.1186/s40168-018-0550-0

Anantharaman K, Brown C, Hug L, Sharon I, Castelle CJ, Probst AJ, Thomas BC, Singh A, Wilkins MJ, Karaoz U, Brodie EL, Williams KH, Hubbard SS, Banfield JF (2016) Thousands of microbial genomes shed light on interconnected biogeochemical processes in an aquifer system. Nat Commun 7:13219. https://doi.org/10.1038/ncomms13219

Ashburner M, Ball CA, Blake JA, Botstein D, Butler H, Cherry JM, Davis AP, Dolinski K, Dwight SS, Eppig JT (2000) Gene ontology: tool for the unification of biology. Nat Genet 25:25. https://doi.org/10.1038/75556

Behrens S, Lösekann T, Pett-Ridge J, Weber PK, Ng WO, Stevenson BS, Hutcheon ID, Relman DA, Spormann AM (2008) Linking microbial phylogeny to metabolic activity at the single-cell level by using enhanced element labeling-catalyzed reporter deposition fluorescence in situ hybridization (EL-FISH) and NanoSIMS. Appl Environ Microbiol 74(10):3143–3150. https://doi.org/10.1128/AEM.00191-08

Bell PJL (2022) Eukaryogenesis: the rise of an emergent superorganism. Front Microbiol 13:858064. https://doi.org/10.3389/fmicb.2022.858064

Brettin T, Davis JJ, Disz T, Edwards RA, Gerdes S, Olsen GJ, Olson R, Overbeek R, Parrello B, Pusch GD, Shukla M, Thomason JA 3rd, Stevens R, Vonstein V, Wattam AR, Xia F (2015) RASTtk: a modular and extensible implementation of the RAST algorithm for building custom annotation pipelines and annotating batches of genomes. Sci Rep 5:8365. https://doi.org/10.1038/srep08365

Chen L, Zheng D, Liu B, Yang J, Jin Q (2016) VFDB 2016: hierarchical and refined dataset for big data analysis-10 years on. Nucleic Acids Res 44:D694–D697. https://doi.org/10.1093/nar/gkv1239

Chen LX, Méndez-García C, Dombrowski N, Servín-Garcidueñas LE, Eloe-Fadrosh EA, Fang BZ, Luo ZH, Tan S, Zhi XY, Hua ZS, Martinez-Romero E, Woyke T, Huang LN, Sánchez J, Peláez AI, Ferrer M, Baker BJ, Shu WS (2018) Metabolic versatility of small archaea Micrarchaeota and Parvarchaeota. ISME J 12(3):756–775. https://doi.org/10.1038/s41396-017-0002-z

Davis JJ, Gerdes S, Olsen GJ, Olson R, Pusch GD, Shukla M, Vonstein V, Wattam AR, Yoo H (2016) PATtyFams: protein families for the microbial genomes in the PATRIC database. Front Microbiol 7:118. https://doi.org/10.3389/fmicb.2016.00118

Dodsworth J, Blainey P, Murugapiran S, Swingley WD, Ross CA, Tringe SG, Chain PSG, Scholz MB, Lo CC, Raymond J, Quake SR, Hedlund BP (2013) Single-cell and metagenomic analyses indicate a fermentative and saccharolytic lifestyle for members of the OP9 lineage. Nat Commun 4:1854. https://doi.org/10.1038/ncomms2884

Dojka MA, Hugenholtz P, Haack SK, Pace NR (1998) Microbial diversity in a hydrocarbon- and chlorinated-solvent-contaminated aquifer undergoing intrinsic bioremediation. Appl Environ Microbiol 64(10):3869–3877. https://doi.org/10.1128/AEM.64.10.3869-3877.1998

Dombrowski N, Donaho JA, Gutierrez T, Seitz KW, Teske AP, Baker BJ (2016) Reconstructing metabolic pathways of hydrocarbon-degrading bacteria from the Deepwater Horizon oil spill. Nat Microbiol 1(7):16057. https://doi.org/10.1038/nmicrobiol.2016.57

Edgar RC (2004) MUSCLE: multiple sequence alignment with high accuracy and high throughput. Nucleic Acids Res 32:1792–1797. https://doi.org/10.1093/nar/gkh340

Eloe-Fadrosh EA, Paez-Espino D, Jarett J, Dunfield PF, Hedlund BP, Dekas AE, Grasby SE, Brady AL, Dong H, Briggs BR, Li WJ, Goudeau D, Malmstrom R, Pati A, Pett-Ridge J, Rubin EM, Woyke T, Kyrpides NC, Ivanova NN (2016) Global metagenomic survey reveals a new bacterial candidate phylum in geothermal springs. Nat Commun 7:10476. https://doi.org/10.1038/ncomms10476

Hedlund BP, Chuvochina M, Hugenholtz P, Konstantinidis KT, Murray AE, Palmer M, Parks DH, Probst AJ, Reysenbach AL, Rodriguez-R LM, Rossello-Mora R, Sutcliffe IC, Venter SN, Whitman WB (2022) SeqCode: a nomenclatural code for prokaryotes described from sequence data. Nat Microbiol 7(10):1702–1708. https://doi.org/10.1038/s41564-022-01214-9

Hess M, Sczyrba A, Egan R, Kim TW, Chokhawala H, Schroth G, Luo S, Clark DS, Chen F, Zhang T, Mackie RI, Pennacchio LA, Tringe SG, Visel A, Woyke T, Wang Z, Rubin EM (2011) Metagenomic discovery of biomass-degrading genes and genomes from cow rumen. Science 331(6016):463–467. https://doi.org/10.1126/science.1200387

Hug L, Baker B, Anantharaman K, Brown CT, Probst AJ, Castelle CJ, Butterfield CN, Hernsdorf AW, Amano Y, Ise K, Suzuki Y, Dudek N, Relman DA, Finstad KM, Amundson R, Thomas BC, Banfield JF (2016) A new view of the tree of life. Nat Microbiol 1:16048. https://doi.org/10.1038/nmicrobiol.2016.48

Imachi H, Nobu MK, Nakahara N, Morono Y, Ogawara M, Takaki Y, Takano Y, Uematsu K, Ikuta T, Ito M, Matsui Y, Miyazaki M, Murata K, Saito Y, Sakai S, Song C, Tasumi E, Yamanaka Y, Yamaguchi T, Kamagata Y, Tamaki H, Takai K (2020) Isolation of an archaeon at the prokaryote-eukaryote interface. Nature 577(7791):519–525. https://doi.org/10.1038/s41586-019-1916-6

Imelfort M, Parks D, Woodcroft BJ, Dennis P, Hugenholtz P, Tyson GW (2014) GroopM: an automated tool for the recovery of population genomes from related metagenomes. PeerJ 2:e603. https://doi.org/10.7717/peerj.603

Jiao JY, Fu L, Jua ZS, Liu L, Salam N, Liu PF, Lv AP, Wu G, Xian WD, Zhu Q, Zhou EM, Fang BZ, Oren A, Hedlund BP, Jiang HC, Knight R, Cheng L, Li WJ (2021b) Insight into the function and evolution of the Wood-Ljungdahl pathway in *Actinobacteria*. ISME J 15:3005–3018. https://doi.org/10.1038/s41396-021-00935-9

Jiao JY, Liu L, Hua ZS, Fang BZ, Zhou EM, Salam N, Hedlund BP, Li WJ (2021a) Microbial dark matter coming to light: challenges and opportunities. Nat Sci Rev 8:nwaa280. https://doi.org/10.1093/nsr/nwaa280

Kallifidas D, Kang HS, Brady SF (2012) Tetarimycin A, an MRSA-active antibiotic identified through induced expression of environmental DNA gene clusters. J Am Chem Soc 134(48):19552–19555. https://doi.org/10.1021/ja3093828

Kanehisa M, Sato Y, Kawashima M, Furumichi M, Tanabe M (2016) KEGG as a reference resource for gene and protein annotation. Nucleic Acids Res 44:D457-462. https://doi.org/10.1093/nar/gkv1070

Kang DD, Froula J, Egan R, Wang Z (2015) MetaBAT, an efficient tool for accurately reconstructing single genomes from complex microbial communities. PeerJ 3:e1165. https://doi.org/10.7717/peerj.1165

Kantor RS, Wrighton KC, Handley KM, Sharon I, Hug LA, Castelle CJ, Thomas BC, Banfield JF (2013) Small genomes and sparse metabolisms of sediment-associated bacteria from four candidate phyla. MBio 4(5):e00708–e00713. https://doi.org/10.1128/mBio.00708-13

Koch R (1882) The etiology of tuberculosis. ZentralblBakteriolMikrobiolHygA Med MikrobiolInfektParasitol 251(3):287–296

Kochetkova TV, Zayulina KS, Zhigarkov VS, Minaev NV, Chichkov BN, Novikov AA, Toshchakov SV, Elcheninov AG, Kublanov IV (2020) *Tepidiformabonchosmolovskayae* gen. nov., sp. nov., a moderately thermophilic *Chloroflexi* bacterium from a Chukotka hot spring (Arctic, Russia), representing a novel class, *Tepidiformia*, which includes the previously uncultivated lineage OLB14. Int J Syst Evol Microbiol 70(2):1192–1202. https://doi.org/10.1099/ijsem.0.003902

Konstantinidis KT, Rosselló-Móra R, Amann R (2017) Uncultivated microbes in need of their own taxonomy. ISME J 11(11):2399–2406. https://doi.org/10.1038/ismej.2017.113

Law V, Knox C, Djoumbou Y, Jewison T, Guo AC, Liu Y, Maciejewski A, Arndt D, Wilson M, Neveu V, Tang A, Gabriel G, Ly C, Adamjee S, Dame ZT, Han B, Zhou Y, Wishart DS (2014) DrugBank 4.0: shedding new light on drug metabolism. Nucleic Acids Res 42:D1091-1097. https://doi.org/10.1093/nar/gkt1068

Lewis K (2018) Antibiotics from the microbial dark matter. 31(S1):257.2-257.2. https://doi.org/10.1096/fasebj.31.1_supplement.257.2

Liu Y, Makarova KS, Huang WC, Wolf YI, Nikolskaya AN, Zhang X, Cai M, Zhang CJ, Xu W, Luo Z, Cheng L, Koonin EV, Li M (2021) Expanded diversity of Asgard archaea and their relationships with eukaryotes. Nature 593(7860):553–557. https://doi.org/10.1038/s41586-021-03494-3

Lloyd KG, Schreiber L, Petersen DG, Kjeldsen KU, Lever MA, Steen AD, Stepanauskas R, Richter M, Kleindienst S, Lenk S, Schramm A, Jørgensen BB (2013) Predominant archaea in marine sediments degrade detrital proteins. Nature 496(7444):215–218. https://doi.org/10.1038/nature12033

Loceya KJ, Lennon JT (2016) Scaling laws predict global microbial diversity. PNAS 113(21):5970–5975. https://doi.org/10.1073/pnas.1521291113

López-García P, Moreira D (2020) Cultured Asgard archaea shed light on Eukaryogenesis. Cell 181(2):232–235. https://doi.org/10.1016/j.cell.2020.03.058

Mao C, Abraham D, Wattam AR, Wilson MJ, Shukla M, Yoo HS, Sobral BW (2015) Curation, integration and visualization of bacterial virulence factors in PATRIC. Bioinformatics 31:252–258. https://doi.org/10.1093/bioinformatics/btu631

Marcy Y, Ouverney C, Bik EM, Lösekann T, Ivanova N, Martin HG, Szeto E, Platt D, Hugenholtz P, Relman DA, Quake SR (2007) Dissecting biological "dark matter" with single-cell genetic analysis of rare and uncultivated TM7 microbes from the human mouth. Proc Natl Acad Sci U S A 104(29):11889–11894. https://doi.org/10.1073/pnas.0704662104

Maruyama T, Mori T, Yamagishi K, Takeyama H (2017) SAG-QC: quality control of single amplified genome information by subtracting non-target sequences based on sequence compositions. BMC Bioinformatics 18:152. https://doi.org/10.1186/s12859-017-1572-5

McArthur AG, Waglechner N, Nizam F, Yan A, Azad MA, Baylay AJ, Bhullar K, Canova MJ, De Pascale G, Ejim L, Kalan L, King AM, Koteva K, Morar M, Mulvey MR, O'Brien JS, Pawlowski AC, Piddock LJ, Spanogiannopoulos P, Sutherland AD, Tang I, Taylor PL, Thaker M, Wang W, Yan M, Yu T, Wright GD (2013) The comprehensive antibiotic resistance database. Antimicrob Agents Chemother 57(7):3348–3357. https://doi.org/10.1128/AAC.00419-13

McKay LJ, Dlakić M, Fields MW, Delmont TO, Eren AM, Jay ZJ, Klingelsmith KB, Rusch DB, Inskeep WP (2019) Co-occurring genomic capacity for anaerobic methane and dissimilatory sulfur metabolisms discovered in the Korarchaeota. Nat Microbiol 4(4):614–622. https://doi.org/10.1038/s41564-019-0362-4

McLean JS, Lombardo MJ, Badger JH, Edlund A, Novotny M, Yee-Greenbaum J, Vyahhi N, Hall AP, Yang Y, Dupont CL, Ziegler MG, Chitsaz M, Allen AE, Yooseph S, Tesler G, Pevzner PA, Friedman RM, Nealson KH, Venter JC, Lasken RS (2013) Candidate phylum TM6 genome recovered from a hospital sink biofilm provides genomic insights into this uncultivated phylum. PNAS 10(26):E2390–E2399. https://doi.org/10.1073/pnas.1219809110

Mehrshad M, Salcher MM, Okazaki Y, Nakano SI, Šimek K, Andrei AS, Ghai R (2018) Hidden in plain sight-highly abundant and diverse planktonic freshwater Chloroflexi. Microbiome 6(1):176. https://doi.org/10.1186/s40168-018-0563-8

Meziti A, Rodriguez-R LM, Hatt JK, Peña-Gonzalez A, Levy K, Konstantinidis KT (2021) The reliability of metagenome-assembled genomes (MAGs) in representing natural populations: insights from comparing MAGs against isolate genomes derived from the same Fecal sample. Appl Environ Microbiol 87(6):e02593-20. https://doi.org/10.1128/AEM.02593-20

Murray AE, Freudenstein J, Gribaldo S, Hatzenpichler R, Hugenholtz P, Kämpfer P, Konstantinidis KT, Lane CE, Papke RT, Parks DH, Rossello-Mora R, Stott MB, Sutcliffe IC, Thrash JC, Venter SN, Whitman WB, Acinas SG, Amann RI, Anantharaman K, Armengaud J, Baker BJ, Barco RA, Bode HB, Boyd ES, Brady CL, Carini P, Chain PSG, Colman DR, DeAngelis KM, de Los Rios MA, Estrada-de Los Santos P, Dunlap CA, Eisen JA, Emerson D, Ettema TJG, Eveillard D, Girguis PR, Hentschel U, Hollibaugh JT, Hug LA, Inskeep WP, Ivanova EP, Klenk HP, Li WJ, Lloyd KG, Löffler FE, Makhalanyane TP, Moser DP, Nunoura T, Palmer M, Parro V, Pedrós-Alió C, Probst AJ, Smits THM, Steen AD, Steenkamp ET, Spang A, Stewart FJ, Tiedje JM, Vandamme P, Wagner M, Wang FP, Yarza P, Hedlund BP, Reysenbach AL (2020) Roadmap for naming uncultivated archaea and bacteria. Nat Microbiol 5(8):987–994. https://doi.org/10.1038/s41564-020-0733-x

Neufeld JD, Murrell JC (2007) Witnessing the last supper of uncultivated microbial cells with Raman-FISH. ISME J 1(4):269–270. https://doi.org/10.1038/ismej.2007.55

Nichols D, Cahoon N, Trakhtenberg EM, Pham L, Mehta A, Belanger A, Kanigan T, Lewis K, Epstein SS (2010) Use of ichip for high-throughput in situ cultivation of "uncultivable" microbial species. Appl Environ Microbiol 76(8):2445–2450. https://doi.org/10.1128/AEM.01754-09

Ondov BD, Treangen TJ, Melsted P, Mallonee AB, Bergman NH, Koren S, Phillippy AM (2016) Mash: fast genome and metagenome distance estimation using MinHash. Genome Biol 17:132. https://doi.org/10.1186/s13059-016-0997-x

Overbeek R, Begley T, Butler RM, Choudhuri JV, Chuang HY, Cohoon M, de Crécy-Lagard V, Diaz N, Disz T, Edwards R, Fonstein M, Frank ED, Gerdes S, Glass EM, Goesmann A, Hanson A, Iwata-Reuyl D, Jensen R, Jamshidi N, Krause L, Kubal M, Larsen N, Linke B, McHardy AC, Meyer F, Neuweger H, Olsen G, Olson R, Osterman A, Portnoy V, Pusch GD, Rodionov DA, Rückert C, Steiner J, Stevens R, Thiele I, Vassieva O, Ye Y, Zagnitko O, Vonstein V (2005) The subsystems approach to genome annotation and its use in the project to annotate 1000 genomes. Nucleic Acids Res 33(17):5691–5702. https://doi.org/10.1093/nar/gki866

Parks DH, Rinke C, Chuvochina M, Chaumeil PA, Woodcroft BJ, Evans PN, Hugenholtz P, Tyson GW (2017) Recovery of nearly 8,000 metagenome-assembled genomes substantially expands the tree of life. Nat Microbiol 2(11):1533–1542. https://doi.org/10.1038/s41564-017-0012-7

Rinke C, Schwientek P, Sczyrba A, Ivanova NN, Anderson IJ, Cheng JF, Darling A, Malfatti S, Swan BK, Gies EA, Dodsworth JA, Hedlund BP, Tsiamis J, Sievert SM, Liu WT, Eisen JA, Hallam SJ, Kyrpides NC, Stepanauskas R, Rubin EM, Hugenholtz P, Woyke T (2013) Insights into the phylogeny and coding potential of microbial dark matter. Nature 499:431–437. https://doi.org/10.1038/nature12352

Saier MH Jr, Reddy VS, Tsu BV, Ahmed MS, Li C, Moreno-Hagelsieb G (2015) The transporter classification database (TCDB): recent advances. Nucleic Acids Res 44:D372–D379. https://doi.org/10.1093/nar/gkv1103

Salam N, Xian WD, Asem MD, Xiao M, Li WJ (2021) From ecophysiology to cultivation methodology: filling the knowledge gap between uncultured and cultured microbes. Mar Life Sci Technol 3:132–147. https://doi.org/10.1007/s42995-020-00064-w

Schomburg I, Chang A, Ebeling C, Gremse M, Heldt C, Huhn G, Schomburg D (2004) BRENDA, the enzyme database: updates and major new developments. Nucleic Acids Res 32:D431–D433. https://doi.org/10.1093/nar/gkh081

Seitz K, Lazar C, Hinrichs KU, Teske AP, Baker BJ (2016) Genomic reconstruction of a novel, deeply branched sediment archaeal phylum with pathways for acetogenesis and sulfur reduction. ISME J 10:1696–1705. https://doi.org/10.1038/ismej.2015.233

Spang A, Saw JH, Jørgensen SL, Zaremba-Niedzwiedzka K, Martijn J, Lind AE, van Eijk R, Schleper C, Guy L, Ettema TJG (2015) Complex archaea that bridge the gap between prokaryotes and eukaryotes. Nature 521(7551):173–179. https://doi.org/10.1038/nature14447

Stamatakis A (2014) RAxML version 8: a tool for phylogenetic analysis and post-analysis of large phylogenies. Bioinformatics 30:1312–1313. https://doi.org/10.1093/bioinformatics/btu033

Stamatakis A, Hoover P, Rougemont J (2008) A rapid bootstrap algorithm for the RAxML web servers. Syst Biol 57:758–771. https://doi.org/10.1080/10635150802429642

Strous M, Kraft B, Bisdorf R, Tegetmeyer HE (2012) The binning of metagenomic contigs for microbial physiology of mixed cultures. Front Microbiol 3:410. https://doi.org/10.3389/fmicb.2012.00410

Vavourakis CD, Andrei AS, Mehrshad M, Ghai R, Sorokin DY, Muyzer G (2018) A metagenomics roadmap to the uncultured genome diversity in hypersaline soda lake sediments. Microbiome 6(1):168. https://doi.org/10.1186/s40168-018-0548-7

Wagner M, Nielsen PH, Loy A, Nielsen JL, Daims H (2006) Linking microbial community structure with function: fluorescence in situ hybridization-microautoradiography and isotope arrays. Curr Opin Biotechnol 17(1):83–91. https://doi.org/10.1016/j.copbio.2005.12.006

Wattam AR, Davis JJ, Assaf R, Boisvert S, Brettin T, Bun C, Conrad N, Dietrich EM, Disz T, Gabbard JL, Gerdes S, Henry CS, Kenyon RW, Machi D, Mao C, Nordberg EK, Olsen GJ, Murphy-Olson DE, Olson R, Overbeek R, Parrello B, Pusch GD, Shukla M, Vonstein V, Warren A, Xia F, Yoo H, Stevens RL (2017) Improvements to PATRIC, the all-bacterial bioinformatics database and analysis resource Center. Nucleic Acids Res 45:D535–D542. https://doi.org/10.1093/nar/gkw1017

Whitman WB (2015) Genome sequences as the type material for taxonomic descriptions of prokaryotes. Syst Appl Microbiol 38(4):217–222. https://doi.org/10.1016/j.syapm.2015.02.003

Wright GD (2017) Opportunities for natural products in 21st century antibiotic discovery. Nat Prod Rep 34(7):694–701. https://doi.org/10.1039/c7np00019g

Wrighton KC, Castelle CJ, Wilkins MJ, Hug LA, Sharon I, Thomas BC, Handley KM, Mullin SW, Nicora CD, Singh A, Lipton MS, Long PE, Williams KH, Banfield JF (2013) Metabolic interdependencies between phylogenetically novel fermenters and respiratory organisms in an unconfined aquifer. ISME J 8(7):1452–1463. https://doi.org/10.1038/ismej.2013.249

Wrighton KC, Thomas BC, Sharon I, Miller CS, Castelle CJ, VerBerkmoes NC, Wilkins MJ, Hettich RL, Lipton MS, Williams KH, Long PE, Banfield JF (2012) Fermentation, hydrogen, and sulfur metabolism in multiple uncultivated bacterial phyla. Science 337(6102):1661–1665. https://doi.org/10.1126/science.1224041

Wu YW, Tang YH, Tringe SG, Simmons BA, Singer SW (2014) MaxBin: an automated binning method to recover individual genomes from metagenomes using an expectation-maximization algorithm. Microbiome 2:26. https://doi.org/10.1186/2049-2618-2-26

Xian WD, Salam N, Li MM, Zhou EM, Yin YR, Liu ZT, Ming YZ, Zhang XT, Wu G, Liu L, Xiao M, Jiang HC, Li WJ (2020) Network-directed efficient isolation of previously uncultivated *Chloroflexi* and related bacteria in hot spring microbial mats. NPJ Biofilm Microbiome 6:20. https://doi.org/10.1038/s41522-020-0131-4

Yarza P, Yilmaz P, Pruesse E, Glöckner FO, Ludwig W, Schleifer KH, Whitman WB, Euzéby J, Amann R, Rosselló-Móra R (2014) Uniting the classification of cultured and uncultured bacteria and archaea using 16S rRNA gene sequences. Nat Rev Microbiol 12(9):635–645. https://doi.org/10.1038/nrmicro3330

Zamkovaya T, Foster JS, de Crécy-Lagard V, Conesa A (2021) A network approach to elucidate and prioritize microbial dark matter in microbial communities. ISME J 15(1):228–244. https://doi.org/10.1038/s41396-020-00777-x

Zaremba-Niedzwiedzka K, Caceres EF, Saw JH, Bäckström D, Juzokaite L, Vancaester E, Seitz KW, Anantharaman K, Starnawski P, Kjeldsen KU, Stott MB, Nunoura T, Banfield JF, Schramm A, Baker BJ, Spang A, Ettema TJ (2017) Asgard archaea illuminate the origin of eukaryotic cellular complexity. Nature 541(7637):353–358. https://doi.org/10.1038/nature21031

Zhu F, Han B, Kumar P, Liu X, Ma X, Wei X, Huang L, Guo Y, Han L, Zheng C (2009) Update of TTD: therapeutic target database. Nucleic Acids Res 38:D787–D791. https://doi.org/10.1093/nar/gkp1014

Zhu Q, Mai U, Pfeiffer W, Janssen S, Asnicar F, Sanders JG, Belda-Ferre P, Al-Ghalith GA, Kopylova E, McDonald D, Kosciolek T, Yin JB, Huang S, Salam N, Jiao JY, Wu Z, Xu ZZ, Cantrell K, Yang Y, Sayyari E, Rabiee M, Morton JT, Podell S, Knights D, Li WJ, Huttenhower C, Segata N, Smarr L, Mirarab S, Knight R (2019) Phylogenomics of 10,575 genomes reveals evolutionary proximity between domains bacteria and archaea. Nat Commun 10(1):5477. https://doi.org/10.1038/s41467-019-13443-4

MIX
Papier aus verantwortungsvollen Quellen
Paper from responsible sources
FSC® C105338

If you have any concerns about our products,
you can contact us on
ProductSafety@springernature.com

In case Publisher is established outside the EU,
the EU authorized representative is:
Springer Nature Customer Service Center GmbH
Europaplatz 3, 69115 Heidelberg, Germany

Printed by Libri Plureos GmbH
in Hamburg, Germany